WITHDRAWN

Sherrie McLeRoy

Roy E. Renfro

Sherrie S. McLeRoy
& Roy E. Renfro, Jr.

WITHDRAWN

D1286500

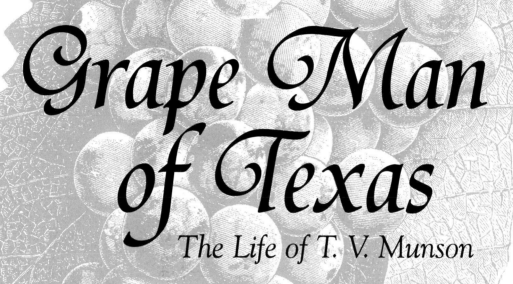

Grape Man
of Texas

The Life of T. V. Munson

Sherrie S. McLeRoy
and
Roy E. Renfro, Jr., Ph.D.

EAKIN PRESS ❦ Austin, Texas

For further information on acquiring this title or the following titles:

Foundations of American Grape Culture
Ten Million Acres: The Life of Wm. Benjamin Munson

Please contact
Grayson County College Foundation
6101 Grayson Drive · Denison, Texas 75020

FIRST EDITION
Copyright © 2004
By Grayson County College Foundation
Published in the United States of America
By Eakin Press
A Division of Sunbelt Media, Inc.
P.O. Drawer 90159 ☏ Austin, Texas 78709-0159
email: sales@eakinpress.com
🖥 website: www.eakinpress.com 🖥
ALL RIGHTS RESERVED.
1 2 3 4 5 6 7 8 9
1-57168-819-6
Library of Congress Control Number: 2003114084

Book Design by Amber Stanfield
Edited by Melissa Locke Roberts

The authors wish to dedicate this book to the

memory of Thomas Volney Munson

and his goal of educating people about

the natural treasures all around them.

Contents

Authors' Preface   vii

Chapter 1: Ancestors and Childhood   1

Chapter 2: The Kentucky Years   7

Chapter 3: On to Nebraska   25

Chapter 4: Gone to Texas   38

Chapter 5: "No Little Work"   53

Chapter 6: "A Great Pest"   68

Chapter 7: Mr. Munson to the Rescue   78

Chapter 8: Back to Work   91

Chapter 9: "A Valuable Treatise"   107

Chapter 10: The World Goes to Chicago   114

Chapter 11: The Waning Century   128

Chapter 12: Meet Me in St. Louis   144

Chapter 13: Munson's Magical Flying Car   156

Chapter 14: "A Great and Good Man"   171

Chapter 15: A New Beginning   189

Epilogue   205

Appendices

        I. Published Works of T. V. Munson   207

        II. Papers and Addresses Given by T. V. Munson   212

        III. Grape Varieties Created by T. V. Munson   213

Chapter Notes   237

Bibliography   269

Index   279

Thomas Volney Munson, 1843-1913. (T.V. Munson Viticulture and Enology Center)

Authors' Preface

Modern research on the life and extraordinary scientific achievements of T. V. Munson began in the 1960s and intensified when this book, the first full-length biography ever written about him, was started in 1996. Though internationally known in his lifetime, Munson has proved an elusive subject because the bulk of his personal archives—correspondence, field and experiment notes, manuscripts—was destroyed after his death. This unfortunately also proved true of many of his closest contemporaries, leaving us to search for him elsewhere.

The pursuit led us all over this country from California to Washington, New York to New Orleans, and across the sea to France. Even collections which contained material were maddeningly incomplete. C. S. Sargent's papers at Harvard, for example, include letters from Sargent to Munson, but none from Volney himself. The index to the Liberty Hyde Bailey collection at Cornell doesn't even list him, and the same was true for the Luther Burbank papers at the Library of Congress and the museum in Santa Rosa, California. One of our greatest disappointments came from the U.S. Department of Agriculture collection at National Archives II in College Park, Maryland. Though we found marvelous and critical letters there, we also discovered that well over a hundred more are now missing, as are Munson's field reports for the Department.

Our great hope is that the publication of this book will result in other relevant papers being brought to light which will allow us to learn more about this scientific pioneer.

Roy and I would like to thank a number of people who helped make this book possible. First and foremost are our families, who tolerated many long absences for research trips, preoccupation, and working vacations. In Montpellier, France, we say thank you to Beatrice Bye, Mr. and Mrs. Henri Bernabe, Dr. Max Rives, Dr. Pierre Galet, Alain Charbonneau, and Dr. Jean-Paul Legros. Dr. George Ray McEachern at Texas A&M University, who kindly read the draft manuscript. Barbara Smith-LaBorde at the Center for American History at the University of Texas. Claire Coates at the Bureau Nationale Interprofessional du Cognac (France). Rusty Russell at the National Herbarium (National Museum of Natural History) of the Smithsonian Institution, who kindly allowed us to search their specimens for ones collected by Munson. Wayne Olsen, reference librarian at the National Agricultural Library in Beltsville, Maryland, who patiently answered my many e-mails. Dorothy Heckmann Shrader in Hermann, Missouri. The archival staff at the Panhandle-Plains Historical Museum in Canyon, Texas. (And a special thank you to the folks there who prepared an index to the W. B. Munson Collection.) Dr. Tom Geraci, a descendant of Herman Jaeger, who tried to help us track down Jaeger's Legion of Honor medal. The family of John and Peggy Brock in Neosho, Missouri, who graciously gave us access to the Jaeger cache of letters. Dixie Foster at the Denison (Texas) Public Library.

A special thank you to the "vine ladies" in Missouri: Kay Hively in Neosho and Linda Walker Stevens in Hermann. They have written biographies of Jaeger and George Husmann, respectively; and we have carried on a lively, at times profane, internet conversation the last few years. Another big thanks to my dear friends Jane and Joe Gleason of St. Louis, who helped me with Isidor Bush material and housed me on several research trips to Missouri.

There are dozens of other people who have answered questions, checked material, and tried to help, and we thank them all.

Without the Munson family and the W. B. Munson Foundation, however, this book would never have been written. Roy and I particularly wish to express our thanks to Joyce and John Maki, who made several long trips from Houston to Denison to bring us marvelous material and answered and sent more e-mails and phone calls than I can recall. The Board of Directors of the Munson Foundation has supported our research and hung in there patiently when the work took much longer than we anticipated. Their devotion to the T. V. Munson Viticulture and Enology Center, the enology program of Grayson County College, and perpetuating the memory of Volney Munson is exemplary.

Roy and I believe that we have discovered at least a part of the man behind all the legends and mistruths that have proliferated over the years. Volney Munson proved to be something of a Renaissance man, with his wide range of interests, and one who thought big even by Texas standards. We hope you will enjoy his story.

Sherrie S. McLeRoy
Roy E. Renfro

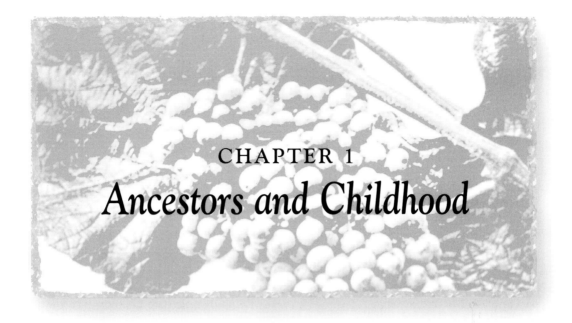

CHAPTER 1
Ancestors and Childhood

He was a man who embraced knowledge all his life, indeed made it a kind of god, more reliable and more comforting to him than the religious dogma so many of his day believed in. Even his family life had to accommodate itself to his dogged pursuit of knowledge. Wife and children were left behind for days and weeks at a time while he searched riverbeds, canyons, and the deep recesses of forests for ever more bits and pieces of information to link together. His son was pulled away from his own interests to run the business and leave the father free to continue his pursuit.

He could be arrogant and vain, this Thomas Volney Munson, sure that his was the only right way and not hesitant to say so. Querulous at times, he liked to complain about his poor finances and his unequal health. And he was unbending in his dislike of organized religion, a stance which earned him suspicion and even enmity.

Yet to see Volney Munson with his hands in the earth, or gently tying up a grapevine, was to see him in the element he was born to enrich. His colleagues respected his achievements and learning, his penetrating vision. Friends held him dear; family loved him and, far from resenting his devotion to work, helped him whenever possible. For they recognized in him a light that is given to few, a desire to *know* at all costs, and the generosity of spirit to share that knowledge for the betterment of others.

European Origins

Handsome, dark-haired, and spare, Volney Munson doubtlessly owed both his long straight nose and his work ethic to a New England heritage. As is true of many with such roots, he ultimately traced his ancestry across the ocean to Northern Europe. That lineage was important to him, and he spent many hours reading and corresponding with oth-

The family crest used by Volney Munson. (T.V. Munson Viticulture and Enology Center)

ers to search it out; in fact, the basis for the next several sections of this text are Volney's own genealogical files.

The Munson name and its several variations, primarily "Manson" and "Monson," can be found in Scandinavia. Derived from "Magnusson," or son of Magnus ("The Great"), the name and the family spread to Britain and particularly to the Shetland Islands of Scotland during the Viking invasions. By the sixteenth century, "Munson," in some form, could be found throughout this region.[1]

To America

The first of the family to venture across the Atlantic was probably Capt. Thomas Monson, who immigrated to Massachusetts about 1635 and eventually helped found New Haven, Connecticut. The next was Richard Monson (ca. 1635–1702), who took up land in the Massachusetts Bay Colony in the early 1660s; it is from Richard that Volney descended. Settling in the Strawberry Banke/Portsmouth area of modern New Hampshire, Monson was a sea captain, a Congregationalist (a Puritan), and a fisherman who did well enough to acquire additional land across from Portsmouth in what is now Kittery, Maine, several sailing vessels, a ship wharf, and a fish "warehouse." His descendants remained in the Portsmouth/Kittery region for many generations. His grandson John, Jr. (1705–1783) changed the family name to Manson, but his great-grandson Richard (1741–1813) reverted to Monson before eventually altering it to Munson. This latter Richard moved to New Hampshire and followed his second son, William, to Vermont in 1810.[2]

Richard's oldest son, Theodore (1775–1839), took his family to Vermont, too, then restlessly pushed on, finally settling in Illinois in 1830. He and second wife Lydia Philbrook had four sons and two daughters, but only one would ever permanently venture out of the state.[3]

Volney's Parents

Theodore's oldest son, William Munson (1808–1890), was born in Wakefield, New Hampshire. By the time the family arrived in Illinois, he was supporting them since Theodore had become an alcoholic. At some point, William met a girl born in Kentucky and, in 1836 in Rushville, Illinois, he married her.

Maria (pronounced Ma-RYE-ah) Linley (1810–1890) was the daughter of Joseph Linley, an English blacksmith and cutlery maker who had immigrated to the United States in 1800, and of Sibilla Benjamin of Pennsylvania. Maria was an intellectual

Maria Linley Munson (1810-1890), Volney's mother.
(T.V. Munson Viticulture and Enology Center)

3

William Munson (1808-1890), Volney's father.
(T.V. Munson Viticulture and Enology Center)

woman, like her mother well read and educated. Another trait the two women shared was a "business capacity of a high order." Maria was "energetic" and "ambitious" and had a fascination for growing things which she would bequeath to her sons, especially Volney. Throughout her life she cultivated extensive gardens and "originated [created] many remarkable varieties" of flowers herself.[4]

In 1836 William and Maria purchased land in Schuyler County, Illinois, on Crooked Creek (now called the La Moine River) and built a log cabin. Daughter Louisa Elizabeth was born there in 1840. T.V. Munson would write: "Wild broad prairies and forests lay about on every side; nightly, wolves howled around and miles intervened between their home and that of the nearest neighbor. They toiled early and late at the plow and loom to make scanty food and clothing for their little family, and for company mother [Maria] sang songs the live-long day, and during the long winter evenings, father [William], while mother sewed, read aloud, such books as they could get."[5]

A Growing Family

It was the author of one of those books, *Les Ruines, ou meditations sur les revolutions des empires (The Ruins, or Meditations on the Revolutions of Empires: and the Law of Nature)*, by Comte de Constantin F. C. Volney, who inspired the name of their child born in 1843. By this time, the Munsons had moved to Fulton County, Illinois, where they had bought a 160-acre farm near Astoria from Maria's brother Isaac Linley and his wife Louisa, her niece's namesake.[6] In another log cabin there, Joseph Theodore (The, or Theo, as his family called him) was born in 1841 and named after his paternal grandfathers. When the second son made his appearance on September 26, 1843, William and Maria recalled the book given them by a friend that had helped beguile many a long evening, "for they loved serious thought." They named the baby Thomas Volney for Maria's uncle, Dr. Thomas Linley, and for the book's author. However, family members always called him Volney or Vol.[7] Third brother William Benjamin (Ben) was also born in the log cabin in 1846.

William Munson then built a frame, four-room house where Triphena Mary, or Trite, was born in January of 1848.[8] The following year he sold the farm and purchased a larger, adjacent property. Unfortunately for Maria, this site came with only a two-room log house, and here she gave birth to their sixth and last child, Mary Ginevra (Jennie or Ginnie) in 1853. William afterward built Maria a fine eight-room frame house in a simple Federal style.

In appearance, William and Maria's six children were much alike, and most of them inherited the Munson nose from their father. Volney stood close to six feet tall, with thick, dark hair which he retained until late in life and a distinguishing characteristic still to be found in his descendants: a slightly quirked, or raised, right eyebrow. As he got older, Volney would grow the obligatory beard and mustache of his generation, though he shaved them off in later years.

A Farm Childhood

Volney and his siblings grew up on the Fulton County farm and attended the local school in Woodland Township. The boys then went on for a last year at the Fulton Academy in nearby Lewistown, a Methodist-related school which catered to more affluent society and had just built a new four-story facility.[9] When not in school, they worked on William's farm, where he raised grains of various sorts, fruits, and livestock. By 1860, when Volney was sixteen, his father's farm and personal property were valued at $20,000, making William Munson the wealthiest man in town and one of the wealthiest in Illinois.

As was typical of the period, all the children helped with the farm chores, but Volney took to it with a passion. In the 1860 census, he is the only one of the siblings listed with an occupation: farmer. He spent time tinkering with the machinery and building new pieces, a talent he would use all his life.

But thanks to his mother and grandfather, horticulture became his true passion during these years.[10] Maria's English-born father, Joseph Linley, moved to Fulton County about 1853, some years after wife Sibilla died. Volney would later write of his grandfather, "I remember him well, as a very particular, neat, kindly old man, who showed me how to lay out a flower-garden in pretty walks and raised beds, in old English style," designs Linley would have known from his upbringing as the son of a Yorkshire nobleman.[11] Maria was more interested in the scientific side of gardening, always reading horticultural journals and writing for new seeds, and exchanging information and ideas with family in Kentucky. She taught Volney all she knew of grafting, budding, and pruning, skills needed to maintain the farm's fruit orchard.

More Schooling

Because William Munson had been his family's principal breadwinner due to his own father's alcoholism, he instilled a strict work principle in his own children. Theo taught school to earn college money, as did Volney and Ben later. William also gave each son a plot within the main property to farm for his own use, allowing them to keep the proceeds.

William and Maria urged their children to obtain as much education as they could and provided each one funds for their first year of college; any additional schooling, the young Munsons had to pay for themselves. Louisa went to Abingdon College, chartered in Abingdon, Illinois, in 1855 by the Church of Christ. Oldest son Theo followed her there for a short time, as did Ben. Trite and Jennie would eventually attend Columbia College in Missouri and Daughter's College near Lexington, Kentucky. Volney taught for two years in Fulton's rural schools and saved his money.[12] In 1861, with civil war engulfing the country, the eighteen-year-old went off to Chicago to enroll in the Bryant & Stratton Business College, one in a chain of forty-six such schools across the United States and Canada.

Thomas Volney Munson in 1870. (T.V. Munson Viticulture and Enology Center)

Housed in an impressive stone building on the corner of State and Washington streets, Bryant & Stratton offered an extensive curriculum covering every aspect of business: commercial law, partnerships, freighting, insurance, bookkeeping, investments, and penmanship. There were German and French classes as well as architectural and mechanical drawing. Volney spent the four years of the war there and received a thorough grounding in the mechanics of operating a business.

In 1865, Volney returned to Fulton County and taught school for a year in Astoria, once more saving his money. He and Ben had decided they wanted to enroll in a new school scheduled to open in Lexington, Kentucky, for the fall 1866 term.

CHAPTER 2

The Kentucky Years

*V*olney and Ben were probably very familiar with northern Kentucky and the Lexington area, for it was full of their Linley relatives. Their mother Maria had grown up on the family farm less than a hundred miles to the northeast, in Lewis County on the banks of the Ohio River.

Lexington and their college experience would have a profound impact on both brothers and, in Volney's case, would directly affect the future course of his life.

Lexington

Lexington was a cultured and genteel town which had emerged from the Civil War relatively unscathed when Volney and Ben arrived. Founded in 1779 and named for the Revolutionary War battle, Lexington was in the heart of Kentucky's bluegrass and horse racing country. The "Great Compromiser," Henry Clay, built his home "Ashland" there in 1806, and Mary Todd Lincoln grew up in Lexington. A compact town filled with graceful Federal and Georgian-style houses, it early became a center of learning on what was then the western frontier. The oldest circulating library west of the Allegheny Mountains opened there in 1795, and the oldest educational institution in the west, Transylvania College, had moved to town a few years earlier. But the college barely survived the Civil War and declined to a high school, its buildings used as Union hospitals. About 1864, it merged with Kentucky University in Harrodsburg. The resulting school, known as Kentucky University, was located on Transylvania's Lexington campus since it had suffered little physical damage from the war.

In the waning months of that war, on February 22, 1865, the state legislature established a new school to be part of the university complex, the Agricultural & Mechanical College of Kentucky. With the financial help of Lexington citizens, officials purchased

Henry Clay's Ashland estate and the adjoining "Woodlands" property of his daughter, Anne Clay Erwin. While the liberal arts and law departments stayed at the former Transylvania campus on Broadway, A&M opened late in September of 1866 at Ashland, a short distance to the southeast.[1] The 433 acres there allowed ample room for additional new buildings and an experimental farm.

All A&M students received instruction in "Military Tactics" and were uniformed cadets. They were required to spend at least two hours per day working in some capacity on the farm, which was credited to their room and board. Academically, students chose from philosophy, English language and literature, mathematics, chemistry and experimental philosophy, natural history, history, modern languages, civil engineering and mining, and fine arts.

Aggies

Ben and Volney, now twenty and twenty-three respectively, arrived on the A&M campus with jobs already lined up to support them. In the 1866–67 college catalogue, they were listed as both tutors and instructors, doubtlessly because of their previous teaching experience. They taught two hours a day for eighty cents, important wages when the tuition for the nine-month session was $30.[2] They also labored on the college farm many Saturdays for another twenty cents an hour, cut firewood over Christmas break and sold it, and kept their boarding costs to a dollar a week by setting up spartan bachelor quarters together. (Students were allowed to live in private homes or in college dormitories.) Money earned from the crops they planted at home over the summer helped, too.

The oldest surviving examples of Volney's writing date from this year. One is an exercise for Prof. Joseph D. Pickett's English and literature class; the other two are articles for the Ashland Institute, a literary society Volney helped found.[3] In "The Origin of Language," he concluded that, since language has clearly evolved from the rudimentary to the complex, it is "nothing more than the consequent outgrowth of man's nature and circumstances," since "if God had given language directly to man, we would expect it to be perfect and unchangeable." This was a theme which appeared in Volney's writing throughout his life, his belief that if God created something, it should be perfect.

In his article for the Institute, Volney declared that the love of glory was a universal principle, that even Jesus Christ begged on the cross to be glorified in heaven. Fame and glory were not bad in themselves, only the "envious effort to obtain [them] at the expense and destruction of the happiness of others." Again, these are themes he retained for decades; but in later years he would use them to refute the existence of God and Jesus as presented by various "organized" religions. At seventeen, Volney had joined the Christian Church, today's Disciples of Christ. He was an earnest church-goer who taught Sunday school, prayed daily, and exhorted joyfully in prayer meetings; he carried with him to Lexington a small cased Bible he'd been given in May of 1862.[4] This strict Campbellite upbringing (see footnote 9) showed itself again in the concluding paragraphs of his exercise: "All true glory and honor belong to God. He is the source of it . . . we[,] the workmanship of His hand[,] should show forth his glory here that He may glorify us in Heaven." But Volney's conviction would change drastically over the four years of college under the influence of his science classes and the changing times about him. In a note appended to this manuscript in 1900, when he was fifty-seven, Volney looked back scornfully at the naive young man he had been: "... how lame does the logic of T.V.M. of 1867, struggling through the cobwebs of religion, appear! The idea of an infinite God creating a puppet—man—in his own image, to flatter himself!"[5]

A characteristic which stood him in good stead throughout his professional career was the subject of the third essay, "Patience," which foreshadowed the bent for research that would eventually bring him fame. "The greater the cause, the more slowly does the effect appear. Man's impatience has not allowed him to await the results but ever restless, he has pushed on to seize the first fruits, which usually make the most conspicuous display, and when the real harvest has arrived, he is far away in the other fields." Once more he turned to religion for examples, reminding his readers that "a day is as a thousand years" to God. "The final secret of success," Volney asserted, "in nearly every calling is patience," necessary to acquire anything from wealth to a young woman's heart.

Cadet Life

Theo joined his younger brothers at Lexington for the 1867–68 school term, when Ben and Volney once again earned extra money as tutors. This is the only year of their stay at Kentucky A&M for which cadet records have survived. The "Reports and Records" of the Military Department show all three Munsons were exemplary cadets. Only in the spring did Ben and Theo slip, being cited for "noise in quarters." Theo was also caught a few weeks later skipping tattoo (the evening call to quarters before taps). Volney enjoyed a position of leadership as captain of Company B.

As students at a military college, the cadets were called on to appear at special events and on solemn occasions. Just before school let out for 1868, the "Brass Button Boys" traveled thirty miles by train to Frankfort, Kentucky, to participate in the funeral of U.S. Ambassador to Guatemala B. L. Clark. Along with cadets from Kentucky Military Institute, the A&M boys led the procession to add "the bones of another of Kentucky's noble sons" to the earth. A solemn, almost obsequious Volney noted in his article for the *Brass Button,* a cadet publication, that the school officials accompanying them exercised strict control over the students that day, helping them to maintain an "excellent deportment." A more gregarious Munson, in 1900, had the final comment, though, noting in the margin, "Rats!"[6]

Setting Goals

Volney wrote many articles and poems for the *Brass Button,* and in "Soliloquy of the Brass Button" he implied that he was the editor if not the founder of that journal.[7] He apparently faced some criticism of his management in the fall of 1868, for "Soliloquy" was a witty attack on his critics and a plea for help to his friends. Another effort, "The Regent's Soliloquy" (undated), was a humorous response to the administration's call for a new and more costly uniform. Volney satirized Hamlet's "to be or not to be" speech when he wrote: "A tail, or not a tail, that is the question / Whether 'tis noble in the boys to suffer / The jeers and taunts of fickle, silly women / Or to take Coatees in place of Bob-tails / And, by wearing, please them?"

Origin of Language.

(Class exercise under Profesor Pickett, May 23 1864)

As no man knows how language originated, all are equally entitled to their opinions in reference to it.

Men, using their reason and imagination upon the subject, have invented three different theories concerning it. These are too well known to need repeating here. I will simply state my views of the matter.

When we examine the history of Gods dealings with man, we find that He has never communicated any knowledge to him which he could obtain from the world about him through the powers he was endowed with in his creation. Now we know that man has the power to conceive and name ideas; thus language is still growing and will so continue as long as new ideas are required.

I can distinguish no difference in the making of a machine and a word. Each represents the concept or thought of the mind. When a person, wishes to convey an idea I can discover no reason for which he has no means already at hand, as one speaking to another of a different language. why man, in a more perfect state, (as he was in the Beginning) to a horse or dog, as well as now, he uses some gesture or sound by which to express would not himself. He would use the most natural means for this so as most readily to be understood by the person or animal addressed purpose. As man is the most imitative, those expressions made by one would be copied by others; thus would take permanent language form and be transmitted from generation to generation. The further back in the history of language we go, the more simple and rudimentary it is,

2

which seems to confirm what I have said.

If God had given language directly to man, we would expect it to be perfect, and ~~that it would degenerate as man became more imperfect~~, which we know is not the case.

If language were to man as song to the birds, it would be perfect as their songs are perfect, and all would talk alike. I conclude that language is nothing more than the development of attributes belonging to the nature of man.

——— ——

The Love of Glory.

(Written for the Ashland Institute, Nov. 5th 1864)

The love of glory,— a principle as universal as intelligence, which exists throughout the entire spiritual universe, belonging to angel as well as to demon.

Before assenting to this proposition let us determine if it is true or not, and if found true then will we seek wherein the good and evil lies, for it is a very fruitful cause of both.

To establish the truth of the universality of this principle it will be necessary to produce examples as manifested in the various stages of existence.

Those who censure, and even pretend to despise fame and honor, if they would examine their own hearts, would be surprised to find their censures

Volney's earliest known writing was this paper written for a class at Kentucky A&M.
(T.V. Munson Viticulture and Enology Center)

Volney had set his sights on a career in some field of science, as evidenced in "The Future of the Agricultural College" (April 1868): "Never was there a greater demand for thoroughly scientific men than exists throughout the Great South and West. A great revolution in agriculture, mechanics, public schools and public improvements in general is inevitable. These improvements should be made in the full light of science."

With these opinions, Volney demonstrated his ability to "see the big picture." For the last three decades of the nineteenth century witnessed an amazing transformation in how humans viewed their world, thanks to science. Concepts and interrelationships between the sciences which we now take for granted were new and marvelous discoveries in Munson's day. Advances in transportation and communication enabled people around the world to share their knowledge with each other in ways that had never been possible before. Science became the new god which would make life easier and better.

But at the time Volney wrote, in 1868, American botany and horticulture were still years away from a true scientific approach. His assertions about the value of science indicate that he was *au courant* with European thoughts and methods which, even there, were in their infancy. Only in the early nineteenth century, for example, had serious work to improve domesticated plants begun, in England. Few bothered to study the heredity of plants; even the great Swedish botanist Linnaeus had believed that the formation of new species could only occur naturally and over time.

Most American botanists studied *about* plants, not the plants themselves; they relied on dried specimens rather than direct observation since few could afford expensive microscopes to study interior structure. Taxonomy, or classification, was the rage. And Charles Darwin and other progressives had not yet issued a call to the new generation of botanists "for investigation anew of the living plant in *all* its relationships, *all* its physical, biological, and environic factors." (Authors' italics)[8] Volney Munson was on the cutting edge of a new age in botany, horticulture, and agriculture; and he was one of the few to recognize it.

But embracing the new scientific viewpoint carried a price for some, and that was a challenge to their religious faith. At Kentucky A&M, Volney and his classmates studied Charles Darwin's explosive 1859 book, *Origin of Species,* which typified the growing trend to explain all life in scientific terms. It was a book that rocked many a mind, both young and old. No wonder then that many people, Volney Munson among them, began to question institutional religion, still mired in the past.

Warring with God

Volney's commitment to the scientific approach was, paradoxically, deeply emotional and tinged with his religious upbringing. "When I say Naturalist," he wrote in the fall of 1868, "do not understand me to speak of him who has merely learned the technicalities of Botany, Zoology, Mineralogy, and Geology. I would have you think of one who loves the communion of the objects of Nature, as they come undeformed and undefiled from their Maker's hand ... He who finds in Nature no spirit of communion ... Such a one is an Atheist, Nature's grocest [grossest] deformity." Volney also deplored the medieval superstitions that still ruled in his day. Why did A&M, of all institutions, he continued, hire a water-witch to find a well on the Ashland grounds when anyone with "a little common-sense" and scientific knowledge could have figured it out as well?

of, excellent deportment among the cadets. (rats! 1900)

At five o'clock in the evening our worthy Governor appeared before us with a very beautiful and complimentary little speech, after which we took to our coaches again, which the old 'Iron-Horse' now brought dashing along. An hour or two of puffing and champing, ending with a loud and furious neigh and the ding-dong of the driver's bell, we stepped out among our Lexington friends. To sweeten our last steps, we again took our way along the neat and pleasant Hill-street to enjoy the tender emotions its lovely faces inspire. Though the temptation was strong to cast an look around on those beaming eyes, Gray-Coats were too soldierly to glance aside. Finally the charming Woodlands welcomed them home and echoed their joyous whoop as their breaking ranks dispersed to quiet rest and pleasant dreams.

——— — —

T.V. Munson was Captain of "Company B" of the Cadets at the time this excursion was made.

Age, 24 yrs. 8 mos. 12 ds.

An older, more cynical Munson commented on the naivete of his youth.
(T.V. Munson Viticulture and Enology Center)

His Campbellite heritage was now seriously at war with his studies.[9] "I was ever honestly seeking truth," he would write many years later; but the exact nature of that truth was unclear to him. The Bible told him that God created the earth in six days, while his geology professor told him it took millions of years, a fact he accepted in his head. Yet Volney could not completely tear himself away from the church's teachings. His 1868 article "A Journey Backward and a Glance Forward" is a good example of the confusion which plagued him. He invited his readers to travel with him back through time to see the rise of civilization, the Great Flood, and the Garden of Eden. Then he switched to a scientific description of earth as it was 50,000 years ago and wrote of geological time periods and fossils. Even more ancient were the "huge reviving crocodiles" and the "great Aligator [sic]."[10] Volney continued his look back to a time when all higher forms of life were simply inorganic matter, and beyond that, when "the entire globe is one mass of igneous fluid and gas." But there he was stymied. For the science of 1868 couldn't explain any more; it knew

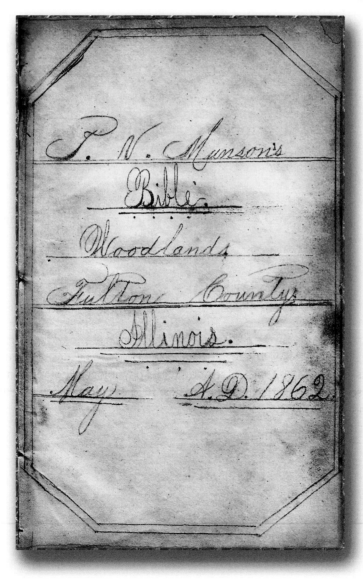

Volney carried this leather-cased Bible, published in 1858, to college. (T.V. Munson Viticulture and Enology Center)

nothing of the "Big Bang" and other theories. Volney had built as much as he could on a scientific base; in the end, he was forced to return to God, "the Omnipresent Spirit," who "called [Earth] into existence."

Both he and Ben, who was majoring in mathematics and civil engineering, struggled with religion versus the new science while at Kentucky A&M. Complicating the matter was the economic impact of putting aside their religious beliefs. In 1906, Volney wrote a pamphlet entitled *The New Revelation* under the pseudonym of Theophilus Philosophius, in which he defended his own beliefs. (See Chapter 13.) In it, he discussed candidly the choice he and Ben had faced in college, when "to disavow my religion meant almost universal ostracism and loss of honorable and lucrative position." Volney was practical enough to know that perceived atheists found it hard to get work in 1870 America. He hoped he could reconcile faith with knowledge but "my whole being revolted against the savagery of ancient heathenish idolatry. . . . I was mortified and ashamed of myself for having been a duped idolator for ten of the best years of my life."

In the end, each of the brothers came to an accommodation which suited him. Both withdrew from organized religion as practiced in the churches of the day, "determined to stand by demonstrated truth." A later biographical sketch would declare Ben an evolutionist, "believing in the unfolding and growth of our finer individualism from our inherited natures to a full stature of man and womanhood."[11] But recognizing the church's role in promoting good moral conduct, which he honored, Ben would prove a generous benefactor to several denominations. And Volney, while loathing the narrow-minded, superstitious strictures that made his scientific outlook seem suspect to many, never lost his belief in the God that had created the wondrous world about him. That viewpoint was, in fact, typical of many of botany's brightest figures, such as Asa Gray of Harvard. Obviously there was a Creator of some kind: otherwise the universe would be chaos instead of order.

But outside the scientific community, reaction to Volney's unorthodox beliefs would dog him all his life.

Rising Faculty Star

Volney entered his third year at A&M with a higher faculty ranking. While Ben remained a tutor, Volney was named an assistant instructor in natural history, even as he continued his own studies. (Theo did not return to Lexington with his brothers but demonstrated his ongoing interest by donating a set of weasels to the new museum.) Natural history students pursued such topics as structural and physiological botany, human anatomy, and geology; seniors were introduced to the new field of paleontology, the "history of fossil animals and plants."

In addition to his science classes, Volney studied French and German and, for at least one year, mathematical mechanics under Prof. H. H. White. A journal he began in college contains notes on two of his earliest "inventions" done "for the benefit of my class in Mechanics" in 1870. "Diagram showing the relation of Time, Space and Velocity, of a falling body under the influence of Gravitation alone" and "Cycloidal Pendulum" illustrate his fascination with engineering that had begun while repairing farm equipment back in Illinois.

In his senior year at A&M, Volney moved up yet another rung on the career ladder when he was named adjunct professor of chemistry and natural history, a position which placed him in the Senate of the University. By this time he was alone at A&M, for Ben had completed the four-year course in just three years and graduated in June 1869, becoming the first graduate of Kentucky A&M (now the University of Kentucky).[12]

A Summer Trip

Ben decided to celebrate the end of his years of hard work by taking a trip. He, Volney, and classmate John Edward Leet of Sedalia, Missouri, headed to the mountains of Kentucky for "the purpose of recuperating our health, gathering botanical and geological specimens and seeing the country."[13] They began their adventure on June 15, traveling by train to the small town of Paris, then walking to Blue Lick. There the young men sampled the spring water and viewed the "celebrated" Blue Lick Battlefield.

The next day they became a bit lost in the high, steep hills and deep hollows south of Elizaville. At Sherburn Mills (modern Sherburne), the explorers met two local geologists. Arrogant as only college students can be, they dismissed the men as knowing "about as much about geology as a jackass." In the hills outside Owingsville, they were delighted to find "lots of fossils" from the Devonian formation. Even more amazing than the fossils, though, was the sight that met their eyes in town: "Saw a man riding a velociped!"[14]

At Mount Sterling the next day, John, Volney, and Ben "shipped a box of fossils" back to the Museum of Natural History at A&M. Near town, they met "an old gent named Spratt," who, John remarked scornfully, was "still" a Christian who "actually believes that the world was made in six days and nights." John apparently suffered from the same dichotomy that plagued the Munson brothers, for the skeptic went on to write the following day that the view from Pilot Knob caused him to send "a hymn of thanks to the Creator for the greatness of his work." Prosaically, he also noted that he and Volney found "no less than 15 new flowers" while climbing the summit. They analyzed all the flowers there, then pushed on to the Red River and State House Rock.

Arriving at the Kentucky River, Leet and the Munsons bought a boat for $5 and navigated the river to Irvine, where they visited the dilapidated White Sulphur Springs Hotel, then continued down the river. Unfortunately, Leet's diary ends in the middle of June 22; the remaining pages are missing.

His Gal

Leet's diary, though regrettably incomplete, provides a clue to Volney's social life. On June 20 John wrote that, while Ben analyzed a flower, "Volney wrote a letter to his gal."

That was Ellen Scott Bell, born in Lexington on December 3, 1849, and the only daughter of Charles Stewart and Margaret Bunyan Smith Bell.[15] She was christened for her maternal grandmother but promptly nicknamed "Nellie" by her father, and Nellie was how she was ever after known to family and friends. Where and when she met Volney is unknown, but it may have been at A&M since her father, C. S. Bell, was the first superintendent of the college's experimental farm and a man who would prove an important influence to the young Munson.

Born in Scotland in 1823, Bell had followed his own father, an estate manager, into

Nellie Munson in her lilac going-away dress, 1870. Oil painting by her niece Linley M. Tonkin from the original daguerreotype. (T.V. Munson Viticulture and Enology Center)

horticulture. He worked his way through several estates and nurseries in Britain before coming to the United States in 1842 and then to Lexington, where he met and married Margaret, also Scottish-born.[16] In 1849 Bell was chosen superintendent of the Lexington Cemetery, a position he would hold most of the rest of his life. Bell became famous for his horticultural work there, planting many unusual varieties and species on its extensive grounds and originating the park system of cemeteries, which made these areas very nearly botanical gardens. In 1866 the new Kentucky A&M hired him to supervise the grounds and farm at Ashland. He stayed only a year before deciding to leave and expand his own nursery and florist business. But cemetery trustees enticed him back there, and he remained superintendent until his death in 1905.

Bell's daughter Nellie was a shy, dark-haired girl with large, expressive eyes who tried to follow her boisterous brothers but frequently had to be rescued from the trees they blithely climbed. Family members recall that her most outrageous expletive was "Oh Sugar."[17] As a young girl, she enrolled in Lexington's prestigious Sayre Female Institute, founded in 1854 and still in operation today. Nellie graduated in 1868 since Sayre continued to operate through the war years.[18] In addition to the usual female accomplishments of music and drawing, Sayre girls studied such hefty subjects as philosophy, ancient languages, math, chemistry, astronomy, French, and rhetoric. So it could have been Nellie's chemistry professor—the second of the Kentucky characters destined to influence Volney—who introduced her to him.

Physician and geologist Robert Peter was born in England in 1805 and came to the United States in 1817. He moved to Lexington in 1832 and began teaching at Transylvania University, where he also earned his own medical degree two years later. He taught everything from geology to pharmacy and lectured regularly at the Sayre Institute, too. He taught at Kentucky University and at A&M after the war, where he became the mentor and friend of Volney Munson. "It was not mainly or merely pure science, the passionless quest of ethereal truth" which motivated Dr. Peter "but Science applied in the service of his kind, to bless and exalt, to dignify and embellish human life, that commanded [his] undivided devotion."[19] This desire to make knowledge truly useful by improving the lives of those around him was an important trait Robert Peter passed on to the young Volney.

Ellen Scott "Nellie" Bell at the time of her wedding to Volney in 1870. (T.V. Munson Viticulture and Enology Center)

Did he introduce the quiet-spoken and intelligent Nellie Bell to his star pupil? Or did the couple meet at Ashland while the Bell family lived there (1866–67)?

Married Life

However they met, Nellie and Volney were married in Lexington's Broadway Christian Church on June 21, 1870, just a few days after Volney became Kentucky A&M's second graduate. The young couple had much to celebrate. Volney had presented the graduation oration (a discussion of Baconian philosophy ponderously entitled "Inductive Reasoning the Only Road to Advancement") and was informed, a few days later, that he had been unanimously selected by the Board of Trustees to continue as adjunct professor of chemistry and natural history. That decided the question of whether he would return to Illinois, and the newly married Munsons settled in with every apparent expectation of living in Lexington.

They probably lived with the Bells in their house on West Main Street, just opposite the entrance to Lexington Cemetery; Nellie's brothers Charlie and George also lived at home. The 1871 Lexington tax rolls list "T. B. Munson" and note that he owned no land or outstanding personal property, but that he was enrolled in the city militia as was required for men of his age. The 1872 rolls show no change nor are there any deed records for Volney in Fayette County, indicating that he and Nellie probably continued to board with her parents.[20]

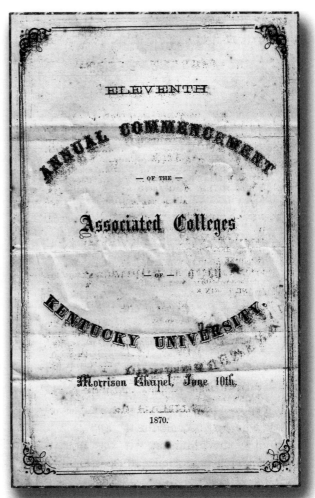

Volney was the second graduate of Kentucky A&M, now the University of Kentucky. (T.V. Munson Viticulture and Enology Center)

What did change was Volney's health. He threw himself into his duties at the university with such zeal that, added to the previous stress and work of his college studies, he became exhausted and ill. It was, he would later recall, "the same sad story told by many a young teacher"; the appointment he'd received with such joy "came near being his death warrant."[21] On his doctor's advice, he retired from the faculty after only a year. "Knowing no other pursuit than the tillage of the soil," he went to work for his father-in-law in Bell's private nursery business, hoping that the physical labor and outdoor life would heal him. Bell was a meticulous and methodical man from whom Volney learned much, since he was drawn to the new field of scientific experimentation.

Volney had likely been sketching for some time, too, and a delightful small book of "Drawings Selected and Executed by T. V. Munson" still survives in the T. V. Munson Viticulture and Enology Center at Grayson County College, just west of Denison, Texas. Within are eighteen sketches, mostly in pencil, which Volney drew in 1872. All kinds of plants and animals caught his eye—a cedar waxwing, American holly, sphinx moths, a rose, and shells among them. On the last page is a flower, a "nymph," marked "originated by T. V. Munson 1872." This may well be the first of the many plants he would breed in his career. As interesting as the attention to detail in the subjects themselves, however, is the way in which he embellished the captions. Every letter of "Hedera helix," or European ivy, for example, is drawn as ivy tendrils. He clearly loved to sketch, though in technique he was only a middling artist; and his aesthetic sense, which appreciated the beauties of nature, led him to try and bring that beauty into his own life wherever possible ... even in a humble drawing book.

Inventing

Volney also continued to dabble with mechanical inventions during the Lexington years. In the fall of 1870 he drew plans for a "Force Pump for general use in wells and cisterns." Soon his journal carried detailed drawings and a narrative description for a cookstove fueled with either wood or coal. The chief attraction of the "Perfection" stove was the firebox, which was made in one piece as part of the back; the whole apparatus was hung so it fell open and dropped the ashes for cleaning. In 1872 came the "Continuous Stream, Chain force Pump," which he considered an improvement on the common chain pump. The next year he designed a "Rotary Dasher Churn" which was attached to an ordinary churn to make it easier to operate and with a better end product. He also designed the "Lightning Weeder and Cultivator" and drew it "at work killing weeds." This may be the prototype of a scuffle hoe, the only invention he actually patented (1879).

Did he build these for the personal use of the Bell-Munson household? Did he sell duplicates? Or was the "inventing" strictly a pleasurable mental exercise? Unfortunately, the journal says nothing of what happened to his machines and tools.

Stay or Leave?

Despite Volney's enjoyment of his work with Charles Bell, he was battered by the personal troubles which struck him and Nellie. Their first child, Huxley, was born dead on May 25, 1871, and buried in the Lexington Cemetery his grandfather cared for.[22] A second son, Elmerwin, born November 26, 1872, lived less than three months and was laid next to his brother under a single tombstone. That summer Nellie's sister-in-law Eudora, married only eleven weeks to George Bell, also died, of consumption.

An economic depression struck central Kentucky during this period—the beginnings of the full-fledged national "Panic of '73"—which made Lexington an increasingly expen-

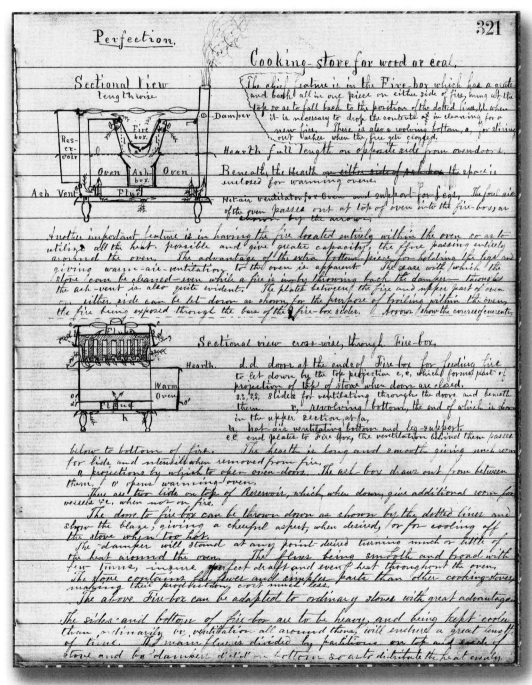

Volney invented the "Perfection Stove" sometime in the early 1870s; the lengthy explanation demonstrates his attention to detail. (T.V. Munson Viticulture and Enology Center)

Nellie Munson with her parents, Margaret and C.S. Bell, and one of her sons at the Bell home in Lexington, Kentucky. (T.V. Munson Viticulture and Enology Center)

sive place to live. Volney, now in his late twenties, yearned to start his own nursery business somewhere. But Kentucky no longer seemed a good location with the economy in poor shape and the city plagued with post-Reconstruction racial tension.

Thinking of Texas

Volney considered following Ben to Texas. In 1871 the youngest Munson brother had moved to Sherman, just fifteen miles from the Red River in northern Texas; there he studied law and set up a practice. When the Missouri, Kansas & Texas Railroad announced it would build into Texas, Ben began speculating in land at the proposed new terminus of Denison, a few miles north of Sherman. It proved an astute move, for Denison "grew like topsy" with the arrival of the "Katy" late in 1872. Ben extolled the advantages of North Texas to his family; eventually, most would join him there.[23]

Volney certainly planned to. In July 1872 he purchased land northeast of Sherman and bought more that fall. He must have visited the region that summer to look at property and study the area.[24] But two events intervened. A Texas summer convinced him that his health was not yet up to the climate. And Nellie was pregnant with Elmer, so the Munsons decided to stay in Lexington and not risk another stillbirth such as had killed Huxley. Yet Volney continued to be fascinated with Texas. He gave his power of attorney to Ben and bought and sold several parcels of land in Grayson County over the next year.

22

Texas was not the only state he considered. During his college years, and perhaps before that, Volney had visited "several of the most famous Western states, with a view of living someday in the one which seemed best adapted to [his] taste in climate, soil, society, and trade."[25] Just which areas he visited, and when, are not entirely known. Kansas, apparently, was one of them since he owned land in Parsons at least two years, 1874 and 1875.

Older sister Louisa had moved in 1865 to Nebraska, where she and her husband had a small farm in Tecumseh, near Lincoln. She urged Volney to consider her area, and he did. Agriculturists were rushing there to try out experimental dry-land farming techniques, so it seemed an up-and-coming spot for someone in Volney's line of work. Liking what he saw, Volney and Nellie decided to move west in the summer of 1873.

Passion for Grapes

But one last critical event would take place in Lexington which shaped Volney Munson's life as much as had his college experience and work with C. S. Bell. Volney decided to pay a final visit to his mentor, Dr. Robert Peter, who had moved outside of town after the war to "Winton," an estate that had been in his wife's family for several generations. There Peter had planted vineyards in such profusion that he made and marketed his own wine.

With interests ranging from archaeology to winemaking, Dr. Peter had profoundly influenced the young Volney, who wanted to follow him into research of some kind. But how to reconcile that with his desire for a nursery business, too? In the Winton vineyards, Peter showed him the way.

Volney made the improvement of America's native grapes his lifelong work.
(Photo by Roy E. Renfro)

23

The warm fall weather had ripened his vines, which included most of the then-available American varieties. In this lush atmosphere, as Volney would later recall, "the Doctor ... discoursed freely upon the character of vine and fruit of the varieties." And suddenly Volney had a revelation. "It seemed to me that there might be numerous combinations, which would naturally occur in such a vineyard, and that one could expect some of the seedlings grown from such crossed seeds to turn out better than any in the vineyard, by combination of excellencies of both parents in the crosses. This reflection aroused within me a strong desire to test the matter"—a prosaic and dry explanation of what was a truly astonishing line of reasoning for the times.[26]

There had been as yet little experimentation in breeding and hybridizing American grapes; the American *vitis* was the "red-headed stepchild" of the grape world, scorned for its foxy taste. The development of the Concord grape in Massachusetts and the Rogers hybrids of the 1840s had been the most exciting work done thus far. What Volney proposed in that sunny Kentucky vineyard was astonishing in its daring and vision.

Dr. Peter obligingly gave his former student clusters of thirty or forty varieties, and Volney carefully carried them home, saving and labeling the seeds. In Nebraska, where agricultural experimentation was encouraged,[27] he knew he could use those seeds to create a new kind of American grape.

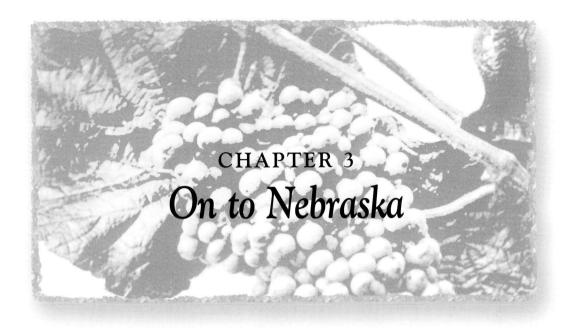

CHAPTER 3
On to Nebraska

$\mathcal{T}$he territory of Nebraska became a state in 1867 when parts of modern Colorado, Idaho, and the Dakotas were carved from it. The Burlington & Missouri and the Union Pacific railroads, which were the first to build through it on the new transcontinental rail route, heavily promoted immigration to the infant state from both Europe and other parts of America. Since Nebraska was the first major semi-arid area in the United States to be settled, and no one knew what would grow there, it attracted men like Volney Munson who were interested in experimentation and dry-land farming. And it seemed particularly ideal for Volney, since grapes were the state's most abundant wild fruit, growing so thickly in some timbered areas as to make them impassable and producing quantities of fruit sufficient for a young winemaking industry.

Lincoln

The capital of Lincoln was established in the eastern half of the state along Salt Creek, where it lies in the bottom of a "bowl" surrounded by low hills which stretch toward the Platte River. By December of 1870, when the Burlington & Missouri reached it, Lincoln boasted a population of 2,500. The town's growth had been so rapid that wild game had only recently stopped roaming the streets when Volney arrived on September 23, 1873. (Nellie stayed behind in Astoria, Illinois, to await the birth of their third child, William Bell, or Willie, on December 15; the two joined Volney later.) Yet the capitol had already been built, classes had begun at the state university, and there was a state asylum and penitentiary. Business boomed with six hardware stores, eight grocers, six dry-goods dealers, mills, packing houses, hotels, an opera house, and half a dozen weekly and daily newspapers.

Volney took lodgings at the Commercial House, a genteel establishment which reminded him of the Old Phoenix on Lexington's Main Street, "excepting the white instead of the colored waiters."[1] For a week he was courted by real estate agents who showed him all over the city and countryside. He ultimately purchased from Oscar Law thirty acres on Vine Street near Wyuka Cemetery for $1,300, and declared the land to be conveniently located and with a view equal to any.[2] It would also increase in value when the University of Nebraska bought a nearby property as the test farm for its new College of Agriculture. Charles Bessey, the college's dean, later described the area which encompassed the Munson farm: ". . . the whole distance from the Antelope Valley to the [College] Farm was filled with cornfields, wheat fields, orchards, and even wild and unbroken prairie land. In muddy weather one had great difficulty traversing the soft dirt roads and it was a bad hour[']s drive from the city."[3]

The Lincoln Farm

Volney frugally continued to use the journal he'd started in college; within its pages is a detailed layout of the Lincoln farm. Two and a half miles from the market square, the property was bisected north/south by Main Avenue and east/west by Orchard Avenue.[4] North of Orchard, Volney partitioned the farm in two large rectangles for fruits and vegetables. The western and southern portions, labeled "Forest," were probably where he grew his nursery trees. To the east was a long, narrow picnic grove bordered by a "Shady Walk," and beyond that was the vineyard. A triple row of evergreens on the north and east served as a windbreak, along with an osage hedge on the south and willows on the west.

Fronting on Vine Street, and just east of Main, was the house, a "neat cottage" roughly thirty-four by nineteen feet. His "Cottage of Two Rooms with Basement," as drawn in his journal, places the kitchen and dining room in the basement of the house along with the cellar. Interestingly, the first floor has a library, living room, front hall, and closet—but no bedrooms. No doubt Nellie pointed out this error, for Volney noted that the house was built according to his drawings but "with some changes" for about $1,000. In front was a half-circle driveway with an extension to the stable, while paths led to Nellie's chicken yard and to the "Vegetable & Fruit House," probably a storage building.

Volney later added a copper lightning conductor—another of his inventions—to protect the house from fierce prairie storms. He twisted copper wire into a rope and ran it up one side of the house, along the roof, and down the other side. On the roof were three or four "pointed projections" which looked like barbed wire gone mad, several strands twisted together with the top cut and spread to expose individual, sharpened wires.[5]

Starting a Business

Volney was quite pleased with his new home, the "broad, rich, rolling prairies covered with high luxuriant grass" which fed so many fat cattle and the "great amount of balmy sunny weather." Really on his own for the first time in his life, and excited about his plans, he saw everything in a rosy light: the soil was deep and rich, the roads dustless and mudless, and the winter mostly pleasant. Even the prairie fires that regularly scoured the landscape excited his aesthetic sense.[6] He wrote several glowing reports on Nebraska to the *Kentucky Gazette.*

He kept a journal of life in Lincoln, the only one of his diaries that has survived, and

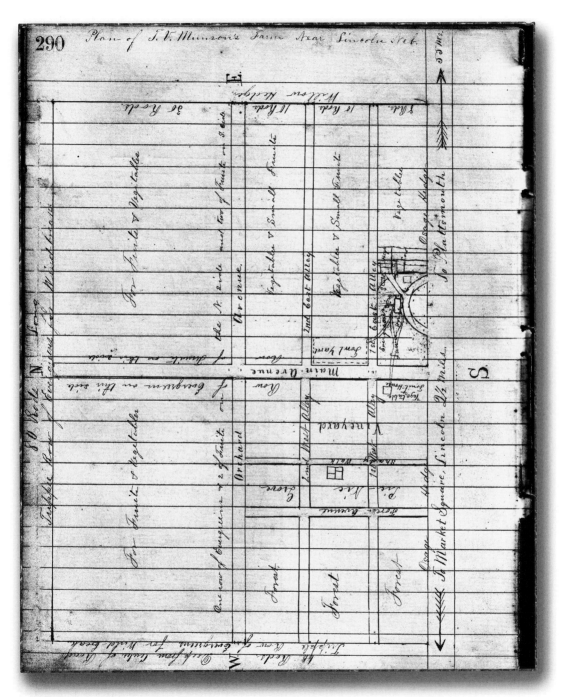

Volney drew this meticulous layout of the Lincoln property in his journal.
(T.V. Munson Viticulture and Enology Center)

Cottage of Two Rooms with Basement.

Front Elevation.

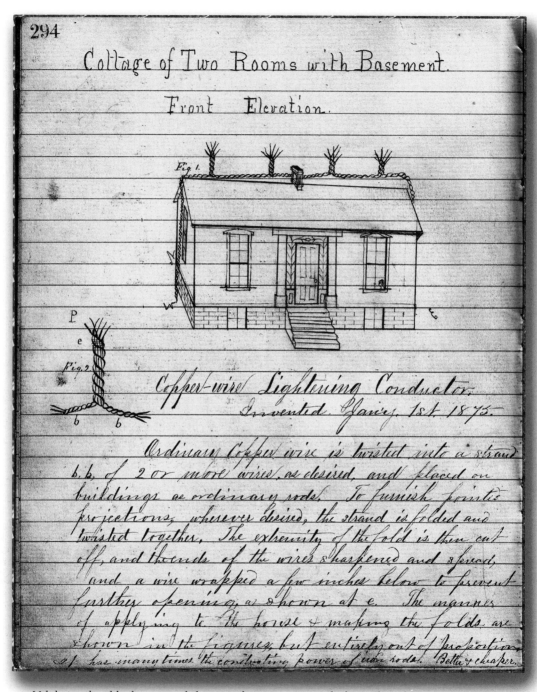

Fig 1.

Fig 2.

P
e
b b

Copper-wire Lightening Conductor.
Invented Janry. 1st 1875.

Ordinary Copper wire is twisted into a strand b.b. of 2 or more wires, as desired, and placed on buildings as ordinary rods. To furnish pointed projections, wherever desired, the strand is folded and twisted together. The extremity of the fold is then cut off, and the ends of the wires sharpened and spread, and a wire wrapped a few inches below to prevent further opening, as shown at e. The manner of applying to the house & making the folds are shown in the figures, but entirely out of proportion, as has many times the conducting power of iron rods. Better & cheaper.

Vol designed and built a copper lightning rod array to protect the house from fierce prairie storms.
(T.V. Munson Viticulture and Enology Center)

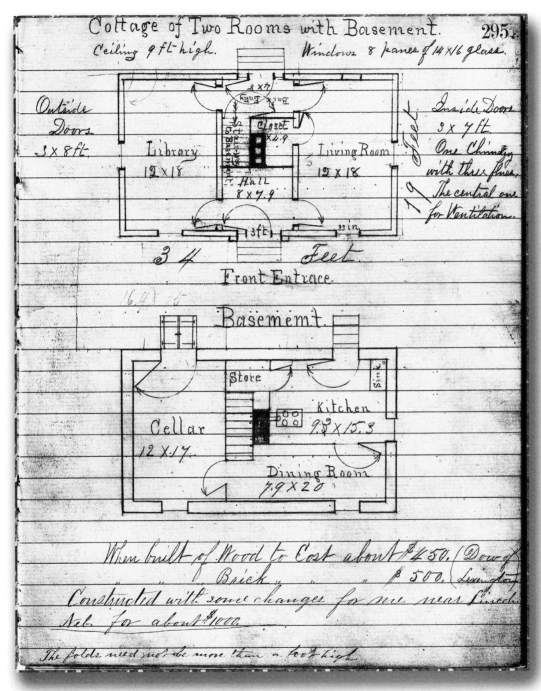

Vol and Nellie's Lincoln home was very similar to this plan he'd drawn.
(T.V. Munson Viticulture and Enology Center)

then only for 1875. It includes a meticulous listing of what he planted, including the trees he would sell to a prairie population hungry for wind- and snow-breaks: "Beginning near Main Drive between 1st & 2nd East Alleys. Raspberry, Gooseberry, Ash, Box-Alder [sic], Barberry Common and Evergreen, Coffee Bean, Catalpa, Cedar, Juniper, Chestnut, Pecan, Walnut, Apple, Peach."[7] Next came a record of the grape seeds—Catawba, Concord, Muscadine, and other varieties. This vineyard included Robert Peter's grape seeds, brought from Kentucky, which he planted the first spring. Volney put in a vegetable and market garden to feed his own household and for surplus to sell, laboriously hauled manure and straw for the beds, trapped and killed rats, and built cold frames. He regularly walked the few miles into Lincoln to pick up mail, seeds, and supplies and to peddle his produce and poultry.

He began to form a circle of friends which included the Laws (from whom he and Nellie bought their land), Prof. S. R. Thompson of the university,[8] and others. He also involved himself in the happenings of the community. When the Independent Literary Society organized in December of 1875, Volney attended and immediately joined the debate group, speaking for the negative view on the "Patent Law question." Earlier that year he helped organize a new school district (No. 87) for his area, was appointed to its governing committee, and drew up plans for the schoolhouse, which featured a ventilation system the *Daily State Journal* of Lincoln considered worthy of mention.

Members of Volney's and Nellie's families made the long journey west to visit at least once. Maria Linley Munson traveled from Illinois in the summer of 1875 to see both Volney and daughter Louisa in nearby Tecumseh. And George Bell came out from Kentucky in the early fall.

Hoppers and Drought and Money

But Volney's rosy opinion of Nebraska waned fast, for he and Nellie had picked the worst possible time to move there.

The hot, dry prairie winds of late July 1874 "cooked" his nursery and garden, including Volney's precious Kentucky grapevines. Then the grasshoppers arrived and devoured almost everything that remained green. The air was filled with swarms of hoppers so thick that they literally hid the sun; within hours they had eaten every plant in their path, leaving behind only the hole in which it had grown. The trains stopped running when the rails grew so slick from hopper carcasses that derailments became a threat. Fish and poultry stuffed themselves until their flesh became inedible to humans. Even after the major swarms moved on, the insects' eggs continued to cause problems for several more years.

The endless wind was so aggravating that Volney blamed it for almost every ill. His health began to fail again in 1875, and he complained often of feeling poorly, possibly from the monumental task of trying to break through the hard, compact prairie soil to plant. The January entries in his journal are filled with references to a "bad day" and his being "not well." By March he was really depressed; "sat around the fire and grumbled at the weather," he noted sullenly.

The national economic depression hit Lincoln hard while it was still reeling from the severe drought of 1873 and the hopper invasion of 1874. Scandal surrounding the state government sent many investors scurrying back east and further contributed to the eco-

Feb. 4. Made a drawing of Cycloidal Pendulum P. 331. Went to town and got cucumber seed from Henderson & Co. also a card from N H Spooner stating he got my order (of 25¢) but no cash. (stolen.) Nellie read Physical Geography at eve.

" 5. Caught a Rat and a Rabbit in traps. Wrote letter to C. Douglas and card to N H. Spooner. Mr. Goodrich borrowed wagon and $5. Killed pig and cut it up. Rec'd papers and pamphlet from com. of lands of A. & P. R. R. St. Louis, which we read at night. Fully in the notion of going to Mo.

" 6. Mr. Goodrich returned wagon and my note given him last year for $185.84. He got a ham and spare-rib. Made 14 lbs. sausage. Wrote two letters to Mo. enquiring into climate &c to Albert Newman of Rolla and Ira Haseltine of Springfield.

S. " 7. Read papers. Grumbled at the weather and climate. Nellie wrote to Bro. G. K. Bell.

" 8. Got out stuff for Cultivator. Hauled 1 load manure. Got Gregorys; Hovey & Co & Dreer's Catalogues. Read papers, Catalogues and Physical Geography in the evening. Nellie Washed.

" 9. Hauled 2 loads manure. Received Briggs & Bro. Catalogue & Fannie's Home Journal which we examined in evening. Put floral pictures in "Scrap Book." Received a letter from Father and Mother (at Astoria Ill.)

" 10. A terrible storm of wind prevailed all day from N.W. Sat by the fire and read papers Nellie read Physical Geography in evening.

" 11. Got Bran. Rec'd letter from Dr. Douglas & wife. Read Geology and Physical Geography.

Volney kept meticulous journals, but this is the only one that has survived.
(T.V. Munson Viticulture and Enology Center)

nomic decline. Volney was having trouble making ends meet. His tax records for 1874 and 1875 show him with a minimum of household furnishings, one or two horses, and a cow. On a March day in 1875, he and his team were almost injured when the horses slid on ice. He'd been careless, he admitted in his journal, and besides, the horses needed new shoes he couldn't afford.

Still he hung on, planting a new orchard in April and scouring Salt Creek for native grapes that he could transplant.[9] He had already made his first discovery, that the wild grapes were much healthier than *viniferas* or varieties from the northern United States. Yet, undaunted by the loss of the Kentucky vines, "the kindled flame of passion for experimentation continued to burn" in him.[10] Volney ransacked "the timbered belts along the streams of that bleak country" for grapes, marking the best for removal in the proper season. Those he transplanted into his vineyard and hybridized with "larger berried kinds," but drought, cold, and more grasshoppers eventually forced him to give up.

More Inventions

Perhaps spurred by economic necessity, Volney invented a number of agricultural tools in Lincoln, detailed drawings of which adorn his journal. A paddle churn and a "Combined Cultivator and Subsoiler" were among the items which occupied him in the early months of 1875.[11] The illustration of his automatic dumping cart featured a pipe-smoking driver presumably modeled on himself, while his revolving shuttle sewing machine was, he declared, lightweight, simple, noiseless, and efficient. His labors on the Lincoln farm also led him to a new "Single-hand Drill, For sowing hot beds, small flower, vegetable & other beds."

Dexterous with his hands, Volney fiddled with anything. He made a baby-chair from a box for little Willie and glass cold-frames for his plants to help them withstand the Nebraska winter. (Volney called them hot-beds.) On one February evening, he even measured himself and "cut shirt pattern for self." He fashioned a set of chessmen and a chess board so he and Nellie could play. Many nights while he worked, Nellie read aloud from such books as Lubbock's *Prehistoric Times,* Alexander Pope's *January and May,* or *Physical Geography.*

He also continued the writing he'd begun in college, sending brief articles and letters to the *Nebraska State Journal* and *Farmers Home Journal.* His first published article appeared in the latter in 1873 and concerned his experiments with potatoes. It shows his growing fascination with research. "Wishing to determine the effects of various methods of Potato Culture, I arranged a plot of ground . . . so as to eliminate every influence except the one whose effect I wished to know." The article also clearly illustrates Volney's confidence in his own methods. "Some ignoramus," he wrote scathingly, had recommended one particular method. To be fair, Volney had tried it even though "I knew it to be directly opposed to vegetable physiology."[12] Sure enough, it hadn't worked.

Looking at Missouri

In his unhappiness with Nebraska, Volney began looking elsewhere for a good place to settle. By the end of January 1875, he and Nellie had decided on the fruit region of cen-

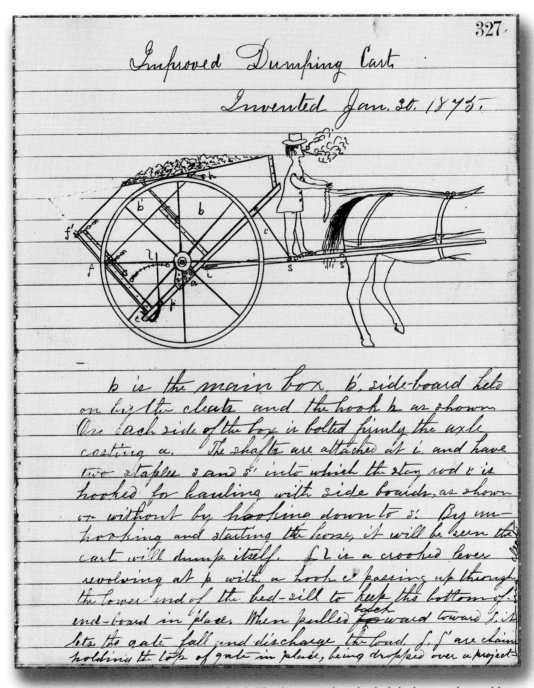

Improved Dumping Cart

Invented Jan. 30. 1875.

b is the main box. b. side-board held on by the cleats and the hook h as shown. On each side of the box is bolted firmly the axle casting a. The shafts are attached at i. and have two staples s and s' into which the stay rod v is hooked for hauling with side boards, as shown or without by hooking down to s'. By un-hooking and starting the horse, it will be seen the cart will dump itself. f l is a crooked lever revolving at p with a hook c passing up through the lower end of the bed-sill to keep the bottom of end-board in place. When pulled forward toward f. it lets the gate fall and discharge the load. f. f' are chains holding the top of gate in place, being dropped over a project-

Volney's cart relied on gravity to dump its load. The driver simply unhooked the horse and moved him out of the trace, and the cart tipped over of its own weight. (T.V. Munson Viticulture and Enology Center)

Combined Cultivator and Subsoiler.

Invented Feby. 1st 1875

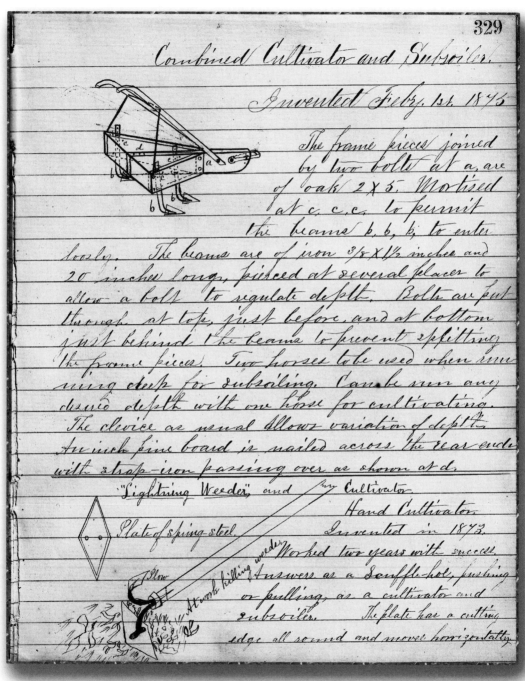

The frame pieces joined by two bolts at a, are of oak 2×5. Mortised at c, c, c, to permit the beams b, b, b, to enter loosly. The beams are of iron 3/8×1½ inches and 20 inches long, pierced at several places to allow a bolt to regulate depth. Bolts are put through at top, just before, and at bottom just behind the beams to prevent splitting the frame pieces. Two horses to be used when running deep for subsoiling. Can be run any desired depth with one horse for cultivating. The clevice as usual allows variation of depth. An inch pine board is nailed across the rear ends with strap-iron passing over as shown at d.

"Lightning Weeder" and dry Cultivator.

Plate of spring-steel.

Plow

At work killing weeds.

Hand Cultivator

Invented in 1873.

Worked two years with success. Answers as a scuffle-hoe, pushing or pulling, as a cultivator and subsoiler. The plate has a cutting edge all round and moves horizontally.

"The Lightning Weeder & Cultivator," another of Volney's inventions.
(T.V. Munson Viticulture and Enology Center)

tral and southwestern Missouri "when we can sell here." In his meticulous way, he wrote acquaintances in Rolla and Springfield for more information on those areas and was pleased with their responses. Rolla, he noted in his journal, seems "a fine place to live pleasantly and economically." Encouraged, he sent for a map of the state and wrote more letters, this time to Carthage, west of Springfield.

But the simple desire to go to Missouri didn't mean he could. Though he showed the Lincoln property several times in 1875, it didn't sell. Volney settled down grimly to make a living and enough money for an eventual move. He tried to hire field hands to help him but was repeatedly disappointed by workers who didn't show up or were, in his opinion, lazy.[13] The sheer labor of operating a market garden and nursery was staggering for a man alone. In the spring of 1875, Volney planted thousands of cabbages, seven hundred tomato plants, and an unknown quantity of beets, onions, peas, salsify, asparagus, turnips, and other vegetables. He "heeled in" chestnut trees (many of which were killed by a subsequent frost), planted shade trees around his own house, and set out apple tree saplings in the orchard. The business was a never-ending round of hoeing, weeding, mulching, and starting all over again. He commented wryly in July, on returning from a brief visit to see Louisa, that the weeds were "growing wonderfully."

Nor were they the only enemy. Scarcely a day passed that Volney didn't write, with seeming satisfaction, of drowning a squirrel and poisoning rats, of running off wolves from the chicken house. During the early summer, his journal is filled with remarks on various pests plaguing him and his crops. Though the great grasshopper swarm had occurred the year before, they had left behind plenty of eggs. "June 11. Hoppers ate up Mr. Bostetter's melons and coming toward my garden ... June 12. Burned old hay to keep back hopper[s] coming from Bostetter's ... did little good ... June 19. Hoppers have eaten off most of my late cabbage." In the fall of 1875 the horses were afflicted with epizoons, an external parasite. Then there were the perennial dangers of prairie fires in the spring and fall months: "Apr. 19. Back-fired against a great fire on the East."

Volney sold both nursery plants and produce, and his market business grew steadily. By the end of May, despite the freezes that had lasted until the first of the month, he was carrying lettuce, radishes, onions, and mustard into town. Sales increased over the summer thanks to an "astonishing crop of potatoes, melons, squashes, and. c [etc.]," and Volney gained "a fine class of customers." Astutely, he displayed his better vegetables in Lincoln shop windows and at the railroad's land office; and he took samples to the *Nebraska State Journal* office, which obligingly wrote up some kind words about the quality of his produce. Sales were good enough that, in October, Volney indulged and bought a stove for the sitting room to make winter evenings there more enjoyable. But the cost of the stove was just a drop in the bucket against his debts.

In January 1874, Volney had taken out a mortgage with Oscar Law for $800. The end of that same year, he sought another one, this time for $922.41, from Eldon Snodgrass. Both, presumably, were to help him build his house, get his nursery business started, and tide him over from the disastrous summer of 1874. Even so, he couldn't borrow enough to keep himself afloat.

Owning a business meant that Volney was constantly dealing with unpaid customer accounts, too. "Aug, 3, 1875 ... Saw Deland and drummed him for money due me for trees."

Family Life

Still, life for the Munsons was not all work. Maria's visit was a pleasant interlude, and Volney, Nellie, and Willie regularly ventured into town to see traveling shows and even a circus and to have a family photograph made, which they hung in the parlour. When the schoolhouse of Volney's design was completed, Nellie baked several cakes and pies for the "sociable" which opened it. "Had a nice supper and pleasant time," Volney reported. He continued to speak at the literary society; his diary entry for November 8, 1875, for example, states laconically: "Gathered 3 large loads corn. Talked on Spiritualism and c. [etc.]."

The family made excursions into the countryside around the penitentiary, where they picnicked on Salt Creek and went on fishing parties. In the early fall they picked wild plums and grapes with friends. Another friend gave Willie a dog but "he tore too many clothes ... to be profitable" and was given to someone else.

Letters from family in Kentucky, Texas, and Illinois were welcome enough to warrant mention in the journal; Volney and Nellie spent many an evening returning correspondence.

However, a string of low-grade illnesses nagged the family all through the summer of 1875. Willie was sick several times, including a bout with dysentery, and Volney finally had to have a wisdom tooth extracted. He was, he wrote, weak from overwork and suffered from a "sore liver." Even Nellie complained of feeling poorly. But by fall, Volney was recovered enough to tackle the seasonal chores: digging cellars to store potatoes and squashes, banking the dependencies such as the chicken house with straw to keep out the wind, and bringing in the last of the harvest, which included more than 200 bushels of winter beets.

Going to Texas

On December 19 the Munsons finally sold their Lincoln property to Henry H. Grimes for $3,000, subject to Grimes assuming the two mortgages Volney still owed.[14] Despite the good summer of sales and the low cost of living in Lincoln, he was undoubtedly happy to get the money, for it allowed the family, at last, to move to Texas.[15] However, there were two big problems. The title to the property was not clear, and Volney spent some weeks trying to get it straight. (He was still wrestling with it when his journal ended on December 22.) And Nellie was pregnant with Fern, so they stayed in Lincoln until their daughter was born on March 1, 1876. It isn't clear whether Grimes allowed them to stay on in the house or if they took lodgings.

Neither is it known how they got to Texas. Fern told her children that they traveled by covered wagon since rail connections were torturous to nonexistent, and the family had no money for train tickets anyway.[16] But that would mean they left as soon as the baby was born, whatever Nellie's physical condition (which doesn't seem likely) for they arrived in Denison on April 6, "the most important date in Southern viticulture," as modern gardening expert William C. Welch declared it.[17]

Why did Volney give up on Missouri? Certainly he retained his interest in that state for many years to come and would, in fact, work closely with several viticulturists there. Likely Ben was the clinching argument for Texas, for the two brothers were very close. By this time, Ben had started a second business in real estate with his law partner and friend,

Jot Gunter, which proved so successful that the two had virtually discontinued their law practice.[18] It appears that he offered Volney a chance to work with them, for there are several deeds from the period 1873 to 1878 which link the three men. Family lore also affirms that Volney worked with Ben to earn enough money to establish his new nursery.

The Nebraska years, though short and mostly unprofitable, demonstrate several of Volney's distinguishing characteristics. He was willing to take risks to further his business and career, or he would have stayed safely in Kentucky working for his father-in-law. He complained a great deal about the ill health and assorted accidents that plagued him most of his life, but he refused to wallow too much in self-pity: he got up and did the work his business required of him.

CHAPTER 4
Gone to Texas

*T*exas may not seem a likely spot to interest a man who wanted to grow grapes; but viticulture, or grape growing, actually has a long history in the state, dating back to the late seventeenth century.

The Vine in Texas

Wild Muscadine, Mustang, and Post Oak grapes were among the species of *vitis*[1] familiar to early Indian tribes of the area. Later European visitors to Texas remarked on the abundance of wild grapes "whose branches climb to the tops of the trees and are covered with clusters of wild grapes, somewhat unpleasant to the taste, from which the natives make an exquisite vinegar."[2] With the aid of irrigation, Spanish missionaries in the El Paso area introduced the first cultivated European grapes, *Vitis vinifera*, in the mid-seventeenth century. Known as El Paso or Mission grapes but probably a Lenoir (black Spanish grape), these were grown to produce sacramental wine and were themselves descendants of vines brought to Mexico from Spain by Cortes. Some brandy was also made. From Chihuahua in old Mexico through Texas and into New Mexico, most of the wine available in the region at the time came from El Paso, which would retain its enological prominence until the late nineteenth century.[3]

Anglo settlers, who began arriving in the 1820s, were pleased with Texas' native bounty; empresario and founding father Stephen F. Austin was moved to write that "Nature seems to have intended Texas for a vineyard to supply America with vines."[4] His cousin Mary Austin Holley also commented on the profusion of grapes: "Almost every variety ... is native in Texas ... but all are of uniform sweetness, and produced in wonderful exuberance."[5]

Other early writers on the state made note of the abundance, too, especially along the

Canadian and Red rivers and in the sandy soils of East Texas. George Wilkins Kendall, in *Narratives of the Texan Santa Fe Expedition,* described trees along the Wichita River bound together by grapevines which had spread from one tree to another. In 1853, *DeBow's Review* waxed lyrical about Texas grapes: "We know there is not a finer country in the world for the cultivation of the grape—not even *la belle France.* ... We are confident that the vine can be cultivated to a far greater advantage in the undulating region of Texas than in any other part of the Union."[6] H. C. Williams, an employee of the U.S. Patent Office (forerunner of the U. S. Department of Agriculture), studied the native grapes of the Southwest in 1858 and prophetically proclaimed West Texas "the Eden of the Grape." Geologist and botanist S. B. Buckley concurred, declaring forthrightly in 1861 that the entire Lone Star State was "emphatically the land of the grape."

By the 1850s winemaking using these native varieties was in full swing. Indeed, El Paso's chief industry was producing wine—200,000 gallons in 1853—from the original Mission vines while French colonists at La Reunion (Dallas) also made wine. The many German immigrants during this period brought a tradition of viticulture and winemaking with them, too. This intensified after the Civil War when Texas experienced a surge of im-

New leaves emerge in the spring from a Munson grapevine in the T.V. Munson Memorial Vineyard. (Photos by Roy E. Renfro)

migration from all across Europe. Scorning the tart, foxy taste of the natives, these new settlers brought cuttings of *vinifera* with them; some vineyardists even attempted to cross the two.

But *vinifera* didn't like the Texas climate and most of its soils. (In fact, *vinifera* disliked all of the United States except California, which is the reason for that state's long wine-making history.) Even the Germans were eventually forced to give up and make Mustang wine with much more sugar than was ordinarily used, to enhance the fermentation process. Texas grapes, they decided, were fine for eating but useless for wine, and their cultivation would never be a major industry.

"My Grape Paradise"

Volney Munson took a different approach, however, for he'd seen those grapes that grew along the Red River.[7] He found "a rough piece of dark limestone, timbered land on the bluffs of Red River. . . . In the woods surrounding, innumerable wild grapes grew." On the higher, sandy lands were Post Oak grapes (*Vitis lincecumii*)[8] which climbed the mighty oaks and hickories that grew there, while Mustangs (*V. candicans*), Sour Winter, or Frost, grapes (*V. cordifolia*), and Sweet Winter grapes (*V. cinerea*) flourished along the river bottoms. There was also the Sand grape (*V. rupestris*) and the Riverside grape (*V. riparia*). "Here were six or eight good species of wild grapes, several of which had not been seen by me previously. I had found my grape paradise! Surely now, thought I, 'this is the place for experimentation with grapes!'"[9]

What Munson had found was something rare, for North Central Texas[10] comprises an ecotone—a transition area between several ecological communities—in this case, the eastern deciduous forests and western grasslands. Modern botanists divide its 40,000 square miles into six vegetational areas: the Blackland Prairies, Eastern Cross Timbers, Fort Worth Prairie, Lampasas Cut Plain, Red River Area, and Western Cross Timbers. This extreme diversity of soils and climates means that a wide variety of both eastern and western plants and their natural hybrids can be found here. And of the fifty counties that make up the region, Grayson, where Volney had decided to live, is among the most complex geologically and botanically.

The major portion of Grayson County lies within the Blacklands, more than 17,000 square miles of grasses and alkaline soils stretching south from Sherman (the Grayson County seat) to San Antonio. Its heavy black, waxy dirt is the result of the weathering of the white Cretaceous rocks underlying it and is excellent for growing cotton. The extreme western edge of the county is part of the Eastern Cross Timbers, a narrow band of hardwood trees extending south nearly to Waco and north across the Red River into Oklahoma. Historically, its understory growth was so impenetrable in places that the Cross Timbers proved a natural barrier to Anglo settlement. Within its acidic and rocky soils are the Woodbine sands, which give rise to many springs and make it a natural area for grapes.

Lastly, the northern edge of Grayson County falls within the sandy riparian soils of the Red River Area, the smallest of the vegetational zones and one that is found in only three counties of North Central Texas (Grayson, Fannin, and Lamar). Here, the ecotone aspect is most pronounced, for this part of Grayson also contains a geological formation known as the Preston Anticline, an ancient fold in the earth which brought deeper strata to the surface and created rugged "canyon-like valleys." These are perfect microhabitats where unusual species and natural hybrids, including several at the limits of their biological ranges, occur.[11]

Seeing the many grape hybrids that had naturally developed here excited Volney and rekindled in him the desire to experiment which he thought had died on the Nebraska prairie. He knew little had been done in the past to improve the American grape, and that little "seemed so inadequate to the needs and capabilities of the country, that I determined to devote a portion of my time and ground to aid the study and development of vine-culture in America."[12]

Staminate.	Pistillate.	Pistillate.	Pistillate.	Staminate.	Staminate. (Upper)
Male.	Female				Pistillate. (Lower)
Vitis Simpsoni.		V. Baileyana.	V. Berlandieri.	V. cinerea.	V. rotundifolia.

This plate from his 1909 book illustrates the flower stage of some of the native grape varieties Volney used in his experimental work. (From *Foundations of American Grape Culture*)

American Hybrids

The science of hybridizing grapes was still in its infancy in the United States and, indeed, in the world when Volney made this resolve.[13] Serious viticulture in this country began in New York State (the Finger Lakes and Hudson River Valley) and Missouri in the early 1800s and proved so successful that soon viticulturists were searching for better seedlings of commercial value.

To growers with any botanical background, it became clear that many of the native grapes being commercially cultivated in the U.S. were actually hybrids themselves which had occurred on their own through natural processes. But how to do the job in a deliberate and conscious manner and under the control of man? The grape is a malleable fruit which reacts well (almost too well, making it difficult to control cross-fertilization) to out-

41

side forces and is easily altered. However, the seeds are notoriously unreliable; offspring seldom inherit the best characteristics of both parents. Consequently, thousands of delicate, time-consuming hybrid crosses may be necessary to achieve one acceptable variety. (Volney himself considered one for every thousand seedlings an excellent return.)

The first important American research into artificial grape breeding came in the 1840s and 1850s with the introduction of Diana (a seedling of Catawba) and of the Concord grape, neither of them true hybrids (that is, a cross between different species) but promising creations nonetheless. Concord proved an especial favorite and marked "the beginning of the popularity of the grape as a dessert and culinary fruit, whereas previously it had been used almost entirely for wine manufacture."[14]

Dr. William W. Valk of Flushing, New York, developed the first authentic hybrid vine and the first one deliberately hybridized in this country in 1852. "Ada" resulted when he crossed Isabella, a *labrusca* introduced in 1816, with Black Hamburg, a *vinifera*. John Fisk Allen of Salem, Massachusetts, developed another successful hybrid two years later by crossing Isabella with Golden Chasselas, which consequently stimulated great new excitement about and interest in grape hybridization. And within another two years, E. S. Rogers, also of Salem, exhibited the first of the hybrid grapes which would make him famous.

These successes accelerated the pace of grape experimentation by both amateurs and professionals to unprecedented heights: it is estimated that 2,000 new American hybrids were introduced before 1900. Much of this hybridizing research was carried out in the eastern United States and relied on *labrusca* varieties, the native grape of that region and thus easily available but known for its tart, foxy taste—not good wine material. It was in the western sections of the country that plant breeders first turned to other native species to develop grapes more suited for their region. Jacob Rommel, who began his work in Missouri in 1860 and would prove to be one of Volney's major influences, was the first to use *riparia* in his crosses and the first to hybridize American grapes on a large scale. He would be followed by two other Missourians, Hermann Jaeger and George Husmann, and by Volney Munson.

A Life Plan

So Volney began to lay out his life's work: to improve the American grape. A short and simple statement but a staggering job. He divided the task into two parts: (1) describe and classify all the grapes, and (2) develop new varieties through scientific cross-breeding of both native grapes and commercial cultivars, including *vinifera*. These creations ideally would ripen at different times, thus lengthening the market season, and would combine resistance to disease with quality of fruit and taste. Moreover, they would adapt well to several climates, especially those of the southern and southwestern United States. And because he had now found his "grape paradise" in Texas, Volney could concentrate on noneastern, non-foxy species, which set him apart from many other American hybridizers of the time.

Saying he would describe and classify American *vitis* was another deceptively simple statement to make. Differentiating between species and varieties of grapes is extremely difficult since they hybridize themselves (cross naturally) in the wild. At the same time, the very definition of "species" was still unclear as it related to *vitis*. As late as 1880, even so authoritative a volume as *The American Cyclopaedia* admitted that "there is at present some confusion about the species." After all, as Volney himself recognized, the distinctions were simply based on arbitrary bounds established by someone in the past and easily redefined many times.[15]

The renowned Swedish botanist Carl von Linnaeus (1707–1778) was the first to attempt a botanical nomenclature of American grapes in 1753. By the mid-nineteenth century, naturalists had become obsessed with such classifications, seeking to make sense of nature by fitting all plants into neat and unchangeable compartments. This strongly conservative approach can be seen in the first grape nomenclature published by George Engelmann of Missouri in 1869. He declared there were ten species of native species and refused for years to budge from that stance. (He grudgingly increased the number to thirteen in 1883.) Volney Munson, then, proposed to blaze new trails in both a strictly botanical aspect and in his philosophical approach to that work.

Making a Home

Before he could begin this lifetime's work, Volney had to build a home for his family. Like most newcomers to the railroad boomtown of Denison, the family probably lived first in a tent or covered wagon since cut lumber for construction was very scarce. Volney selected a forty-three-acre tract just north of Denison on what is now U.S. Highway 69[16] and, atop a rocky sandstone bluff, built a frame cottage very similar to their home in Lincoln. He and Nellie named it "Scarlet Oaks." Fruit orchards eventually embraced the house on

Denison was a rough-and-ready young railroad town when the Munsons arrived. Note the pig hanging on posts in the foreground. (Red River Historical Museum, Sherman, Texas)

two sides, with vineyards and nursery below. The several outbuildings included a shop, large barn, and an office, all close to the house.

And while the Munsons were newcomers, they were not without local, familial support. Trite and Jennie, Volney's sisters, were already in Denison, where they kept house for Ben, still a bachelor. Theo had also joined his siblings and started a land abstract business, Tone & Munson. After Ben married Ella Newton on September 3, 1876, Jennie moved with them to nearby Sherman, while Trite spent most of her time in the country with Volney and his family. Their parents, William and Maria, arrived in the fall of 1878; the three unmarried children then moved with them to the homeplace William built at what is now Houston and Parnell streets.[17] Volney had once owned this property across the road from Scarlet Oaks but had sold it in 1874 to his father.

Soon after Volney, Nellie, and the children settled into their new home, they all became ill, a situation which continued for nearly two years. Part of the problem proved to be drinking tainted spring water before a new well was completed. Another cause was working throughout the day, as they'd been accustomed to doing in cooler climates, without allowing for the more intense Texas heat. All eventually recovered.

Volney and his young family stand on the porch of "Scarlet Oaks," their first home in Denison.
(T.V. Munson Viticulture and Enology Center)

To Business

Volney dabbled in land sales with Gunter & Munson, but his real business lay within the earth and, despite his illness, he began improving the north Denison property. The first summer it seemed as if the grasshoppers had followed them from Nebraska; Volney finally saved his garden by using a homemade concoction of Paris Green (a bright green powder containing an arsenic compound) and flour. (In 1877, after further experimentation, he developed a spray for grasshoppers and sold the recipe to customers for fifty cents each.) The market garden soon began to yield produce he sold in town, regularly walking the mile each way with his baskets since he had no money for a horse. This was the humble beginning of what would become the world-famous Denison (later Munson) Nursery, the largest in the South. For at least one year, 1877, Volney ran the nursery with C. L. Edward of Denison, but it appears to have been a short-lived partnership.[18]

Volney kept meticulous records on his nursery to see which plants were successful and which were not.
(T.V. Munson Viticulture and Enology Center)

Like a father watching his children grow, Volney kept meticulous lists in his journal, noting what flourished and what didn't in this new climate, since there was, as yet, no reliable horticultural information on North Texas to guide him. "Peaches Budded 1877" listed the varieties he tested in the Texas sun, Alexander, Early Rivers, Cooper's Mammoth, and Stump the World among them. His fruit orchard contained well over five hundred plum, cherry, crabapple, mulberry, peach, and pear trees; they, too, were platted out in the journal. He grew about three dozen varieties of grapes and noted them, row by row, in "Record of Vineyard Planted in 1878 & 1879." (This first site was two acres in size; in the 1881–82 season, he nearly doubled it.) Also in 1878 he planted dozens of different types of strawberries, all recommended by his friend and colleague Samuel Miller of Sedalia, Missouri.[19]

For the next fifteen years, Volney devoted late summer and early fall to his wild grape studies. On foot and horseback, he criss-crossed Texas and the Indian Territory, using the "cowpaths" that ran through the woods and seeking the finest examples of native grapes. He tasted "thousands of different varieties, saving seeds of a few of the best, and marking their vines to be later transplanted into the vineyard for experimental purposes."[20] After

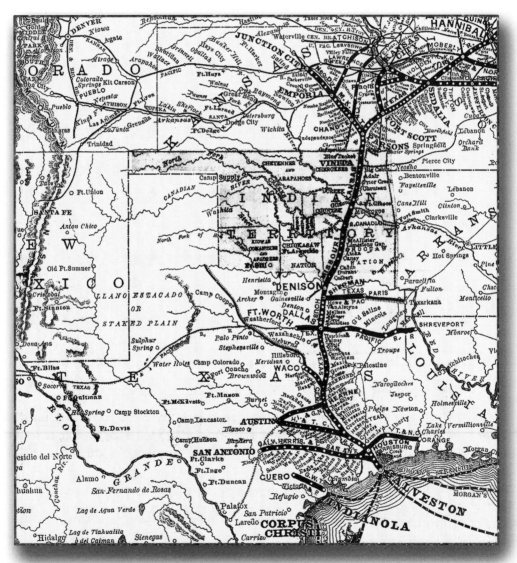

Denison's prominent location on several railroad lines made it easy for Volney to travel and to ship his nursery stock. An 1877 map of the M-K-T Route. (Rand McNally map from collection of University of North Texas)

those vines bloomed and were cross-pollinated, Volney planted the seed of the mixture and culled the resulting plants to find the best "parent" stocks, a process which took many patient years. Often he crossed first-generation plants with native cultivars rather than other new seedlings to maintain fruit quality. Many of his creations were never sold but were kept to breed even more varieties. He also studied the original plants in their native habitats to determine which climatic and soil conditions were most favorable for each species, characteristics he meticulously recorded and used in his breeding program.

As the nursery business and, consequently, his finances improved, Volney expanded his search to other parts of Texas and then to other states. Writing a retrospective of his work many years later, he declared that he had traveled, either purposely on his studies or incidentally while on other trips, "some 75,000 miles, reaching into every state and territory in the United States ... except six, and considerably into Mexico, I hunted grapes from train-car windows, jumping off to collect specimens at every stop in the wood to water, or coal, or cool a hot-box, or wait for other trains."[21] As a result of these travels, he personally observed almost every species of American *vitis*.

Seeking Others

This field work was the first phase of his goal to describe and classify American grapes. A second step, begun in the late 1870s, involved corresponding with others in the field. This in turn led to his acquaintance, friendship, and even collaboration with leading horticulturists and viticulturists in America and France. All these people would play major roles in Volney's career over the next decade and were of invaluable help to him in developing his own theories.

First he began to weave a network of correspondents, writing hundreds of both amateur and regular botanists and grape-growers throughout the country and requesting specimens of the grapes in their regions; the response was generous. Then he turned to the better known horticulturists of the day.

Gilbert Onderdonk (1829–1920) was a native of New York who moved to Texas in 1851 for his health and settled in Victoria County on the Gulf Coast, where he started the Mission Valley Nursery (1858). Onderdonk studied Texas' native fruits and flowers and became famous for his research on peach breeding. He eventually originated more than eighteen varieties of peaches and plums, including one he named for Volney, as well as other fruits.

It was likely at this time, too, that Volney met a reclusive Swiss-born viticulturist in Neosho, Missouri. Hermann Jaeger (pronounced YAY-ger) (1844–1895?) moved to southwest Missouri in 1865 and planted his first vineyard the next year. As Munson would do later in Texas and Oklahoma, Jaeger combed his region for the best native grapes, especially the Post Oak, to cross with Concord and Virginia from the eastern United States; in this way, he created some one hundred new varieties. He would name his No. 70 for Volney Munson, and Volney, in turn, would use it to create his "America" family of hybrids. Jaeger is thought to have been the first viticulturist in the western United States to spray for fungal diseases on grapes, in 1874.

George Husmann (1827–1902) was also a Missourian whose German father settled in Hermann in 1840; the two experimented over the years with grapes and viticulture.

George founded Bluffton Wine Company (1869) and, later, Husmann Nurseries in Sedalia. He also edited *The Grape Culturist,* the first American journal devoted to viticulture. With Parker Earle, he later helped organize the Mississippi Valley Horticultural Society.[22]

Isidor Bush (1822–1898) was another European who served American horticulture well. His American Wine Depot in St. Louis began making wine from native grapes as early as 1869 and won a medal at the Vienna Exposition of 1873. On the banks of the Mississippi River south of St. Louis, Bush later established a vineyard with the help of George Husmann. The Bushberg Nursery was among the first to send cuttings of American vines to France in the 1870s to help fight the phylloxera epidemic that raged through Europe (see Chapter 6), and Isidor Bush's catalog was one of the most important viticultural publications available. By 1875 he had taken into partnership Gustave Edward Meissner, who would also send vines to France in 1877 from his own vineyard.

A Plague in France

Volney was intrigued by the research surrounding the phylloxera epidemic that had now devastated the vineyards of France and Europe for a decade. As early as 1870 he began corresponding with Missouri state entomologist Charles Valentine Riley (1843–1895), who was honored by the French in 1874 for his research in identifying phylloxera. Their letters were published in Riley's sporadic journal, *The American Entomologist.*

Volney also wrote the leading French viticulturists. Jules-Emile Planchon, a pharmacologist and entomologist who, in collaboration with Riley, identified the life cycle of the phylloxera louse, led a mission to the United States in 1877 seeking help in eradicating or at least controlling the pest (again, see Chapter 6 for more information).[23] Pierre Viala of the French agricultural station at Montpellier near Nimes was another who corresponded with Munson.

Volney's interest in phylloxera was threefold. First, he was intrigued by the ravages wrought by an insect that native grapevines of the eastern United States routinely ignored. What lessons could be learned and applied to his work in hybridization? Could he make his American hybrid vines phylloxera-resistant? Second, in 1880, he lost one of his own vineyards to mildew and black rot, which also plagued the French, and he hoped their work would aid him, too. And last, how could the extensive research being done on grapevine botany as a result of phylloxera help him clarify the classification of American grapes?

Challenging the Experts

Volney had now begun the third phase of his work, a reevaluation of grape botany. "It was at once apparent to me," he wrote retrospectively in 1890, "that a thorough botanical investigation of all species of our wild grapes must be made before much valuable work in this field could be done." In other words, understanding the botanical structure of the grape—its components and their functions—would help him determine the best methods of hybridization and cultivation.

He turned to previous works on the subject, including the standard classification of

American *vitis* by Dr. George Engelmann of Missouri, to identify the species he found on his grape rambles.[24] He even "ransacked" Millardet's and Planchon's work in French on their grapes. But in all of them Volney found "many imperfections and errors." So he decided to develop "a new, more natural and thorough classification."[25] This, he believed, would provide him the "necessary foundation material on which to base American Grape Culture, to fill its sphere fully as it by Nature is capable."[26]

Setting the Stage

In 1880 Volney turned his attention to this project, which would consume the next five years. The nursery business now supported him financially, and he had even published his first catalog in 1879, a one- or two-page listing of his offerings and their prices. That same year he had received a patent for the "Diamond Scuffler Hoe," which he began marketing in May of 1880—another source of revenue.[27]

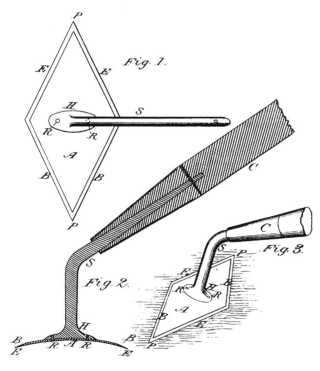

T. V. MUNSON.
Souffle-Hoe.

No. 213,584.   Patented Mar. 25, 1879.

Attest:

Inventor.
T. V. Munson

Volney's hoe remained popular into the 1920s.
(From U.S. Patent Office records)

During the course of this initial botanical research, Volney's family would grow as well as his business. His second son, Warder, named for American Forestry Association president Dr. John A. Warder, was born in the north Denison house in 1878. Neva, who inherited her father's artistic interests, followed in 1881, and Olita two years later.[28]

With this financial and familial stability, Volney could devote more time to his studies. In 1879 he helped organize the North Texas Horticultural Society and served as its first president.[29] He also joined the Mississippi Valley Horticultural Society (later the American Horticultural Society) and was elected a vice-president, representing Texas, the first of many offices he would hold. In addition to the voluminous correspondence which Nellie helped him maintain, Volney visited several key herbariums in the United States: Harvard, the Academy of Sciences in Philadelphia, the National Museum (now known as the Smithsonian), and the Missouri Botanical Gardens in St. Louis. In the latter city, he visited and studied with Engelmann himself and established a relationship with William Trelease, then professor of botany at Washington University and, later, director of the Gardens.[30] In all these locations, Volney studied live specimens in their test beds as well as herbarium specimens (examples of leaves, seeds, fruits, etc. which had been pressed, dried, and mounted on paper). He was pleased with the results as they allowed him to correct old classification errors, discover new species, and find wild varieties unknown to him that looked promising to develop "new families of market, table and wine grapes."

To expand his knowledge and his circle of regional correspondents, Volney also joined the fledgling Academy of Science of Texas, which had formed in Austin in 1880 and moved to Palestine in East Texas shortly after. He maintained his membership after it reorganized in 1892 as the Texas Academy of Science.[31]

Formulating His Theories

Volney spent several years simply accumulating information. As ideas occurred to him, he "tested" their validity with some of his horticultural colleagues. Unfortunately, much of this correspondence has been lost. George Husmann's papers were burned after his death, Samuel Miller's are also gone, as are most of Hermann Jaeger's. However, a recently discovered cache of Jaeger correspondence[32] included twelve complete or partial letters from Volney to his friend that make it clear this small circle of men exchanged information and critiqued theories in their quest to understand the science of the grape.

One of the most fascinating letters, that of May 30, 1884, is seven pages long; in it, Volney expressed his thoughts on the evolution of American grapes. In the very first paragraph, he referred to the "Scientific Viticulturist," a striking illustration of the difference he discerned between the forward-thinking professional such as himself and the amateur who was content to simply muddle along as his predecessors had for centuries.[33] The evolution of grapes interested him, he wrote Jaeger, because it was obvious that, over time, some species must have evolved more closely than others. "If we can discover this specific sequence in time and region, we have [the] key to aid us in 'remarrying' or hybridizing to best advantage these various families." He then drew on geological and biological background to theorize how the varieties likely evolved from east to west; even migratory birds and seeds drifting across the Gulf of Mexico were taken into account. In its most basic form, Volney's theory of American grape evolution was that the older varieties were to be found in the geologically more ancient northeastern states, the younger ones in the South and Southwest. This was not an original idea, however. Sir Joseph Hooker had proposed roughly the same evolution in his 1879 article, "The Distribution of the North American Flora," which Volney had no doubt read.[34] But few others were yet thinking in these terms.

As if unsure of himself—and he was indeed breaking new scientific ground—Volney paused toward the end of the letter to ask if Jaeger had "ever studied the specific development in this way? I know much of such conclusions must be theoretical." But then he became more certain: "The theory I have outlined for the grape in this letter, roughly to be sure, seems the most natural to me. The more I observe and collect on the subject, the more support does the theory obtain." He acknowledged his ignorance of Central American and West Indian species but believed nonetheless that "the US owes much to the Gulf stream for its peculiar flora."

Speaking Out

To pave the way for the eventual dissemination of his new classification, Volney began to give papers at horticultural society meetings.[35] His first major presentation of which there is any record was at the 1883 meeting of the Mississippi Valley Horticultural Society in New Orleans,[36] where Volney's paper on "Systematic Horticultural Progress" challenged members to think beyond the status quo. Cannily, he appealed to the profit motive and to humor to grab his audience and get across his message. "Gentlemen, as much as we all love a grand display and forty feet columns of Ben Davis apples, yet what horticulturist would not rather come here to see and learn of the origination of a single variety of grape, apple, pear, plum, or potato, superior in every point to anything yet produced, than to behold mountains of old and well known kinds? 'Tis well we have planted largely of the standard varieties for market, but must we never try to elevate the standard? This Society can do us no good by piling great tables full of these same old kinds from year to year for us to stare at vacantly, or attend with languid minds the old tale of their growing. We want something new here, yet something better than the old; if possible, something which will *profitably* produce in soils and localities where others fail. In a word, we want this Society to be one of systematic, *profitable* progress." (Authors' italics)[37]

Volney urged members to "choose a field for systematic experimentation"[38] and report their results back to the Society in the form of papers at the annual meeting. The MVHS should reward worthy papers, he declared, since they represented the authors' "triumphing over the elements in exhorting from them some new secret or luscious product." In his rabble-rousing eagerness, Volney envisioned compiling these papers into books, selling them to an eagerly waiting audience, and raising enough money to enable the Society to compensate its "over-zealous, faithful officers." The Society, he suggested, might even become "the University of American Horticulture," large and powerful enough to push for and obtain government appropriations. It should establish experiment stations to test new products and protect the public from horticultural fraud.

No one would ever accuse Volney Munson of thinking small.

All this was necessary, he assured his audience, because "in fifty years more of peaceful growth[,] this land will swarm with people of every clime and nation, and the struggle for the necessities and comforts of life will wax hot. We hope it may not come down to a mere question of 'survival of the fittest' . . . but (by then) . . . we may have learned how to grow two blades, with double the nutriment in each, where but one grew before." He mourned the loss of forests to "the ruthless axe," the rising cost of fresh food, and the loss of fertile soil to erosion and overplanting. Volney concluded by pushing the competitive button of every Southerner in those postwar years: he praised "our Northern brethren" for their foresight in establishing well-regulated horticultural societies.

It's a remarkable speech for its day and clearly illustrates the breadth of Volney's interests and his progressive, scientific outlook. In less than four printed pages, he accurately forecast the population explosion, the need for experiment stations (Texas did not open its first station until 1887), and major famines in Third World countries such as India, which would all surely result from simply maintaining "the standard" in horticulture.

Making a Name

In 1883 and 1884, Volney continued making his name known in horticultural circles. This recognition, he felt, would give him credibility when he published his new classification. His thesis on "Forests and Trees of Texas" earned him the master's degree from his alma mater, Kentucky A&M. The research for it had consumed a good part of 1882 and more than a thousand miles of travel throughout the state. Volney then put the material to good use, revising and publishing it in the *American Journal of Forestry* (1883) and giving a paper on "Trees Peculiar to Texas" to the Mississippi Valley Horticultural Society, which published it in their Society's *Transactions*.[39] He continued his involvement with the MVHS, serving as vice-president and on several committees. He also traveled to St. Louis in June of 1883 to attend the annual meeting of the American Association of Nurserymen and Florists, where he presented a paper—thriftily using his thesis material once more—on ornamental trees and shrubs of North Texas.

He also devoted much time to expanding his business, which was outgrowing the forty-three-acre north Denison location. In the spring of 1883, he and Theo swapped several pieces of land, giving Volney something over 100 acres on the south side of town, where he intended to move the entire operation within a few years. He also bought a small tract on Mirick Avenue from friend and fellow horticulturist James Nimon; Nimon, in turn, bought the adjoining property so the two could work together.[40]

The Denison Nursery kept Volney quite busy. He had developed his first hybridized grape seeds in 1880–81; now the fruits from those began to appear in the vineyard. Murray's Steam Printing House in Denison produced 10,000 copies of the nursery's sales catalog, which Volney mailed out across the country.[41] It paid off, for in November of 1883 the Denison *Sunday Gazetteer*—which maintained an almost breathless round of news reports on Volney and his activities—announced that he had shipped 1,000 peach trees to a large New Jersey orchardist.[42] A carload of 13,000 trees went to Missouri, a large consignment to the State Agricultural College of Mississippi, and other orders to nearly every state in the Union as well as Mexico and Prince Edward's Island (Canada). The scuffler hoe continued to bring in good business, too. According to the *Gazetteer,* "nearly every mail carries one or more of these useful instruments to some enterprising farmer or gardener." (April 29, 1883)

Despite this schedule, Volney also managed a July trip to Del Rio, Texas, to collect grape specimens and another to Neosho, Missouri, to visit Hermann Jaeger. Also in 1884, on one of his Denison grape expeditions, he found what would prove to be one of his best "mother" stocks, a vine he named Lucky. It was, he wrote, "a beautiful sight, that made me leap with joy." Three miles west of Denison was another favorite spot, "a little region three or four miles across in which I found more good wild grapes than in all other of my ramblings."[43] These rambles yielded several good mother stocks for about ten years, but by the mid-1880s, the land increasingly was being cleared for pasture.[44] Volney reminisced about those early days in his 1909 book, *Foundations of American Grape Culture,* and modestly declared that "the finding and bringing into cultivation and using these vines as breeders marks an epoch in American viticulture that will go down in history with more value and renown than even the Norton and Concord and Herbemont."

Perhaps he can be forgiven this bit of braggadocio in light of the achievements which would result from this work of the early- to mid-1880s.

CHAPTER 5
"No Little Work"

𝒯he year 1885 proved a critical one for Volney for it was then that his "investigations began to find expression," as friend and renowned horticulturist Liberty
Hyde Bailey later wrote.

Looking to New Orleans

In January 1884, Volney was appointed to the Mississippi Valley Horticultural
Society's committee on the upcoming World's Industrial and Cotton Exposition in New
Orleans (December 1, 1884–May 31, 1885). Parker Earle, the Society's president and a
personal friend, had been named chief of the Horticultural Department for the Expo, and
he asked Volney to prepare a display on the American grape.

Volney immediately enlisted Hermann Jaeger to help him. Since each man was fluent
in German and/or French, they set about contacting European viticulturists, especially in
France, to send specimens. They decided to illustrate both wild and cultivated varieties
and to feature live specimens growing in pots as well as mounted and framed herbarium
sheets. It was to be the largest exhibit ever done on the genus *vitis*, occupying Volney for
the ensuing year and costing him "considerable outlay of money." No photograph or detailed description of it is known to have survived, but it was likely similar to, though
smaller than, the one he would prepare for the World's Columbian Exposition in 1893 (see
Chapter 10).

The Horticultural Hall which housed the exhibit was the largest conservatory ever
built in the world up to that time, resembling a huge Greek cross 600 feet by 194 feet and
with a center tower. An avenue of magnificent oaks lined the approach to the Hall, and
its glass roof covered more than three acres; the entire structure cost $100,000. (Actually,
it was the least expensive of the Exposition's five main buildings, although the third

largest.) Inside, beneath the tower, was a "magnificent fountain," while tables covered with 20,000 plates of specimen fruits ran the entire length of the building, illuminated by "brilliant electric light," the latest in technology.[1] The most eminent American horticulturists gathered at the Exposition and competed fiercely for an astonishing $32,000 in prizes.

In addition to his own exhibit on grapes, Volney also contributed to the Texas state display, which occupied the largest space in the Government Building. In September 1884 he sent to New Orleans for installation a "mammoth grape vine," fifty feet long, which he had found growing on a sycamore tree along PawPaw Creek near the Red River.

Putting It on Paper

To supplement his and Jaeger's grape exhibit, Volney presented a paper to the Mississippi Valley Horticultural Society, which met in conjunction with the New Orleans Exposition in January 1885. This paper set forth in concrete form the theories he had developed over the past several years. "Native Grapes of the United States" set the horticultural world on its ears by sweeping away all past classifications and proposing a new one based on geological changes.

In his presentation, Volney was quick to praise George Engelmann, who had died just the previous year, as a man who had done "so much for useful botany" and who had developed "the ablest and most accurate [classification] yet published." But he also felt Engelmann's work had several problems. Asa Gray had earlier classified only four species of American grapes, and Engelmann thirteen; Volney believed there were more. Second, Engelmann and others, including French ampelographer Jules-Emile Planchon, took the view that "'honest nature abhors hybridization,' as much as to say there is a dishonest nature, or else there are no hybrids." If that was truly the case, Volney reasoned, then Engelmann's own thirteen species were wrong for there should be only one species. The root of the problem was the definition of "species," and he offered his own: "a type embodying peculiar and uniform general characteristics . . . which continually occur by natural distribution over a large area, or in a great number of individuals, and which have great antiquity, and may be supposed to relate purely to a common parent in the remote past."[2]

Based on that, and "knowing well how I expose myself to scathing criticism and the charge of pedantry in suggesting a different order," he proceeded to do just that "because I believe it to be the most natural and useful to the viticulturist." His classification, he declared, was based on geological changes to the North American continent that had "violated" the natural development of plants and which, in turn, had forced the plants to adapt to different localities and soils.

Volney theorized that Canada was the first part of the continent to rise above "the great universal ocean" of paleontological times. One by one, other mountain ranges uplifted in a roughly U-shaped pattern to eventually form the Mississippi River basin. The first grape, a *riparia*, lodged on the rim of this basin in Canada and along the Atlantic Seaboard; with no mountains to climb, *riparia* and its "various tribes" of *rupestris*, *nuevo mexicana*, and *arizonica* filled the entire area. The southern rim "caught new importations from the warmer southern seas" of the Caribbean, "landing the *Cordifolia, Aestivalis,* and *Cinerea*" from Virginia to Texas, from whence they had spread. As the Gulf of Mexico retreated over the millennia, its winds brought other, fully developed grapes—Scuppernong, *labrusca*, and Mustang—probably from an ancient and long extinct continent east of the Caribbean Islands in the Atlantic Ocean.[3]

With this explanation, Volney declared, one could easily see that northern grapes were older and southern ones younger. It was also patently clear that not all grapes would grow in all climates since each had adapted to a specific environment; a northern vineyardist, therefore, needn't bother with most southern grapes. Volney proposed a classification "according to their Natural Affinities and Distribution" and sorted American grapes into seven groups: Riparian, Cordifolian, Cinerean, Aestivalian, Vulpina, Meaty-fruited and Soft-rooted (which included *vinifera*), and Rotundifolia.

Volney then took up the "knotty question" of the origin of the southern hybrids Warren, Herbemont, Le Noir, Pauline, Louisiana, and Harwood. French ampelographer A. Millardet had deduced from botanical analysis that they were a cross of *cinerea* and *aestivalis*; after years of his own research and that of Hermann Jaeger, Volney agreed. "I humbly claim the honor of the solution. ... I am happy to have the privilege ... to declare this discovery before the largest horticultural society in the world ... under the span of the greatest horticultural hall, and within the grounds of the greatest exposition the world ever saw."

Like his vision and drive, Volney Munson's ego was also not limited.

Why hybridize with disease-prone *vinifera*, he asked his audience, "when we have such examples of pure American blood with freedom from disease?" The only obstacle standing in the way of American viticulturists was the lack of researchers. If so many wonderful

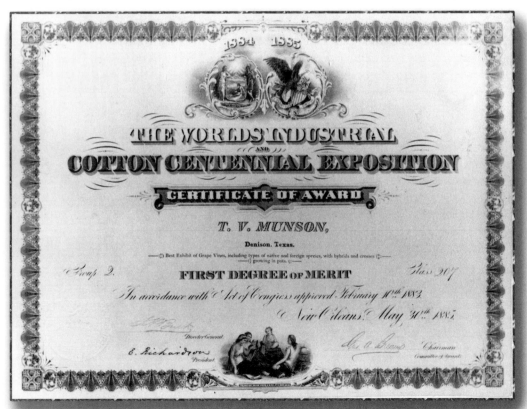

Volney won the first of his many major awards for grapes at the New Orleans Cotton Exposition in 1885.
(T.V. Munson Viticulture and Enology Center)

apple varieties could be produced from just one or two parents, what might not they accomplish with "no less than ten species of grapes … possessing great capabilities in the hands of skillful experimenters?"

Volney's "able paper" drew a great response from the horticulturists gathered to hear him, and they gave him an official vote of thanks "for his exhaustive labors in systematizing our American grapes." One listener applauded his "nerve" in making the topic "practical as well as scientific," while another declared Volney's paper alone worth far more than the Society's annual membership fee.

The MVHS meeting and the Cotton Exposition were major successes for Volney. His grape display won the highest premium: a handsome gold medal and $50 in cash.[4] (After the Exposition closed, he donated the exhibit's herbarium specimens to the University of Missouri at Columbia, but they were later destroyed in a fire.) Parker Earle appointed him to the Committee on Nomenclature to confer with the American Pomological Society on classification and asked him to serve as an exhibit judge. Volney was reelected first vice-president of the Society, which changed its name at this meeting to the American Horticultural Society.[5]

His new classification was now bringing him the attention he'd hoped for, since his goal was to educate both vineyardists and the public alike. But it was, he sighed, "no little work."

Revisions

The classification paper was in great demand after the New Orleans presentation. In quick succession, Volney revised and published it in the *Pacific Rural Press, The Wine and Fruit Grower,* and the *Annual Report of the Ohio State Horticultural Society.* George Husmann stopped off in Denison to see Volney after his own visit to the Exposition, and "he was so well pleased with [my paper]," Volney wrote Hermann Jaeger, "that he asked permission to take it and have it published in the 'Wine Merchant' of San Francisco."[6] And through the auspices of Volney's French correspondents, Planchon and Viala, the work was translated and published in *Le Progres Agricole et Viticole,* bringing him his first international audience.

Through the summer of 1885, he continued to work on the classification for his next big presentation at the American Pomological Society meeting in Grand Rapids, Michigan, in September.[7]

Michigan

In preparation for this trip, Volney designed and had printed a form "for recording notes of grapes" which he passed out to many of the attendees. His goal was "to get at the exact starting [point] of every Am[erican] grape soon," and he hoped the result would be "worthy of [an] article."[8] He also sent Jaeger enough forms to cover 200 varieties and asked his friend's help "in this large work in the interest of viticulture." Besides, he added with his usual practicality, it would be good advertising for both of them since he planned to write several articles based on the information.

To Grand Rapids he carried what he modestly described as "a very small show of my late seedling grapes" and was pleased to win the Wilder Award for it.[9] He was sorry, he wrote Jaeger, that his friend hadn't done likewise: "it would have paid you well, as the display of grapes was meager, and samples from great distance attracted unusual attention."[10]

At this meeting, he met another horticulturist destined to become a lifelong friend and one of his few chroniclers. Liberty Hyde Bailey (1858–1954) had begun his career at Harvard in 1882, working for Asa Gray.[11] It was Bailey who would really transform American horticulture into a profession over the next few decades as he, like Munson, stressed the scientific approach.

"A Timely Paper"

Volney's presentation to the American Pomological Society was quite different from the one he had given in New Orleans just eight months earlier. Little more than half the length of the first paper, "American Grapes: Importance of Botanical and Other Scientific Knowledge to the Progressive Horticulturist, and Especially to the Viticulturist" was supplemented with a lengthy table labeled "the tabulated facts of part of my experiments," which described the characteristics of seventy varieties of American grapes.[12] Volney opened his talk by reminding his listeners of the many facts that a horticulturist had to know about plants: structure, growth patterns, reproductive cycles, origin, etc. He flirted with comparing the plant originator and experimenter to a deity when he declared that "the thought almost inspires him with the feeling that creative power is, to some degree, within his control." And he evoked Charles Darwin by praising his "great work," *Darwin's Variation of Animals and Plants*. Lest any of that annoy the more religious members of his audience, Volney added that "it is far from my desire to intrude a thought or insinuation upon theology." He was, he assured them, referring only to the same principles that had long been recognized in stock breeding. Pomologists were just taking the first steps toward selection and hybridizing that successful stockmen had been using for years, and Volney felt that plant breeders could do no better than emulate them.

He skipped the extensive geological explanation he had given in New Orleans, possibly in the belief that enough people had since read his material. Then came his big announcement: he had increased the number of species he believed existed in the United States. He set forth his new classification of "no less than sixteen distinct species of grapes in the United States … classified according to natural distribution, development and adaptability to climate and soil, and by physical markings."[13] This had been extensively revised since January, reducing his previous seven groups to six: (1) Riparian, (2) Frost Grape, (3) Meaty-fruited, Fleshy-rooted, (4) Hard-wooded summer grape or Aestivalis, (5) Wooly-leaved, and (6) Warty-wood groups. The latter three categories showed the most evidence of revision. He also added a new variety, Munsoniana, "recently identified by myself from specimens found in S. Fla. [South Florida] by Mr. J. H. Simpson … and by him named for me."

Volney next discussed erect versus reflexed stamens and their importance in hybridization, since the failure of the vines to cross-fertilize "is the cause of much loss to fruit-growers." In his opinion, given the fact that "over half of all wild vines are males … if only one variety is planted in a vineyard, it must have erect stamens, in order to make a crop." He presented some general facts based on his fertilizing experiments but declined to discuss diseases as it was "entirely too large and difficult a subject for a paper like this." He referred the interested listener to the "able and beautifully illustrated French work of Pierre Viala, which should by all means be translated for the use of American vine-growers."

Volney urged growers to work slowly. "Great jumps in hybridizing widely divergent species should be avoided, as we are liable to get only a mule or mulatto—bad things to breed from." His beliefs on this subject may have been influenced by Charles Darwin, who wrote in the revised edition of *The Descent of Man* (page 187) that "the inferior vitality of mulattoes is spoken of in a trustworthy work as a well-known phenomenon."

Praise

Liberty Hyde Bailey called the Michigan presentation "a timely paper, with a plea for better breeding in harmony with accumulating knowledge and recognition of evolutionary principles." Horticulturists and viticulturists were also quick to sing Munson's praises for the breadth of his experiments. Soon his revised classification became the accepted and standard one for American *vitis* even though Volney himself considered it "yet very imperfect."

Publicity about Volney increased, too. The *Rural New Yorker*, for example, published a flattering page-length biography and portrait of him in October. The Agricultural College of Mississippi, anxious to attract this bright new star, offered him a position as professor of horticulture. Volney declined, wishing to continue his experiments.

Volney was particularly pleased at Hermann Jaeger's response to his paper. "It is very flattering to my pride to have you express such warm support of my 'Am. Grapes.'" [14]

A Gentleman of Ability

Following the success of these papers, Volney spent what was for him a fairly quiet 1886. Fifth child Volney Earle Munson was born at "Scarlet Oaks" early in the year; three days later, Volney was elected a director of Denison's First National Bank. The bank was a family affair, with brother Ben as president of the board of directors and Theo the vice-president.

In March 1886, thanks to the "thoughtfulness" of J. W. Throckmorton, former Texas governor and current U.S. congressman from McKinney (in adjoining Collin County), Volney received "a large box of very rare house and foliage plants, evergreens, etc. from the botanical gardens [in] Washington, D.C." Most of these were foreign plants "sent out to a few of the most experienced nurserymen" for testing. [15] A second illustration of Volney's growing reputation was the personal visit paid him in August by H. E. Van Deman, who had just been named head of the U.S. Department of Agriculture's new Division of Pomology. [16] Van Deman was intrigued by the *Prunus Simonii* trees he saw growing on the Denison Nursery grounds and declared they were the largest he'd ever seen. A variety of plum, the trees had fruited the year before, and Volney believed it was the first time that had happened in this country.

In his first divisional report, written after this Texas visit, Van Deman expressed his dismay that so little research had been done previously to improve the American *vitis* and wrote glowingly of Volney's work. "But now we have a gentleman of the most eminent ability, both scientific and practical, in the person of Prof. T. V. Munson, who has taken the

T.V. Munson in the late 1880s. (T.V. Munson Viticulture and Enology Center)

matter in hand. He has not only transplanted to his vineyards vines of the best wild varieties he could find, but he has with wondrous care cross-fertilized these with some of our finest cultivated varieties, and has grown seedlings from this fruit. I might say that Professor Munson is raising a new race of grapes, and with the most promising results already. ... It is really worth a trip to Texas to see Mr. Munson's rows of young seedlings and to enjoy the benefit of his intelligent explanations."[17] Van Deman also noted Jaeger's similar experiments in Missouri.

Letters between Volney and Van Deman in the USDA collection in National Archives II document the long-lasting friendship and professional respect begun on this first trip. Volney was quick to support Van Deman against the "carping criticisms which are so often hurled at the Dept. by persons who speak without knowledge of the work done and con-

Henry E. Van Deman was not only a personal friend but an important influence in Volney's professional life. (National Agricultural Library, Beltsville, Maryland)

demn without a hearing, until the very name of 'Agricultural Department' is a signal for ridicule by a thousand one-horse [news]papers (envious because they haven't their finger in the governmental pie), till the people actually believe the Comr. [Commissioner] and all his aids [aides] go to Washington to have a good time in riotous living at Uncle Sam's expense." [18] Volney proposed a series of articles in major journals to better publicize the work of the USDA.

During 1886, he continued as president of the North Texas Horticultural Society, one of several around Texas which helped rejuvenate the mostly defunct state group that summer. [19] Sponsoring clubs met in Dallas late in October to reorganize the Texas State Horticultural Society and elected Volney as chair. Shortly afterward, the delegates also elected him state president for the 1886–87 term and selected Tyler as the site of their first meeting in July of 1887.

Volney wrote several articles for national gardening magazines and continued his grape experiments even though the summer of '86 was marked by a severe drought. He was particularly interested in Hermann Jaeger's No. 43, one of his favorites, which he was anxious to cross with several of his own varieties; it would prove to be one of his most reliable parent stocks. More and more, he was being called upon for seeds and cuttings, requests he tried to honor though they brought him little revenue, since his primary goal was education and dissemination of the latest information.

Late in the winter of 1886–87, he began negotiations with Jules-Emile Planchon to sell the Frenchman a collection of American grape specimens, presumably for l'Institut National Agronomique in Montpellier. He asked Hermann Jaeger to "consummate the sale" in exchange for "a part of the proceeds." The initial shipment to Planchon was to consist mainly of leaves but would be supplemented in the summer with flowers and "such things as are wanting [now] ... so as [to] finally make the collection the most complete in existence of American Vitis." Volney thought 75 francs for every 100 specimens a reasonable price though it didn't cover his costs, but as they "would go to where it would do the most good, would let it go at that." [20]

He had also been selling cuttings and seeds for several years to commercial French vineyardists. During the decade in which these transactions occurred (1884–1895), Volney shipped most of his products through the Richter family, still an important name today in the French wine industry.

The Richter family in southern France shipped many of Volney's products overseas.
(Courtesy of Henri Bernebe, Montpellier, France)

The workforce at Denison Nursery prepares a shipment to Victor Vermorel, a prominent viticulturist in Le Havre, France. Warder sits on the packing box, Will is in the wagon, and Volney stands on the extreme right. (T.V. Munson Viticulture and Enology Center)

Vinita Home

Another of Volney's goals that winter was to move the Denison Nursery to the southern property he'd obtained from Theo several years earlier. With that in mind, he planted a vineyard of two acres there in the 1886–87 season, the first of seven over the next two decades that eventually covered almost fifteen acres of the sandy soil.[21] The main part of the nursery he planned to locate at the "south end of [the] street car line on Myrick [Mirick] avenue."

He also began drawing plans for a grand new house less than a mile southeast of the vineyard at the modern intersection of Hanna and Mirick streets. A brick structure in the Italianate style, later painted white, the house bore little resemblance to the Gothic fantasies he'd once sketched in his journal.[22] Square and simple in style, it had a main living area of two stories and a single-story kitchen on the back, plus a dormered attic and a basement (very unusual in that part of Texas). Overhanging eaves with brackets and tall, slender windows marked the Italianate influence, while long, covered porches on the northern and eastern sides of the house made the Texas climate a bit easier to bear atop what was then a treeless knoll.

The Munsons pose on the porch of "Vinita Home" on July 19, 1890. From left: Will and his cousin W.H. Munson in their Denison Zouave Guards uniforms, Fern, Viala Laussel, Nellie, Volney, Warder, Olita, and Neva. (T.V. Munson Viticulture and Enology Center)

Three of the Munson children pose in the Denison Nursery grounds at its new southern location, circa 1887.
(T.V. Munson Viticulture and Enology Center)

On the first floor was a sitting room/library which was, as grandson Marcus Acheson later recalled, Volney's "domain."[23] Here he entertained friends and visitors who debated every topic from religions of the world (the library had many books on that subject) to whether Francis Bacon wrote Shakespeare's plays. An oversize glass-front bookcase held part of Volney's book collection, which was both eclectic and extensive. Many a cutthroat game of chess was played in the library, for Volney believed that it helped improve the mind; he enjoyed a regular Sunday game with Ben and others. There were no novels, either. Volney felt no need for fantasy when truth was so much more interesting. He kept his drawing instruments here to work on geological charts and even drew the Munson coat of arms. A wall-mounted, hand-cranked telephone was located in this room.

Also on the first floor were a spacious foyer and staircase with landing, a dining room large enough to seat all the family, an office, and a parlour which included a piano most often played by daughter Viala.[24] In the kitchen, a large wood-burning stove supplied all the hot water in the house for many years. A tank on the kitchen roof was kept filled with water pumped from the windmill. And south of the kitchen was a "two-holer" outhouse, built of brick and painted white to match the house.

All the bedrooms were upstairs, simply furnished with beds, dressers, washstands, and chamber pots. On hot nights, the family often slept outside on the roof of the east porch. Open space in the attic above this floor was in the shape of a cross. Half the area was filled with shelves holding Volney's correspondence; the other half was for storage.

A center support divided the basement of native sandstone. On one side were stored fruits for both home consumption and shipping while the other housed small machinery and Volney's experimental work. There was a winemaking area as well, and Volney and Nellie regularly hosted both grape and wine tasting parties.

The lawn was eventually lined on two sides with native and imported trees including magnolias, which Volney planted in defiance of the belief that they would not survive a North Texas summer. (The children had the task of bucket brigade to keep them watered.) About a hundred feet south of the house, the barn housed several mules to pull equipment and a horse for the carriage. One well and a windmill supplied all the water for both house and nursery buildings, which meant that water sometimes ran short. There was a packing house and several storage buildings for chemicals and machinery as well as the ubiquitous manure pile so necessary for the health of the nursery stock. As many as a dozen workers were required for the nursery at the height of its season.

Volney and Nellie named their new home "Vinita" and moved the family there from Scarlet Oaks early in 1887.[25] But shortly after, baby Volney Earle died of cholera infantum. The disease must have struck him very suddenly, for Volney wrote Hermann Jaeger on April 11 and mentioned the "grave illness of one of my brothers" but said nothing of the boy, who died on April 20. The funeral took place at Vinita Home; Theo read the eulogy, and Volney spoke at the graveside about his "bright little child."

Now three of Volney and Nellie's eight children had died at tragically early ages.

The Government and Mr. Munson

On July 1, 1886, recognizing the growing economic importance of fruit production in this country, the United States Department of Agriculture had established a Pomology Division headed by Volney's new friend, H. E. Van Deman. (Prior to this time, fruits fell under the USDA's Superintendent of Gardens and Grounds.) Identifying and describing American fruits became the first major task of the fledgling department. Grapes were a special focus since viticulture had become a major industry in the eastern part of the country, thanks to the development of such stalwart varieties as Catawba, Delaware, and Concord. The natural immunity of eastern and southern grapes to phylloxera was another factor in the plant's importance to the government, given the still deteriorating situation in France and Europe (see Chapter 6).

A year earlier, in 1885, French vineyards had suffered yet another catastrophe when American grape cuttings, shipped abroad for grafting to fight phylloxera, introduced the additional scourge of black rot. Perhaps feeling guilty over this unfortunate result, the Department decided to undertake experiments to find a cure.

Frank Lamson-Scribner[26] directed the project as chief of the Department of Vegetable Pathology; at a time when few people had even heard of fungal diseases of plants, much less understood how to treat them, he was already an authority. Working with Van Deman, Lamson-Scribner selected four southern vineyards for the work, the first in Fayetteville, North Carolina, and the second in Charlottesville, Virginia. For the third he chose Volney Munson and the Denison Nursery; at Volney's suggestion, he selected Hermann Jaeger in Missouri as the last site.[27] All four owners were to spray their grapevines in the spring with Bordeaux mixture—a combination of lime, water, and copper sulphate which stained everything it touched blue—and with several other compounds, observe the reactions to the various treatments, and report the results.[28] Lamson-Scribner furnished each man with a Vermorel pump and sprayer from France and a sulphuring bellows, which were to be returned to the Department after the experiment was completed.

Volney had complained only in February to his friend Van Deman that he'd not been well and had "more work than I am able to endure in my [nursery] business," yet he'd been intrigued enough by the project to agree to participate, apparently at Van Deman's urging. Another consideration may have been the stipend he was to be paid for the work.[29]

Following his usual procedure, Volney kept meticulous experiment notes in his diary, which has not survived. He reproduced the entries in his report, however, which was published in Lamson-Scribner's *Report on the Experiments Made in 1887 in the Treatment of the Downy Mildew and the Black Rot of the Grape Vine*. Volney not only recommended the Bordeaux mixture but added his opinions on the soil types that favored rot and the varieties of grapes most susceptible to it, as well as a detailed weather chart of the study period. As a result of this test and a second one the following year,[30] the USDA enthusiastically recommended Bordeaux mixture, which proved an invaluable tool: "more than any other one thing [it] influenced and shaped the development of the science of plant pathology during the quarter century following its discovery."[31] Within a few decades, its use became so common that "any school boy," as Samuel Fraser wrote in 1931, knew its efficacy in treating both black rot and mildew. (*American Fruits: Their Propagation, Cultivation, Harvesting and Distribution*)

And Volney Munson had played an important role in bringing that about.

Writing the Book on Grapes

In June of 1887, USDA Commissioner Norman J. Coleman approached Volney about writing the definitive publication on the American grape, to be produced and disseminated by the Department.[32] Because the initial correspondence regarding this project has not survived (or hasn't yet been found), it's difficult to ascertain the exact outline of the project presented to Volney. From one extant letter, however, it appears that he was to do a short monograph first followed by a full-size book with color illustrations: "in so short a space of time as I must devote to the treatise under the *first commission,* I can do little more than to give an outline classification, with viticultural aspects of each species touched upon." (Authors' italics)[33]

Munson's job was daunting: to observe and collect vines from as many states and territories as possible, sort out the different names that had been given to each variety, describe the characteristics of each variety for identification purposes, examine specimens in major American herbaria, and grow representative vines in his own Denison vineyards—then to make sense of the whole matter and prepare a new classification system based upon this data. Not only that, but the material was to be presented in such a way that any viticulturist, amateur or professional, could understand and use it.

The commission carried with it status as a special agent of the USDA and provided some travel expenses as well as a modest stipend. Volney politely pointed out in a letter to Van Deman that the pay "for the first work" was inadequate to cover the costs of his correspondence, experimentation, and travel.

So why did he agree to undertake the project?

Just the year before, Volney had written in *Texas Farm and Ranch* (June 1, 1886) that he was considering such a book himself, with the "results published for [the] general good (almost entirely at my individual expense). . . ." So Coleman's offer came at a propitious time. Volney was already collecting data, and the Department would both save him the cost of publication and give the book added stature. Nobly he vowed not to allow the pay issue "to interfere with the thoroughness of the work"; just so no one would miss the point, however, he reiterated the Department's "recognizing my labors in a most <u>useful</u>, <u>original</u>, but most likely unprofitable (to me) direction."[34] Volney accepted the commission and, believing that his "dear Friend Van Deman" had directed it to him, wrote that "I must thank you for honoring me with this exceedingly important work, and one which requires such close discrimination and large collection of data to arrive at correct conclusions."[35]

Volney felt he needed more information on *Vitis Novo Mexicana* and on "a new and entirely distinct and peculiar species identified by me" which he had named *Vitis Texana*. And since there was nothing he considered "more exhilarating and entertaining than wild grape hunting and testing," he set off early in August—one of the hottest months in Texas and the Southwest—as soon as he returned from the state horticultural meeting in Tyler.[36] Unfortunately, few records have survived of his itinerary, which concluded in Denison late in the month. The local newspaper reported only that he went on "an extended tour of the west and south"; Volney wrote later that he visited the Bradshaw Mountains of Arizona, among other sites. He did report to Van Deman that his first venture covered seven hundred miles in ten days, from the "headwaters" of the Red River in the Panhandle to the Brazos and Colorado rivers in southwestern Texas, where he searched for species of wild grapes "as yet little known to botanists."[37] On this route he discovered an entirely new grape, which he named *Vitis Doanii* in honor of Judge J. Doan of Wilbarger County, who made a tolerably good wine from it.[38]

This specimen, collected by Volney in Lampasas County, Texas, is now in the National Herbarium at the Smithsonian. (Photo by Roy E. Renfro)

Van Deman had great plans for Volney's book, hoping to capitalize on the growing American grape industry and the phylloxera catastrophe in Europe. "I mean to have that thing done well if it takes two years," he wrote Volney in mid-August, for he wished the publication to "do the country and all of us credit."[39] On August 1, he

Volney found Vitis Doaniana, or Doan's Grape, in the Texas Panhandle.
(From *Foundations of American Grape Culture*)

had appointed Wilhelm (William) Heinrich Prestele of Iowa as the Pomology Division's staff artist.[40] Prestele was to prepare life-size watercolor paintings of each grape variety "with a view to the whole being published in the highest style of art." Van Deman asked Volney to send him both fresh and pressed samples of the fruit, branches, and leaves of each variety that Prestele might work from. "Don't stop for expense or time," he adjured Munson.

But Van Deman's desire to hasten the book to the printers met with several obstacles. Preparation of the artwork, for one, proved slow and painstaking. Volney sent samples to Washington, where they were photographed for working use, then examined by the artist for color and texture. Prestele did rough sketches and sent them to Volney for correcting, which he had to sandwich in between travels and work. The whole process was then repeated at least once more. Inadequate samples slowed Prestele down, too, and required that Volney take the time to find him better ones. "I shall be very careful to send pure typical specimens," he pledged to Van Deman, "so that the blindest person may know what I am talking about, when they see a good picture of them."[41] In all, the production took several years longer than anticipated, and the book eventually appeared only in truncated form (see Chapter 9).

These two USDA projects—the black-rot experiments and the grape monograph—brought Volney Munson to the attention of even more people. His meticulous approach to research and the scope of the work he had already accomplished would now bring him his greatest challenge as he took on the fight against the bug that was eating France.

CHAPTER 6

"A Great Pest"

*F*rance, of course, has been known for centuries for its fine wines: bordeaux, champagne, cognac, and many others. Each wine is made from a particular variety of *Vitis vinifera*, and the grapes are grown primarily in specific areas of the country—the Burgundy District, and so on. It has always been a hazardous business, for *vinifera* grapes are delicate creatures that require just the right soil and climate to flourish. By the time Volney Munson was a young boy, the production of wine from these grapes had become a cornerstone of the French economy.

The growing popularity of wine over ale and liquor in the 1700s and early 1800s, and an expanding European railroad system that allowed them to more easily market their product, had led French vineyardists to dramatically expand their acreage planted in grapes from 3.3 million acres in 1788 to 5.5 million in just six decades. By 1850 the vineyards in the Languedoc and Herault regions of southern France had doubled in size; over the next 25 years, another half million acres were added generally throughout the country. All told, France had more than half of all the wine-producing acreage in the entire world. This new reliance on one crop caused the abandonment of the traditional agricultural system that had been in place for centuries but seemed justified by numbers of staggering proportions. Estimates of wine production in France in the 1840s range from 45 million hectoliters to more than 55 million hectoliters; that's a *billion* to a *billion and a half* gallons to those of us not on the metric system. And the bumper crop of 1875—83 million hectoliters, or well over two billion gallons—has never been repeated. Obviously, those quantities meant equally big money. Indeed, that 1875 harvest brought in enough taxes to make up more than fifteen percent of the *entire* French governmental budget.

Odious Oidium

The first hint of trouble came around 1845, when oidium was discovered on grapevines

The wine-growing region of St-Emilion, France, in the Valley of the Dordogne. (Photo by Roy E. Renfro)

in a hothouse in Margate, England. Oidium is a parasite that to the naked eye resembles white dust and, like most grape pests, is of North American origin. From England, it crossed the Channel and attacked the vineyards of Europe. Sulphur proved a great remedy, though an equally popular solution was to import oidium-resistant American vines and graft the *vinifera* varieties on them. Wine production in France alone fell to as little as 10.7 million hectoliters (278.2 million gallons) before vineyardists were able to turn the tide in the late 1850s. Oddly enough, over the fifteen years or so that followed, there were several remarkable vintage years, such as the clarets of 1864 and 1869.

The Devastator

Interest in American grape varieties continued even after the oidium epidemic, particularly among wealthy landowners interested in exotic new plants of the Americas, botanists, and some religious communities.[1] They hoped that grafting *vinifera* onto the foreign stocks—especially *labruscas* such as "Isabella," a black grape with a raspberry taste—might strengthen the vine and make it more disease-resistant without changing the taste of the grape, and these tests were conducted on a very small, experimental scale. But in hindsight, French and European vineyardists probably wished they'd never had the idea, because some of those varieties, especially "Isabella," brought another little visitor on their roots—phylloxera, which is native to the eastern United States.

Just as with oidium, this new bug was first discovered in England in 1863 and christened "*peritymbia vitisana.*" Viticulturists soon realized its full destructive powers. By the time growers noticed anything wrong, their vines were already dying and beyond help. French ampelographer Jules-Emile Planchon eventually bestowed on it the more appropriate name of "*phylloxera vastatrix,*" or "devastating leaf-dryer"—a reference to the first visible symptom, the leaves curling up and dying. (Today it is known as "*Dactylasphaera viti-*

The tiny devastator, phylloxera. From Dossier: Il y 100 Ans ... Le Phylloxera.
(Courtesy Station Viticole du B.N.I.C., Cognac, France)

foliae.") On vines native to the eastern and southern United States, which had acquired an immunity to phylloxera centuries ago, the aphid merely produced small galls on the underside of leaves.

But in Europe, on *vinifera* vines, the damage was much greater because the cambium layer of the roots is thinner than on American stock and, consequently, easier to chew through.

Unlike oidium, which attacks the grape, phylloxera destroys the very vine the fruit grows on, and it lives only on *vitis.* A yellow to yellow-brown aphid no more than a millimeter (.039 inches) in length (not much bigger than a pinprick or a seed of grass), phylloxera sucks the life-giving sap from the vine's roots which causes the roots to swell with galls. It then leaves behind a saliva that infects the wound and prevents it from healing, a process which can take several years to kill the vine. A yellow dust appears around the roots that, if touched, stains the fingers. The pest has a complicated life cycle and multiplies rapidly; the female lays eggs on the vine stem in the fall which hatch out in spring. Phylloxera travels from plant to plant borne by wind, tools, machinery, clothing, and other means. Most often, it travels on the roots of young dormant plants. "Cute little Yankees as they are," the aphids move on to a new plant after destroying the previous one.[2]

Growers saw their vines dying but couldn't understand why. Moreover, wine made from the affected grapes was of poor, thin quality. Some thought that the cause was a variation of oidium, but by 1867 it was clear that, instead, they were facing a new problem. A year later, the Vaucluse Agricultural Society appointed a commission of experts (including Planchon) to investigate the problem. With the aid of American entomologist C. V. Riley, the commission soon linked the destruction to phylloxera. At the same time, some growers proposed that the diseased vineyards be torn up and replaced with American vines. Given the origin of the problem, this was not a popular solution.

But even as men studied it, the pest spread at an unbelievable rate.

Evidence of phylloxera first appeared in the Gard region of southern France in the mid-1860s, though the louse itself may have arrived several years earlier.[3] From the Gard, the plague mostly traveled east at first, toward the mountains that separate France from Italy. A second major infestation hit the western coast in the Charente district around Cognac and destroyed a third of the vine acreage within a few years. Cognac, wrote one observer in 1880, was a "fair district, [but one] whose vineyards are being rapidly rendered desolate."[4] By 1874 the Vaucluse region was ruined, its production down to almost nothing. Within five more years, the aphid had invaded the Gironde, the Loire Valley, and the Cote d'Or, the speed of its attack (more than twelve miles per year) enhanced by the near continuous vineyards which had been planted earlier in the century to meet market demand. In some areas, this damage was exacerbated by a drought in 1875, but others suffered through several years of equally devastating heavy rains in the late 1870s. By the end of 1879, more than one and a half million acres of vines in France were infected and almost half of them completely destroyed. C. V. Riley reported one example in *The American Entomologist*: "Chateau Lafitte for which Baron Charles Rothschild paid $830,000 two years ago, is nearly ruined." (February 1880)

The infestation spread to Corsica, Switzerland (first detected 1873–74), and Germany (late 1870s), among other countries. Phylloxera liked the warm winters of the Mediterranean, and in most of Spain (1879), Portugal (1867), and Italy (1879), the destruction was astounding; by 1883, for example, Portugal's wine production had dropped 75 percent. Other warm climates, including Australia, New Zealand, and South Africa, also suffered devastating losses. However, the sandy soils of several districts in Portugal and France proved inhospitable, as did northern Italy, where there were no large vineyards, the vines hung on trees, and the soil was too compacted from tree roots for the bug to penetrate. One observer, writing in 1879, estimated that more than 200 million gallons of European wine had been lost from production in the previous two years alone.[5] He continued with a pessimistic forecast: "Once established in a wine-growing district, the disease always ends by invading every portion [of the vine], so that unless some really efficacious remedy be speedily found, the larger part of the French vineyards will ere long have ceased to exist."

Nor was the problem confined to Old World countries. Phylloxera finally reached the Sonoma region of California, too, in the 1870s, where planting had already increased to meet the growing international demand for new wine sources. The culprits were probably infected *vinifera* vines imported from Europe despite early warnings from authorities such as C. V. Riley that such a disaster was likely if growers weren't careful. However, there was a difference. Riley reported that, in California, "the insect has, to all appearance, there undergone a considerable modification in habit which very much limits its destructiveness."[6] This difference was the almost non-existent number of winged females, which limited both reproduction and travel capabilities. Nevertheless, nearly 20,000 acres of vineyards would eventually be ruined.

Why? And What to Do?

No one knew initially if the aphid Riley and Planchon had discovered was the cause of the problem or if it appeared after the vine was already diseased, and that confusion delayed any helpful action. In a day when the god of science was only beginning to show its face, many people naively continued to place their trust in old remedies and beliefs; they were certain that phylloxera could not cross mountains and that those some distance from the Gard were safe. Science, after all, had little to do with agriculture.

But while slow to develop solutions, everyone was quick to find someone or something to blame for the whole mess. It was due to a blight in the atmosphere. Overproduction. The weather. The lack of laws protecting insect-eating birds from sportsmen. "God's Punishment for the Country's Abandonment of the Emperor," Napoleon III.[7]

Politics obstructed any concerted effort to find a remedy. Italy was undergoing unification in 1870, the Second Empire of Napoleon III was overthrown in France the same year and the Third Republic established, and France and Prussia decided to go to war. The latter was a real mistake, for phylloxera would eventually cost much more than the war: an estimated twelve billion francs (old style currency) as opposed to the five billion France would give Prussia in the conflict's final settlement. Scientists who realized the more pressing danger of phylloxera bemoaned the conflict. "But the blighting effects of the war have not only entailed untold misery and woe to millions in France, but have either paralyzed or effectually balked scientific investigation within her borders."[8]

Nevertheless, the French government did take several steps to deal with the situation, most of them after the fighting was over when it became plain they faced a more serious problem than a mere war. Acting on the advice of the 1869 commission, the government offered a reward of 20,000 francs in 1870 for a remedy. Four years later, with no help in sight, they frantically raised it to 300,000 francs. Together with the private awards also of-

The Universite de Montpellier opened an agricultural school in 1872 to deal with the phylloxera problem; this is one of the original buildings still in use today. (Photo by Roy E. Renfro)

fered, the successful inventor of a cure stood to receive at least 600,000 francs. That prize would never be claimed.

The Universite de Montpellier, on the Mediterranean coast of France and near the devastated Gard region, created a separate agricultural school in 1872 to respond to the various crises besetting the wine and grape industries. (L'Ecole Nationale Superieure Agronomique continues in operation today; its collection still includes Munson Nursery catalogs, herbarium specimens sent them by Volney, and, in a coastal vineyard, some of his hybrids.) Gustave Foex, the first professor of viticulture, quickly proceeded to amass what became the largest collection in the world of French and American rootstocks for experimentation.

Foex, L'Ecole, and France's national agricultural department also set aside an infected vineyard there to test proposed remedies. For just as everyone had selected a target to blame, so did everyone have a solution since a catastrophe such as phylloxera brings out all the crackpots. Suggestions ranged from the clichéd "sublime to the ridiculous": toad venom, exorcism, electricity, human urine, sand, flooding the fields, sulphur compounds, beating the ground to drive the louse away, and more. In all, nearly seven hundred remedies were submitted for consideration and about half of them were actually tested in the field. All had problems. Phylloxera didn't like sand, but the cost of transporting the tons needed was too high. Submerging the vines in water to drown the little pests worked—if the vineyard was on level land. Many were on the sides of steep hills.[9] Carbon disulphide also worked well but was expensive, dangerously flammable, and emitted toxic fumes; besides, it didn't keep the aphid from returning. Sulphides in general were also too expensive and drained the soil so that heavy fertilization was required to keep the exhausted vines alive, another cost.

Those Pesky Americans

Soon vineyardists had fallen into two camps: those who advocated chemicals and those who supported grafting onto American vines. A grape variety can grow on any rootstock; it's the upper part of the vine, not the root, which determines the type of grape. (Grafting is an ancient technique in which a cutting, or scion, of one variety or species is literally bound to the rootstock of another.)

But for centuries, *vinifera* had been grown from the original roots; grafting was an heretical idea to all save the experimenters. Moreover, the foxy taste of wine made from American grapes—and the continual reminder of where the phylloxera had come from in the first place—made grafting a tough sell. Some districts, such as Burgundy, actually passed laws forbidding the importation of any foreign vines. One chemical advocate

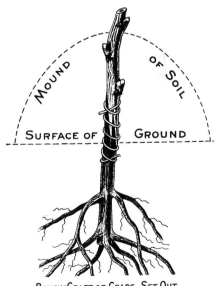

BENCH-GRAFT OF GRAPE, SET OUT

"Grafting the grape," Munson wrote, *"is a simple and easy matter, requiring sharp tools, a good eye and judgment."* (From *Foundations of American Grape Culture*)

summed up his side's feelings very plainly: "O, profanation! ... are the glorious productions of the sun of France to be sacrificed for frightful beverages, of which the odour of the black currant and the fox are the dominant perfumes?"[10]

Americanists such as Planchon argued that the foreign vines were "a means sent by God to permit the reconstitution of those beautiful vineyards which were the pride and wealth of France." Grafting, they believed, would make the vines stronger because of the thicker outer skin of American vine roots.[11] But opponents continued to harp on the same question. Even if growers eliminated the foxy element in the bouquet, how could the grapes, and consequently the wine, possibly taste the same?[12] Besides, the High Commission on Phylloxera was composed mostly of chemists, who naturally leaned toward their own experience.

Jules-Emile Planchon was a major figure in the fight against phylloxera. The base of this statue was erected in his honor, Montpellier, 1894. (Photo by Roy E. Renfro)

Complicating the entire problem was distrust. Many French vineyards were peasant-owned, "small in extent and cultivated by the owners, who as a class are honest industrious men, but narrow-minded and miserly. Of no education, they receive with scorn or suspicion the advice of scientific men."[13] Uproot their vines simply because some unknown professor in a far-off town said so? Hardly. And who would pay for it if they did? In their obstinacy, these growers even refused to report new outbreaks of phylloxera for fear of official interference.

The results were both obvious and to be expected, and, at the same time, wildly indirect. Phylloxera continued to spread unchecked at a terrifying rate. Many small vignerons, unable to afford the solutions being bandied about, packed up and left for new areas such as Australia or South America, where they established viticultural industries but often spread phylloxera via the infected vines they'd carried with them.[14] (In fact, many viticulturists found work in Chile and Argentina, both of which had a wealthy upper class interested in emulating Europe and, even today, no phylloxera.) Some of their land was snapped up by speculators with enough capital to build new vineyards, yet acreage planted completely in vines dropped from an 1852 high of 2.2 million hectares (5.5 million acres) to a low by 1892 of 1.4 million hectares (3.4 million acres). The total effect of this was to significantly alter the French social structure since thousands of small farmers had been involved in the grape industry in some way. Meanwhile, the farmers who stayed in France tended to replant with new crops such as olives, which in turn had an economic impact on other regions and countries that had historically grown and depended on those crops.

Supply and demand also went to work; with less wine available, it fetched much higher prices. That led to fraudulent wines made from raisins or beets and marketed as authentic; grape growers in Greece, for example, enjoyed a vast new demand for raisins to supply this black market.[15] In France itself, some unscrupulous dealers bought immature wines from other regions and sold them as burgundies when supplies in that district were low. Others added sugar before the fermentation process to increase the alcohol content. In Cognac, where farmers traditionally held back part of each year's yield and sold the matured wine later at a higher price, the brandy was diluted with beet juice to increase supply. "Years ago," mourned one writer, "such practices were never attempted but are now common."[16] Blended wines and liquors such as vermouth also became popular at this time. Regions that had escaped or only lightly suffered from phylloxera (e.g., parts of Portugal and Italy) stepped up their wine production and flooded France with imports.

In the United States, grape growers found a new market with the growing demand for grafts, which raised the price of a good American root to a franc. And France's loss was California's gain. Some European vintners moved their operations to the "Golden State" to take advantage of the new markets. Among them, for example, would be Paul Masson, born to a family of vintners in the Cote d'Or, who attended college in California and returned later to open his own winery there.

Setting the Stage

This lengthy background is necessary to understand the truly global impact of the phylloxera epidemic and the consequent importance of Volney Munson's role in it. The French and everyone else caught up in the nightmare were powerless to stop it. Every remedy they tried had an equally disastrous side effect or simply didn't work. One Frenchman, recalling the disaster in 1898, wrote that it appeared at the time to be "a scourge without remedy" which resisted all the "years of inane attempts to cope with it."[17] Indeed, such predictions of woe had begun to appear quite early; the phylloxera, declared one writer in 1880, had already "defied every effort suggested by science and every device prompted by human ingenuity."[18] And in the end, that one tiny bug would destroy two-thirds of Europe's vineyards, including all of those in southern France. In the Gironde district alone, the economic loss to viticulturists was estimated to be 32 million livres by the end of the century.

Finally bowing to the inevitable in the chemical-versus-grafting controversy, the French sent Professor Planchon to the United States in 1873 and charged him with several goals. He was to learn more about phylloxera itself, and he was to determine the best American vines for grafting. Planchon met with Charles Valentine Riley, the former state entomologist of Missouri who had made the first serious study of phylloxera.[19] However, because of Planchon's elderly age, he did not want to travel out of the northeastern U.S., so he relied on Riley's suggestions as to the best phylloxera-resistant vines. The French scientist returned home with a shipment of vines sure to solve the problem and encouraged American vineyardists and nurserymen to send more of the same as quickly as possible.

Help from America

The response came primarily from Texas and from Missouri. In San Antonio, Texas, for example, French-born Francis Guilbeau, a wine and liquor importer, went into business with nurseryman Matthew Knox to ship Mustang grapevines overseas.[20] Between 1876 and 1878, it is estimated they sold the French several hundred tons of cuttings. There were others who dealt with French growers on a similar but smaller basis.

The major players, however, were Isidor Bush and his colleagues in Missouri, C. V. Riley's home and, at that time, a major center of wine manufacture and distribution in this country. (In 1869, Missouri produced 42 percent of all American wine.)[21] Bush's interest in phylloxera extended back to 1867, when he first imported vines from Austria to study the problem. A few years later, after the louse was identified, he invited Riley to experiment in his Bushberg vineyards. The entomologist later persuaded Bush to donate cuttings of native grapevines to France for the first experimental grafts. Between 1875 and 1885, Bushberg Nursery sold several hundred thousand cuttings to French vignerons alone; worldwide, that trade amounted to at least 5,000,000 vines sold around the world each

Rear view of statue in Montpellier dedicated to J.-E. Planchon and American assistance in the phylloxera epidemic. (Photo by Roy E. Renfro)

A statue dedicated to Gustave Foex at L'Ecole Nationale Superieure Agronomique in Montpellier metaphorically depicts the wild grapes of America saving France. (Photo by Roy E. Renfro)

year, from Serbia to Australia to California. Because of this huge demand, Bush scoured the state and called for help in supplying the quantities needed. Volney Munson's friends and colleagues Hermann Jaeger and George Husmann were among those who responded, reportedly sending altogether several dozen train car loads of cuttings overseas from their Neosho and Sedalia operations.

The French were ecstatic. Their vineyards were saved at last, thanks to "a group of deserving and devoted men who refused to yield to despair."[22] Vineyardists set to with a passion, grafting as quickly as they could[23] and waiting for their fields to blossom forth with fruit once more.

And they waited.

But there was just one little problem.

The grafted vines were dying.

CHAPTER 7
Mr. Munson to the Rescue

*T*he vines, by and large, were dying, whether they were direct producers planted to replace diseased vines or had been used as grafts. The French and Americans alike were flabbergasted. The solution *should* have worked. What had gone wrong?

Several things. Many of the vineyardists who chose to graft their stock onto American vines rushed into the process without sufficient knowledge or experience and often used the wrong rootstock, consequently losing many vines and adding new problems such as chlorosis. Then, too, a "lively speculators' market" took advantage of the situation and drove up the cost of the cuttings, making it economically difficult to persevere. And even when the *producteurs directs* managed to survive, they resulted in poor quality wine. Moreover, American *labruscas*, which formed the bulk of the vines sent to France, proved tender and difficult to graft.

But the real problem proved to be, literally, right under everyone's noses—and feet.

A Matter of Dirt

What no one had taken into account was the very basic matter of soil. There is an old maxim in viticulture that says the poorer the land, the better the quality of the grape and, consequently, the wine. So French vineyards, particularly those in the hard-hit regions of southern France such as the Gard or Charente, tended to be planted in chalky, limey soil, unlike that in most of the eastern and northeastern United States.[1] Even the hardy Texas Mustang grapes (*Vitis candicans, Engelmann*) didn't much care for those conditions. Consequently, *labruscas* and others developed chlorosis, turned yellow, withered, and died. The French tried different varieties, but the results were mostly the same, with more time and money expended.

When it seemed that matters couldn't become bleaker, they, of course, became bleaker.

Many French vineyards are planted in chalky soils produced from limestone outcroppings such as these along the Rhone River near Avignon. (Photo by Roy E. Renfro)

The new vines had brought with them still another plague, that of downy mildew (discovered 1878), which destroyed the few leaves and grapes phylloxera had left and soured the wine upon bottling. The French were so irked that they finally responded to this latest crisis with legislation. Despite the fact that growers in the Herault region continued to favor grafting, the government chose to heed the advice of the Academy of Sciences, which declared the American rootstock "costly and doubtful."[2] In 1878 a new law was passed forbidding the importation of any foreign plant material into France. Having fired this salvo, the government effectively threw up its hands and retired from the lists, leaving further initiatives to individual growers.

But politics continued to be part of the problem. Most wine historians consider the 1880s the peak of the phylloxera plague. That was also a decade of major upheavals around the globe which distracted attention from dying grapes and focused it on noisier disasters. The French still had an unsteady Republic and were occupied with securing Indochina, while various European nations, including France, were practicing some imperialistic chest-beating in Africa. The British and Russians were squabbling over Afghanistan and India, and a financial depression that began in 1882 deflated prices just as it had become very expensive to produce wine, thus frightening off investors.

Nor had phylloxera stood still. By 1880, almost two million acres of French vines were utterly destroyed; some experts predicted that, without a solution, and soon, there would be no healthy vines left in the country within sixteen years. The aphid appeared as far away as Australia and continued to wreak havoc in already infected areas. A Phylloxera

Congress in Saragosse, Spain, in the fall of 1880, an International Phylloxera Conference held in Bordeaux in 1881, and several other such gatherings proved ineffective at producing reasonable, usable solutions. While the amount of land planted in vines continued to drop, the French threw more money at the problem: nearly a million francs (old style) in just one year (1880) to treat diseased vines and encourage investigative research. Meanwhile, a few individuals began to cross-breed American and French vines in the hope of finding a hybrid with the vigor of its American father and the bouquet of its French mother.

Monsieur Viala Goes to America

Still, the seemingly insoluble problem continued to be soil the American vines didn't like.

Merchants and viticulturists in the Charente region around Cognac began discussing a new delegation to America to search for different and better vines. Eventually, La Societe de la Charente-Inferieure (i.e., Lower Charente) combined with a similar organization in the Herault department (Montpellier) to offer the government financial support for such a mission.[3] On March 16, 1887, the French minister of agriculture appointed M. Pierre Viala, professor of viticulture at Montpellier's l'Institut National Agronomique and a correspondent of Volney Munson's, to undertake a second fact-finding mission to the United States "for the purpose of searching for varieties of grape stocks found to grow in the marly and chalky formations."[4]

Viala arrived in the U.S. on June 5 for a six-month visit and went directly to Washington, D.C. to confer with scientists and officials at the U.S. Department of Agriculture and the U.S. Geological Survey. Both gave him "small hope of finding vines growing in purely chalky earth." The Frenchman lamented, politely, the Americans' ignorance of the geographical distribution of their own native grapes, though he allowed that

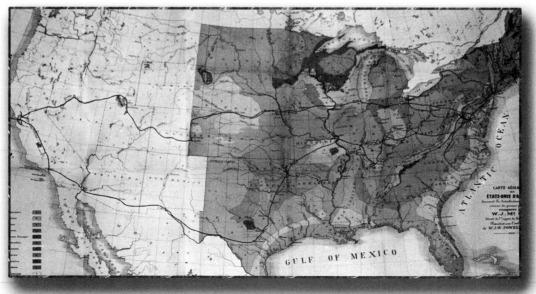

The map accompanying Pierre Viala's 1888 report shows his route (black line) and the various geographic zones of central and eastern United States. (Photo by Roy E. Renfro)

80

it was an "immense country." Unfortunately for his goals, much of America's western and middle sections, extending even into Tennessee and Pennsylvania, were of Upper Cretaceous origin and didn't contain chalk formations comparable to those in southern France. What he needed was scattered irregularly over the country, and he initially feared that he couldn't begin to cover all the possible locations within the span of his visit. No surprise there: his total journey, looping around the United States, would prove to be about 10,000 miles under conditions that were, at best, barely civilized to a European and, at worst, downright hostile. To find the grapes that would save his country meant traveling "mostly in the virgin forests and in uncultivated land." It was, he later wrote with commendable understatement, "adventuresome."

Viala first traveled along the Eastern Seaboard, mostly in its upper reaches.[5] Nothing suitable appeared, so he planned to move westward to Tennessee, Missouri, the Indian Territory (Oklahoma), and California. The first two brought him "some important indications." *Vitis rupestris* and its natural hybrids very much interested Viala; he'd learned of them through reports and specimens from Hermann Jaeger, whom he visited on this trip. But the Lower Cretaceous limestone formations of Texas[6] seemed to hold the best hope, so "it was necessary to go to Texas," too.

To Denison

Before leaving France, Viala made arrangements to see Volney Munson, though his report reads as if Texas was an afterthought. However, Frank Lamson-Scribner of the USDA had written Volney on April 9 about Viala's trip and mentioned that "he intends to visit Texas."[7] A *Sunday Gazetteer* article of May 22, while Viala was still en route to the United States, stated that "Denison will be honored by a visit from several distinguished citizens of France next month. The French government has appointed a committee on vitaculture [*sic*] to visit the United State [*sic*] and study the diseases peculiar to the grape and investigate the native grape. Mr. T. N. [*sic*] Munson says there will be a distinguished geologist, botonist [*sic*] and horticulturist in the party that will visit Denison. They will be the guests of Mr. Munson while here."

Why Munson? Volney and Pierre Viala had been corresponding for some time on the subject of phylloxera, and the French horticulturist knew of his groundbreaking classification and research. After all, as the *Gazetteer* boasted, Volney was "recognized as one of the best authorities on the grape and its culture in the United States." Who better to help Viala? No doubt Frank Lamson-Scribner, who accompanied Viala on his American trip and had worked with Volney on the Bordeaux mixture experiments, also recommended the visit.

The French party spent several weeks in September in the Denison area. (Volney didn't return home from his USDA fact-finding trip until late in August, as discussed in Chapter 5, and he wrote H. E. Van Deman on September 23 that "Prof. Scribner & Viala left me yesterday for Dallas & S.W.") On horseback, Volney, Viala, and Lamson-Scribner explored the grape paradise of the Red River, collected many specimens, and consulted Volney's extensive library at Vinita Home. There the French viticulturist took "copious notes" from Volney's research, refining and expanding his thoughts on American *vitis* in search of an answer and agreeing with his host on most points of his grape classification.

As a result of these days and weeks of discussion, Volney was able to write Van Deman with some satisfaction that he had enjoyed the days "riding in their most enjoyable company" and hoped he'd been "of some service to them."[8]

During those warm September weeks, the relationship between Volney and Viala also seems to have ripened from a professional one into a real friendship. Nellie was in the early stages of pregnancy with their ninth child when the French party arrived; and the daughter born seven months later, on March 29, 1888, would be named Viala Laussel, or "Vee," in honor of Pierre and his wife.

Finding the Answer

After completing their research in Denison, Viala and Lamson-Scribner left and headed south. Volney apparently did not accompany them, though it's difficult to be certain since there was almost no coverage of the mission in Texas newspapers. Even the Denison journal had no more to say about Viala's visit after its initial report. Likely, Volney remained at home, for he was seriously behind in both preparation of his grape monograph and his nursery business as a result of the USDA trip. He wrote Van Deman on October 2 that "I have been pushed beyond my ability . . . the past few weeks." Though he was not physically with the French party, he was still directing them. He passed them along to his friend J. R. Johnson of Dallas,[9] who took them on short jaunts east and west of that city.

But their real goal, as laid out by Volney, was further south. The years of wandering and grape-hunting throughout the region stood Volney in good stead now, and he knew exactly where to direct his French visitor—to the Texas Hill Country, a twenty-five-county region in the geographic heart of the state. Geologically an ancient plateau dissected by creeks and rivers, the Hill Country varied in elevation up to 1,300 feet above sea level. Most importantly for the French, its soils were limey and marly.

In the English translation of his report to the Minister of Agriculture, Pierre Viala spent a great deal of time describing the Cretaceous soils of Texas and the native grapes that grew there, establishing the similarities and differences between them and those in the afflicted areas of France. This section is replete with references to Volney: ". . . a new form, considered as a distinct species by Mr. T. V. Munson . . . Mr. Munson has observed . . . according to Mr. Munson . . . ," etc.

Viala described specific observations of grapes he made in Johnson, Bell, and Lampasas counties, and he mentioned other locations he visited: Tarrant, Ellis, Travis, Hays, and Hill counties, and the towns of Austin, New Braunfels, and Waxahachie, among them. The map accompanying his much lengthier French report shows that he traveled west out of Austin up the Colorado River almost to Marble Falls and made a half-loop to the east as well. From San Antonio, Viala traveled a few miles west to look at grapes before boarding the train for the long journey to El Paso, Tucson, and on to Los Angeles.[10]

In the "poor soils" of the Temple-Belton area (Bell County), Viala found abundant examples of *Vitis berlandieri* that both he and Volney thought would be most useful in France. Bell County lies on the Balcones Fault that bisects Texas, its eastern half in the Blackland Prairies. It was the western half of Bell that had first attracted Volney Munson, however. The broken country there is typical of the Balcones, and its "wild chalky limestone hills covered with thickets of cedar and scrub oak"[11] were a perfect match to the Gard region

This enlarged section of the map from the French edition of Viala's final report shows his route through Texas from Denison south to San Antonio, then west to El Paso. (Photo by Roy E. Renfro)

of France, where the streams can actually run white from limestone deposits. In these hills, for example, Volney had earlier found a natural hybrid of *Rupestris-Candicans* and *Berlandieri,* which he named "Dog Ridge" for the nearby mountains of the same name.[12]

Now, at Volney's suggestion, Viala followed a line south of and parallel to the Leon River, traversing the site of modern Lake Belton (dammed 1954) and across land that today is part of the U. S. Army's Fort Hood installation.

Viala had first observed *Berlandieri* in Johnson County, south of Fort Worth, but that was a different soil. The Bell County examples confirmed that this variety would not yellow due to chlorosis in limestone soils and that "it bears very well a graft in chalky soil. I have seen many conclusive proofs of this fact," he continued in his English report, "at Belton, where certain two-year-old grafts of western vines upon four-year-old vines of *Berlandiere* were perfectly green, and of a beautiful growth in those poor soils."

In Tennessee and Missouri, Viala had also seen splendid examples of *Vitis cinerea* and *V. cordifolia.* What excited him, though, was to see them growing just as well in their southernmost range in the "exhausted white limestone" of Texas. Their success in such poor soil made them perfect for his needs.

Viala had found what he came for. His final report recommended three species—*Vitis berlandieri, Planchon; V. cinerea, Engleman;* and *V. cordifolia, Michaux*—"all Texans," he noted with enthusiasm if not strict accuracy.[13] Even though two of the three could be found elsewhere, he was interested only in the Texas finds, for the soils there were a better match. In fact, the English translation of his report is subtitled in part "showing that

Dog Ridge vines still grow on their namesake hills in Bell County, Texas. (Photo by Roy E. Renfro)

A medallion commemorates Pierre Viala at Montpellier's agricultural school. (Photo by Roy E. Renfro)

this [saving the French vines from phylloxera] can only be done by grafting upon native Texas vines."

Volney Munson had sold the products of his state well.

Berlandieri, cinerea, and *cordifolia* were, in Viala's opinion, "the grafts which are most likely to thrive" in France; but he also added candidly that all three "recover with difficulty from cuttings." It would be better, he admitted, to find varieties not susceptible to this problem, but none of them would grow in Cretaceous soils. The three Texas varieties, all in all, had fewer problems and appeared to offer more benefits than any others. Viala returned to France "more convinced [than before] that ... [with some exceptions] ... we must depend entirely upon American grafts bearing our native varieties to ensure the reconstruction of our vineyards, and to maintain for French vines their legitimate reputation."[14]

He concluded the English version of his report by thanking the U.S. Geological Survey and the Agriculture Department. But chiefly, he wrote, he thanked "Mr. T. V. Munson, of Denison, a modest Texas scientist, whose investigations of the native American grapes have brought much honor, save in his own country." Munson and Hermann Jaeger, he declared, "have afforded me, in the chief part, a solution of the questions upon which I have reported, and I am happy to say that they have recognized my conclusions in presence of facts which would have gone against their commercial interests." This last remark is a bit obscure; perhaps he meant simply that the two men were willing to recommend wild varieties over their own hybrids, an honorable stance but not a financially lucrative one.[15]

Volney was well satisfied, too, with the results of Viala's visit. As he wrote H. E. Van Deman, "The great importance of this work [grape botany and hybridization] is now realized by but few[.] If we had 15 or 20 valuable species of wild apples not yet known, even by most botanists, as is the case with our grapes, everyone [every species] would be well illustrated. I believe our native grapes are yet to occupy a more important position among cultivated fruits than the apple[.]" Reflecting Pierre Viala's excitement on his Texas finds, Volney continued enthusiastically and prophetically, "It now looks a little as though the balance of the world will have to rehabilitate its vineyards chiefly from the native species of the U.S."[16]

But Did It Work?

Pierre Viala returned to France in early December after traveling back to Washington across the Upper Midwest. Behind him he left a number of American nurserymen and horticulturists eager to supply him with cuttings and seeds of vines, among them Volney Munson. Unconfirmed reports state that Volney actually spent some time in the Temple-Belton area in the fall of 1887 overseeing the cutting of fifteen wagonloads of dormant rootstock and its shipment to France. He did provide tens of thousands of cuttings from his own vineyards, where he had long used *Berlandieri* and other wild varieties as breeding stock for his hybrids. In fact, the *National Cyclopaedia of American Biography* (1922) noted that "until the French vineyardists began propagating to supply themselves, he [Munson] had an extensive trade in resistant grape stocks in Texas." Volney also sold the French some of his hybrids (especially *labrusca* blends), but these were not particularly successful. As he had for years, Volney shipped these overseas with the Richter family.

The French hurried to graft and plant, especially with *Berlandieri*, certain that their

85

phylloxera problems were at last over. But the difficulties in rooting the cuttings, which Viala had pointed out, continued. Once successfully grafted, *Berlandieri* proved a valuable stock, but those successes were few and far between. Desperate viticulturists tried once more planting the American stock directly, rather than grafting *vinifera* to it, but the wines that resulted were, as before, not up to French standards and tastes.[17] Pierre Viala was "the high priest of *Berlandieri*," yet the sheer force of his enthusiasm for the species, and the respect the French had for him, could carry the grape only so far.

Within ten years of the arrival of the first cuttings, and in response to these grafting difficulties, the French began genetically developing their own rootstock varieties. No troublesome grafting was involved. Phylloxera-resistant American grapes were cross-pollinated with European (preferably French) ones, meaning that the resulting plant could be called French rather than American.

But what has proved enduring in this long war against phylloxera is grafting, particularly grafting with rootstock of ultimately American origin. The French used the *Berlandieri* and other Texas grapes recommended by Volney Munson

Berlandieri from the Texas Hill Country proved invaluable in developing today's hybrid rootstocks. (Photo by Roy E. Renfro)

and the *rupestris* which Hermann Jaeger supported, as well as *riparia*, to create hybrid rootstocks for grafting. Virtually every grapevine in France, Italy, and most of Europe now grows on grafted rootstocks which can trace their heritage back to the United States.

So did *Berlandieri* fail?

Yes and no.

As direct producing and grafting stock, yes, it did fail. And that is still a disappointment to modern ampelographers such as Dr. Pierre Galet, who notes sadly that it "should have played a great role in replanting," save for the difficulties in rooting it. But as the basis of cross-breeding, *Berlandieri* was a success, after Viala's support brought it to the attention of French and other hybridizers.[18] *Berlandieri* crossed with Chasselas eventually yielded the invaluable rootstock 41B, which gave high fruit quality in chlorose ground; 41B was used to replant both the Champagne and Cognac regions. (Its origins actually date back to 1882, but the publicity surrounding Viala's report gave impetus to the first experimental plantings in 1888.) A *Berlandieri/Cabernet-Sauvignon* cross produced Number 333 School of Montpellier, superior even to 41B in its chlorosis resistance. Both inherited *Berlandieri's* resistance to chlorosis and phylloxera and were instrumental in rebuilding the Charente and Champagne regions.[19] Later crosses included Number 99 Richter, 110 Richter, 1103

At Chateau Pavie in Bordeaux, Merlot still grows on 100-year-old rootstocks provided by T.V. Munson.
(Photo by Roy E. Renfro)

Paulsen, and 140 Ruggeri, the latter two originating in Sicily. *Berlandieri* seeds sent to Hungary in 1896 resulted in many new rootstocks there, as well. In essence, these European hybridizers had begun work that Volney Munson had been doing for a decade: crossing *vinifera* with rootstocks such as *Berlandieri* from the American Southwest.

Other American stocks, including Viala's Texas finds, proved easier to manage as grafting stock, but it took years to find ones suitable for all soil types.[20] Before the battle was over—or at least stalemated, since phylloxera continues to be active today—one-third of France's vineyards were destroyed and never replaced. The cost of grafting or replanting was so high that producers simply gave up, too discouraged and financially beaten to wait for the final outcome.

Experiments on phylloxera-tolerant rootstocks are still ongoing. Tests in the late 1990s by the Bavarian State Institute for Viticulture and Horticulture in Germany, for example, confirmed the continued success of *Berlandieri* and *Riparia* crosses. In fact, one of the institute's recommendations, Wu 83-32-4, is a Salt Creek hybrid. Salt Creek, discovered in Oklahoma by Volney Munson, also remains a popular rootstock in such diverse locations as India, Israel, and Mexico.

Phylloxera continues to be a problem around the world even today. The *Oxford Companion to Wine* reported that 85 percent of the world's vineyards in 1990 were "grafted on to rootstocks presumed to be resistant to phylloxera." Despite that, several vine-growing regions, such as California and Australia, are experiencing a resurgence of the pest—including a new and immune strain, Phylloxera biotype B—due to the many ungrafted vines which have been carelessly planted. It will be interesting to see if Volney Munson's research once more comes into play, more than a century after it first proved its worth.

Chevalier Munson

On arriving home from his American mission, Pierre Viala began searching for a way to honor his friends Munson and Jaeger, who had done so much to make his trip a success. Thus he recommended both men, as well as Frank Lamson-Scribner, for the French Legion of Honor.

In his sales catalogs, Volney proudly displayed his Wilder Award medal (top), the Legion of Honor (middle), and Paris' 1889 Exposition Universelle medal (bottom). (T.V. Munson Viticulture and Enology Center)

The Legion was established in 1802 by Napoleon Bonaparte, since all previous orders of chivalry and knighthood had been abolished during the French Revolution. It was open to members of the military and civilians who fulfilled the Legion's slogan of "Honneur et Patrie." Within the order was a sub-category for "Merite Agricole." (Needless to say, its nominees were lengthy during the phylloxera crisis.) The honor carried with it a medal and a small annual stipend. While the majority of honorees were French, the Legion was also awarded to foreigners who had accomplished great deeds for the country; by 1874, there were already 4,000 foreign recipients.

The announcement of awards for Munson, Jaeger, and Lamson-Scribner was not made until December 30, 1888, a year after Pierre Viala returned to France; the actual presentation of medals took place in April of the following year. American newspapers reported the story right after the announcement, in early January of 1889. The *Dallas (Texas) Morning News* was the first, followed a few days later by the *New York Times,* which carried a small note on the front page that "the only foreigners awarded Legion of Honor decorations yesterday" were Lamson-Scribner, Munson (described dismissively as "a vine grower of Texas"), and Jaeger. (January 3, 1889) Nor did the *Times* comment on why they were selected. The *Fort Worth (Texas) Gazette* was much more brazen. "A Texan Leads the World" blared its headline on January 6.

The three men were named Chevaliers du Merite Agricole in the Legion of Honor. The chevalier, or knight, was the low man on the award pyramid; honorees received a munificent sum equaling £10 per year, and they constituted the largest group within the Legion. (In 1883 the number of chevaliers that could be awarded had been raised to 2,000, while the higher-ranking officers were limited to 300.) In Munson's class, for example, there were fifty other Chevaliers du Merite Agricole—all French—in addition to the three Americans.

Still, it was an honor not to be sniffed at, for it carried a certain cachet. The medal itself was a six-pointed star resting on a wreath of laurel leaves. On the front center was a profile of Napoleon, and on the obverse were the words "Merite Agricole" and the year in which this particular medal was struck, 1883. The whole was made of gold and richly enameled in blue, hung from a ribbon of green silk with red stripes, and cost over 100 francs. Accompanying the medal was a certificate from "Le Ministre de l'Agriculture" of "Republique Francaise," which featured an etching of the award at the top and a charming border of grapevines, wheat sheaves, agricultural implements, and farm animals.

The significance of the award may be seen in the fact that only one other living American at that time, Thomas Alva Edison, had been awarded the Legion of Honor. But a number of questions about this particular presentation remain unresolved because many Legionary records from this period were destroyed in the fighting that led up to the 1944 liberation of Paris.

For example, how were the awards presented? The *Fort Worth Gazette* stated that the decorations were presented at the Exposition Universelle in Paris. But Munson family lore asserts that Viala himself came to Denison to give Volney his medal.[21] If so, no newspaper in Texas—including Denison—covered the event. It's also unclear whether Viala, on the same trip, personally presented Jaeger and Lamson-Scribner with theirs as well.

Second, why were these three men the only Americans honored for their efforts in the phylloxera crisis? Many other nurserymen and horticulturists had worked on the phylloxera question, including C. V. Riley, who first helped identify the louse.[22] George Husmann was another excluded from French honors; a colleague of Munson's and Jaeger's, he was also among the first to send cuttings to France and was widely known for his grape publications. (Husmann biographer Linda Stevens believes the answer, in this case, lies in Husmann's Prussian ancestry. The French had just concluded a costly war with Prussia and weren't likely to honor anyone from there.) In the end, it may have come down simply to no one's thinking of including the others and Viala's strong desire to honor special colleagues and friends.

The Missing Medals

A third puzzle surrounds the current locations of the medals themselves. Jaeger's Legion of Honor star appears to be with a descendant in the southwestern United States, but the authors were unable to confirm that. Lamson-Scribner's medal may also be with family but, again, that's unclear. Only the status of Munson's is known. It was destroyed some six decades ago.

The medal was sold for less than $40, the value of its gold content, in the grim, depressed days of the 1930s. This took place after the death of Volney's son Will in 1931, when Will's second wife, Minnie Secoy Munson, assumed control of the nursery. It is believed that Minnie sold the medal to a Mr. Karchner, a Denison junkman and antiques dealer, along with a number of pieces of Volney's equipment, his patent records, and other items. Fortunately, photographs of the medal survive, for Volney was quite proud of it and featured it in his catalogs. In 1992, in Burgundy, Dr. Max Rives presented a replica to Munson family and Grayson County College representatives. In 1998, Dr. Roy Renfro, director of the T. V. Munson Viticulture and Enology Center, found an original Chevalier du Merite Agricole medal from the same period so that the Center at least has an example to show visitors what Volney's medal looked like.

CHAPTER 8
Back to Work

$\mathcal{A}$fter the frenzy of 1887, Volney settled down once more to his business, his family, and his writing. Now that the phylloxera solution seemed to be in hand, he returned to the educational work that was so important to him.

Still Teaching

Volney had never really left behind his teaching days at Kentucky. As family members could attest, he led them through a wide range of topics every noon at their main meal and expected all present, no matter his or her age, to participate, a practice which his own parents had initiated. "You didn't realize you were being taught," recalled grandson Marcus Acheson, who lived at Vinita Home toward the end of Volney's life.[1]

The USDA monograph on grapes was, in many ways, simply another form of teaching. Volney might—and did—grumble at the cost and time involved, yet he was determined to write it so that even a "15 year old school boy can read and study with perfect understanding, and be able to point out on sight any one of our species."[2] But it was not the only method he had in mind for educating Americans about the native treasures in their backyards.

To reach more adults, Volney knew the best course was to snare them when younger, so, in the mid-1880s, he began making available to U.S. colleges specimen sets of grapes. In October 1887, for example, he wrote Charles Bessey of the University of Nebraska's agricultural school in Lincoln that he had mostly complete sets still available (to be finished out at no charge the following season) for $50. He also sold smaller, species-specific sets of such material to private individuals who approached him in the wake of his growing national and international fame.

In July 1888, Dr. George Vasey, a botanist with what is now the Smithsonian Institution, wrote Volney about purchasing approximately forty large sets through the

Volney used photographs such as this one of "Mature One Year Wood" in the educational sets he produced. From left: V. cordifolia, cordifolia var. sempervirens, rubra, Monticola, Blancoii, cinerea, cinerea var. Floridana, Berlandieri, and Baileyana. (*From* Foundations of American Grape Culture)

"National Museum" to distribute to the country's agricultural colleges. Flattered at the request, and eager to have such a large cash order, Volney began "collecting and pressing various specimens of grapes." Much of their subsequent correspondence dealt with the specifications of what would be included in the sets, uniformity of the contents being important. "It seems to me," Volney wrote, "that each college should have above all other specimens a set of vines of all species which will grow in their localities. The study of the habits of growth, the peculiar diseases, the times of flowering, ripening, &c. [etc.] certainly would afford students valuable practical knowledge, superior to anything they could derive from herbarium specimens only."[3] However, the season was already advanced, and his own vineyard still too young to fill the entire order. Moreover, Volney added, his time to collect was limited due to the press of business, "as it will require my personal selection in every case." He would have to do the work "in pieces."

He asked Vasey for an idea of how much the museum could pay per set. "I prepared a set of leaves, growing shoots, annual wood and seeds, with little fruit for the French Government last year, for its School of Agriculture at Montpellier, for $75, delivered in N.Y. [New York]. But I put in numerous specimens of same species from different localities [to show how variations in climate and soil affected the vines]. Of course I could not

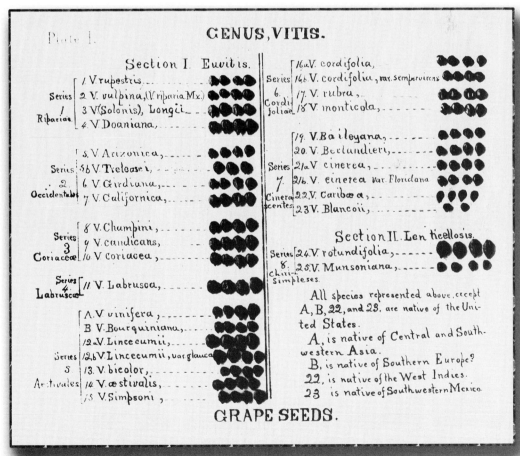

These photographic plates were all life-size to help the reader more easily identify different varieties.
(From *Foundations of American Grape Culture*)

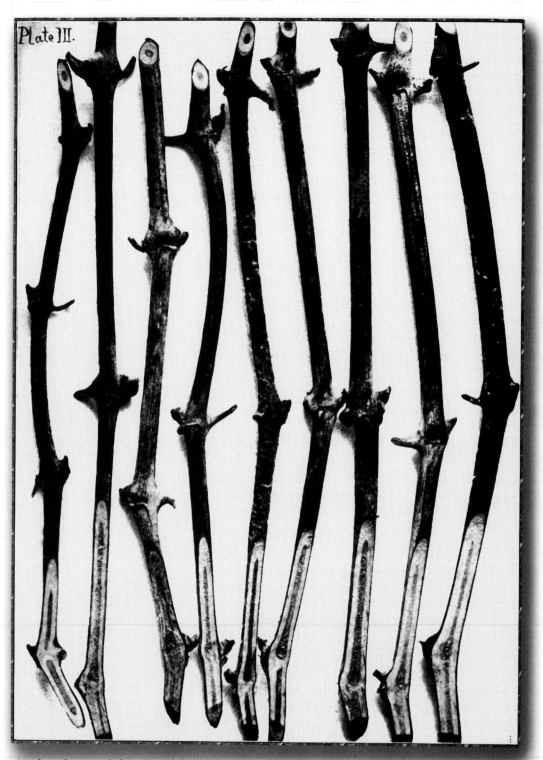

Another educational plate. From left: V. labrusca, vinifera, Bourquiniana [Herbemont], Bourquiniana [Devereux], Lincecumii, Lincecumii var. glauca, bicolor, aestivalis, and Simpsoni. (From *Foundations of American Grape Culture*)

expect such a price per set, yet sufficient to enable me to procure through my correspondents in different parts of the U.S., any I may not have here."[4] He was, in fact, being generous to the museum for these "first sets I collected cost me hundreds of dollars and the procuring and growing of vines of nearly all in great variety, much more."[5]

But George Vasey's superiors at the National Museum choked on the price. In October he wrote back to Volney that he could order only a set for the Herbarium. "A full series for every Agricultural College would be of great value, but the funds at our disposal would not admit of the expense necessary for the full and complete series of specimens of foliage, buds, flowers, fruits. etc." He expressed the hope that the colleges would reimburse and/or purchase sets themselves from Munson.[6]

Volney was annoyed, since he had already put together another twenty sets during the intervening months. These contained specimens of twenty-two species as well as one from Asia and a new grape "just received from Guadalajara, Mexico, found by one of my correspondents, growing among the Sierra Madre Mts. [Mountains]. ... You now can see just what I have done to further your suggestions of July last. You well know," he reproached Vasey, "what it costs to make carefully a collection of such series."[7] Volney then went on to repeat a litany heard frequently in his correspondence over the years with various governmental entities. "My regular business is suffering for my undivided attention, and I must now lay aside this work to another season, and probably entirely as my health is becoming feeble and the demands of a growing family of 6 children calls for *more business and less play.*" (Authors' italics) Nevertheless, he continued, he would be proud and honored to supply the colleges with sets and, "like many other simple beings, for the honor and pleasure of doing others good, would work cheaply."

Vasey apparently wrote back (letter missing) to inquire about the cost of smaller sets rather than the full-blown ones Volney had prepared. "I hardly know what to charge," the latter replied, "not having regularly made such collections before, but would say that Dr. E. Lewis Sturtevant has ordered a set of me, for which I charge $1.00 per species. I agree to supply a growing tip, full grown leaf, 3 to 5 joints [of] mature annual wood and 6 seeds of each species."[8] But since Vasey was interested in forty of these species sets, Volney would discount them to fifty cents each and agree "to supplement with specimen clusters of flowers and fruits, as I may be enabled to procure them, without extra charge. Were it not," he admitted, "that I have plants growing here of nearly every one of our species, I could not furnish sets for many times that am't [amount], if at all."

Thus, his sincere desire to educate overcame personal financial wants and needs.

But once more, the National Museum's budgetary woes stymied even this modest proposal. Vasey had wanted forty sets and Volney had based his price on that; but in the end, the museum ordered only ten, thus bringing Volney not even 10 percent of the several thousand dollars he had anticipated from the original discussions. Moreover, Vasey wanted them immediately, not giving Volney a chance to include blooms. Grudgingly, he agreed to "work away at it at such times as I can get."[9]

Donating His Work

In his later recollections on this project in *Foundations of American Grape Culture* (1909), Volney wrote as if he donated many of these sets; the surviving correspondence between him and Vasey seems to confirm that. Early in January of 1889, he sent the museum ten sets of growing tips and mature leaves from twenty-seven species and varieties, which were to be distributed among unnamed agricultural colleges.[10] (In late February, he shipped the specimen woods to go with them.) In the order was another forty sets "put up for your Division specially," as well as thirty-two packets of grape seeds. He estimated the cost of the entire collection at only $145, "which is far less than the same time and care in my business brings me." Based on his original figures to Vasey, then, it appears he donated roughly $2,000 (1888 prices) to the museum for the college project.

"I have not exactly followed," he continued, "my arrangement of species in the work I prepared for Prof. Van Deman [the projected USDA grape book], as I find my new matter requires some changes in that. The classification I send you is my latest and more nearly accommodates all the facts than any previous arrangement."

For the next six years, he prepared sets of botanical specimens on the various American grape species and sent them to a number of schools. There were three different types of sets. To institutions such as Montpellier in France, Harvard, and Cornell, he sent herbaria specimens of "young and mature wood, leaves, flowers and fruit" of each species. Altogether he prepared eighteen of these larger collections.[11] Volney also donated life-size study photographs of the various components—wood, seeds, fruit, etc.—"to about a dozen colleges and botanists in the United States and Europe." Third, in 1889, he sent live specimens of all the species to state experiment stations and to individual botanists and experimenters here and abroad.[12]

Still, Volney was not finished. "To stimulate experimentation," he recalled in *Foundations*, "over 1000 packets of hybridized grape seeds of select varieties were distributed, gratis, to some 500 grape-growers, located in all parts of the Union." He added with some satisfaction that a number of those had since produced bearing vines.[13]

How much all this cost him personally is unknown. Yet Volney considered it money well spent if the end result was to educate more of the public about the value and potential of the American grape. "This country [should] lead the way, and not wait for France to develop and supply the world with American grapes," he wrote Henry Van Deman. "Hence any plan ... to educate our people concerning all our native species and what [which] of them to chiefly experiment with, for the purpose of improving our viticulture in every direction, is very proper for Government work." And for private citizens as well, Volney would have added.[14]

He also took his educational push to the general public, sacrificing some profits from the nursery to do so. In the 1888–89 sales catalog, he offered customers a real deal: one vine each of twenty-one different species, for a total price of $10. He did this, he wrote, "for the benefit of future grape breeders."

Highlights of the Year

Volney started 1888 with another trip. The American Horticultural Society was to meet in San Jose, California, and officials staged an elaborate cross-country rail excursion with stops at different points to view the scenery. On January 13 the train arrived in Denison, home of AHS vice-president Volney Munson, and the ladies of that town tendered the visitors a "magnificent banquet." The *Dallas Morning News* enthused that this was a great opportunity to fully set forth the horticultural advantages of Texas. Volney joined the excursion in Denison and stayed nearly a month in the western United States, working in a side trip to collect grape specimens in Nevada.

In June he traveled south to College Station, Texas, to give the graduating address at Texas A&M College. "What Shall My Profession Be?," which he had privately printed afterwards, was a mixture of humor, flights of rhetorical fancy, bitter personal feelings, and down-to-earth advice that only Volney could have written. He congratulated the graduates on selecting "a practical cultivator of the soil" such as himself to address them, then waggishly warned that it was their misfortune for doing so if he disappointed. He spoke briefly of his own youth and feelings upon leaving a similar institution—Kentucky A&M—and of his ultimate decision to embrace horticulture, a "laborious pursuit" which he nonetheless could "never abandon."[15]

Volney had harsh words for America's many "so called" farmers, who were in his opinion "the veriest drudges," their families "worse than slaves." Why? Because they had been forced to it, had no love for the work, and considered it "galling servitude" rather than a profession. A farm, he pointed out, was very complex; its management called for an understanding of geology, chemistry, botany, and animal anatomy and physiology. One needed to be a mechanic, too, to deal with equipment. Volney conjured up a vision of "rows of tools correctly classified, clean and in place, like files of soldiers on dress parade."[16] A comment on overproduction harkened back to his first paper before the Mississippi Valley Horticultural Society a few years earlier. "In agricultural pursuits, overproduction to a destructive degree can never come. Who ever heard of a nation, or community, or farmer starving, or freezing on account of having raised too much bread, meat, fruit, wool, and cotton!" Obviously, he gave little credence to the idea of oversupply dropping prices and crippling farmers.

Volney touched on several very emotional topics, too, ones that could be considered questionable in a graduation address but which allow us a glimpse of a darker and embittered side of him. He chastised those people who pried into the private affairs of others as well as those bold enough to define an "incomprehensible God and then pronounce eternal torture against the immortal soul who is brave enough not to subscribe to the definition." Plainly, his unorthodox religious views still rankled with many, who did not hesitate to scourge him about them. Immigrants—"the offal of foreign nations"—also came in for the sharp edge of his tongue. In a tirade reminiscent of his comments on mulattoes, he showed himself a proponent of isolationism. "The world has taken advantage of our too abundant freedom. . . . Homogeneity and elevation of the masses must be secured in any democratic or republican government to keep the people patriotic, and to make itself strong and enduring. We need not go to China to enlighten the heathen, while the heathen are coming to this country so rapidly as to set up and spread their idolatry with amazing success."[17]

Volney had a very emphatic but not too surprising opinion on who should be running the country: the agriculturalists. "The cultivators of the soil, the fundamental producing class, should also be the <u>ruling</u> class," he declared. They, "the largest, most homogenous and patriotic class of people we have," were the only ones who could bring about the "homogeneity and reformation" he believed necessary.

Volney concluded with his "12 Lamps to guide the students through life:

- Sacrifice, which burns the oil of patience and perseverance

- Truth, permitting no deception
- Power, giving strength and endurance
- Beauty, imparting agreeableness
- Utility, always serving a good purpose
- Love, the cement of society
- Memory, which preserves the past
- Progress, which discovers the new
- Obedience to Law, the preserver of all
- Knowledge, the storehouse of facts, of which is born Confidence, or
- Living Faith, by which comes
- Life, or Action." [18]

"You may complain," he then acknowledged, "that I have taught a selfish philosophy. So I have, but it is a far-sighted selfishness, which sees self as a very small part of the whole; and that the best condition of the whole, is that which preserves and develops most fully the greatest number of parts."

This speech well illustrates some of the complexities and vagaries of Volney Munson's character. He envisioned himself as working for the common good, putting food on the tables of America, and enlightening the world with "the gift of useful knowledge" which would lead people "into paths of peace and pleasantness." These were, to him, the highest and greatest callings that a man could have, and every right-thinking person should work toward them as a common goal. Homogeneity.

Yet Volney fiercely respected and insisted on the right of an individual to be just that—an individual. "Choose that pursuit in which you most delight," he urged the A&M graduates. "Have decision. ... Never allow yourselves to become driftwood in a freshet, to be landed anywhere that the tide may cast you." His own religious views were an example of that, for he persisted in maintaining a belief that was far from the ordinary and did so even in the face of persecution. In other words, homogeneity was fine up to the point that it trampled on Volney Munson's individuality.

He had many high-flown words and phrases to describe his goals and beliefs—and sincerely believed every one of them—yet he could descend rapidly and childishly into vituperation when these were called into question. Volney was naive in his sense of what was appropriate to the occasion. Bitter animadversions against personal enemies were hardly the stuff of a graduation oration, yet he went right ahead and spoke them because he felt them, declaring them unpalatable truths that the students—regardless of the circumstances—should be aware of.

All his life Volney would be plagued by this inability to control himself in public when emotionally stressed, whether in speech or writing. Truth was the ideal, he would have said; one should speak the truth no matter when or where or whom one hurt. The unfortunate result was that many people remembered him as much for his outbursts as for his great body of work.

The Ladies

Volney demonstrated his lack of conventionality in yet another area: his attitude toward women.

Perhaps because he had been raised by a strong woman of Scot ancestry, one with her own interests and passions, he encouraged other women to stretch their proverbial wings. His pamphlet with Mrs. J. R. Johnson on organizing a local horticultural society is one example. It could be argued that, as the secretary of the state association, she was the logical choice to work with him; but most men of this period would have selected a male co-author. In only one sentence does Volney show the Victorian mindset: "Take your *wives* with you and have a good social time and interesting meetings," intimating that all the society's members would, of course, be male. (Authors' italics)

But in *Texas Farm and Ranch* (July 1, 1891), he urged the state society to organize local groups and recruit women for their membership. And at the 1888 annual meeting of the Texas State Horticultural Society in Denison, Volney donated one of the only two $10 prizes offered, this one for the "best display of pot plants grown and displayed by a girl under 15 years old."

Nor did he make a distinction between his public utterances and private life. A woman was not a man's slave, in Volney's opinion, but "an encouraging and delightful mate in every field, when *equally educated and not abused.*"[19] (Authors' italics) Nellie Munson was very well schooled for her day and often helped Volney with the voluminous correspondence he maintained, though her real career was raising nine children. (The family joked that Volney ran the outside nursery and Nellie the inside one.) All of his children, the girls included, were encouraged to pursue higher education. Volney and Nellie's five daughters finished Denison High School, four went on to college, and two of them graduated and had professional careers.[20]

A New Strawberry

In contrast to the last, relatively quiet year, 1889 proved to be full of bustle and activity, beginning with Volney's receiving the Legion of Honor and concluding with yet another gold medal from France.

One of his major accomplishments during the year was the commercial release of the Parker Earle strawberry. Though best known for his work with grapes, he actually experimented with many other plant hybrids, particularly the plum, persimmon, and pecan. His new strawberry had its origins in 1881, when he found "very robust seedlings" of another type growing among some strawberries.[21] For two years, he transplanted and encouraged these "No. 3" plants with satisfying results. Then, in 1883, he was offered $500 for "exclusive control" of the berry, provided it continued to show promise. Volney and the buyer agreed it would be named after Parker Earle, longtime president of the American Horticultural Society and a friend of his.

The new berry was exhibited at the New Orleans Cotton Exposition, where Volney also showed his landmark display on the American grape, but that same spring (1885) the plant began to change characteristics. Volney recalled it and returned the $50 advance he'd been given on the deal. At that point, his neighbor, friend, and fellow horticulturist James Nimon stepped in, intrigued by the berry's possibilities. Nimon began crossing the new seeds and eventually came up with one quite close to the original No. 3. By 1887 he had "an enormous crop," and he and Volney estimated that the hybrid could produce as much as 15,000 quarts of berries from an acre plot. After another year of continued suc-

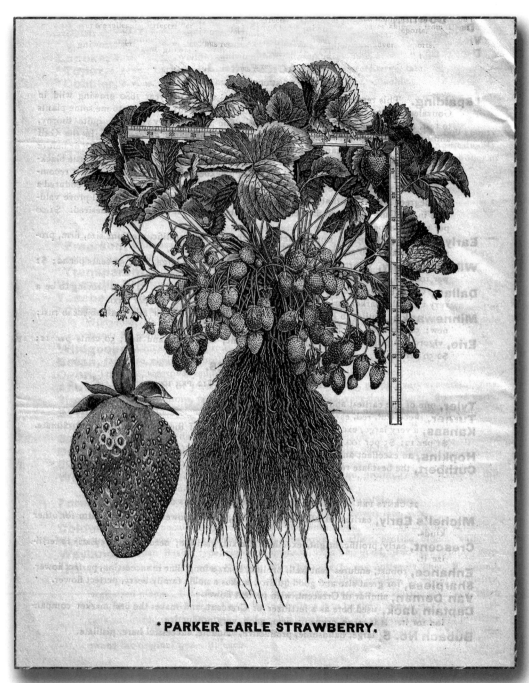

*PARKER EARLE STRAWBERRY.

Munson and Nimon's strawberry was named in honor of the longtime president of the American Horticultural Society. (Courtesy T.V. Munson Viticulture and Enology Center)

cess, the pair agreed to finally bestow on it the name Parker Earle, which had been withdrawn from the earlier plant; and it was released through the Denison Nursery.

Volney also introduced three new hybrid grapes in 1889. Brilliant, a red grape, was the result of a Lindley/Delaware cross which he'd planted six years earlier; it proved one of his more popular hybrids and grew as well in the North as it did in the humid South. Jaeger, named for his good friend in Missouri, was a Post Oak hybrid better suited to the South.[22] Rommel, too, was a southern grape and named in honor of Missouri viticulturist Jacob Rommel, a mentor of Volney's. The latter two plants were later placed on the American Pomological Society's recommended list.

Putting His Name on the Map

Volney was also busy with real estate transactions in the spring of 1889, buying and selling several pieces of property in Denison and the surrounding country. Part of his strategy was to consolidate parcels on Mirick Avenue and in the vicinity of Vinita Home. Like most of his siblings, Volney was attracted by real estate dealings; in late June, following Ben and Theo's lead, he filed a plat at the Grayson County courthouse which laid out "T. V. Munson's First Addition to the City of Denison, Texas." The Houston & Texas Central Railroad cut a slashing boundary across it to the north and west, forming a trian-

gle bounded on the south by Hanna Street and on the east by Fannin, which was a highway in Volney's day. This was all centered on Mirick Avenue and to the north of Vinita Home, which Volney had marked out as Block 12. Within the addition were 148 lots (including the ones on which Vinita Home was situated), the majority of them a standard 50 feet wide and 150 feet long. Most were priced at $400, though a few went as high as $1,000.

However, the addition was never a major success in Volney's lifetime. Only Block 1 (bounded by Heron, Murray, Mirick, and Fannin) and Block A (approximately one-fourth of a full block, just across Heron from 1) sold before 1913. Sisters Trite and Jennie Munson bought half of Block 5 (just south of 1), and Volney and Nellie eventually gave their daughters half of Block 6, while Will received part of Block 4 and all of 2 (which was a fractional block because

The magnolia trees planted by Volney a century ago still bloom on Acheson Street in Denison. (Photo by Roy E. Renfro)

In 1891, when this Bird's Eye View of Denison *was published,* Vinita Home *and the* Denison Nursery *were still in an undeveloped area near the Cotton Mill in the upper right-hand corner.* (Red River Historical Museum, Sherman, Texas)

of the railroad).[23] Another fractional block was sold in 1910 to Denison Light & Power Company; TXU Electric still maintains a transformer station there. Apparently some of the unsold acreage was used for the nursery; the 1896–97 *Denison City Directory* listed Munson's Nurseries on South Mirick at the corner of West Murray, land eventually given to Will.

Today this is a quiet area, full of modest homes, a few of them Victorian but most built from the 1920s and 1930s through the middle of the century. The magnolia trees which Fern helped her father plant and water well over a century ago still line Acheson Street and fill the neighborhood with a divine sweet fragrance in the early summer. The "highway" that Volney noted on the plat is now a simple lane which dead-ends into a modern four-lane di-

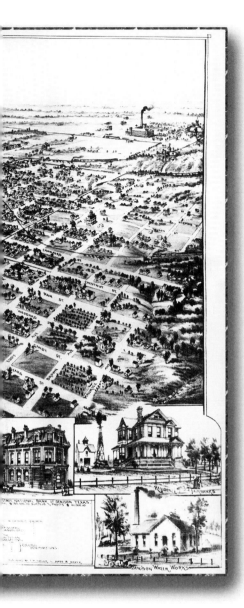

vided highway, but the railroad tracks are still in use. Barrett Street was closed off years ago and reverted to lawns, while Vinita Home, once alone on the crown of the hill, is surrounded by houses.

West to California

The highlight of 1889 was the exploratory trip of almost three months and 13,000 miles which Volney made for the Department of Agriculture. Van Deman first proposed the tour in December of 1888 to collect data for the grape book. But he also wished Volney to examine and bring back specimens of other fruits and nuts as well as statistics and information on wild and cultivated plants of the Southwest and Pacific Coast, an area "as yet but poorly explored" by the Department.[26] To assist him, Van Deman sent along C. L. Hopkins, then a clerk in the Pomology Division. The Department also provided or helped Volney purchase a telescope, microscope, and camera, all of which remain in the possession of the family.[27]

Unfortunately, because of the poor condition of the USDA collection in National Archives II, none of Munson's and Hopkins' field reports could be found; indeed, they appear to have been lost. Only one Denison newspaper article and a few scattered letters and division reports are left to document the journey. In addition to these sources, however, the authors examined grape specimens used in Volney's 1893 Columbian Exposition exhibit and now in the collection of the National Herbarium at the Smithsonian Institution; most are labeled as to the date and/or place of collection. They indicate that Volney and C. L. Hopkins first collected within Texas, around Denison and the Red River, then went south into the central part of the state. During the first two weeks of July, they visited Calvert, Navasota, Waelder, and Gonzales before moving on to Lampasas County and eventually Del Rio on the Rio Grande. Volney and Hopkins gathered the majority of the plants themselves, of course, but also requested people they met along the way to send specimens to Van Deman in Washington. From Waelder, for example, at their behest, J. H. Lewis sent leaves and fruit of *Vitis Spaldingi.*

The pair turned north to the Canadian River then west into New Mexico. By July 23 they were in Albuquerque, where they collected specimens at Bear Canyon on the west slope of the Sandia Mountains. The men stopped briefly at McCarty's Station on the Acoma Indian Reservation and at Cliff Dwellers Gulch near Flagstaff, Arizona.

Their main goal, however, was California, and they arrived in Los Angeles late in July.

San Diego was the next stop, then Santa Rosa, where they briefly visited Luther Burbank. The San Bernardino Mountains, Ojai Valley, and National City were also on the itinerary. While visiting a Mr. Gird there, the trio spotted a mountain lion, "fully 9 feet from tip to tip."

Volney and Hopkins were detained in San Francisco for fifteen days, waiting "for our transportation orders," so they took advantage of the sojourn to explore and collect in the Brown and Napa valleys. Volney also found time to write an article entitled "Hints to Southern Fruit Growers" for the *Southern Horticultural Journal.* They finally left on August 26 and headed north to Redding to collect along the Sacramento River. "It would do your soul good," Volney wrote Nellie and the children, "to see such a river running swiftly, as clear as crystal ... with plenty of trout and other fine fishes." The family, back home enduring the hottest period of a North Texas summer, probably agreed.[28]

Volney collected this specimen somewhere on the line between Floyd and Crosby counties in Texas (northeast of Lubbock). It is in the collection of the "United States National Museum," which we now call the Smithsonian. (Photo by Roy E. Renfro)

To the Summit

At Sissons, California, on August 28, Volney and Hopkins prepared to ascend Mount Shasta. Starting early in the afternoon with horses, a guide, and one other hiker, they arrived at timber line about 8:00 P.M. and struck camp, enjoying "a rude supper" and a few hours sleep. At 1:30 A.M., the party "laid aside every needless trapping" and headed up another 1,500 feet to Horse Camp, where they left the animals. By 9:00 that morning, the men were standing on Red Cliff on the South Crater. "Here," the forty-six-year-old Volney noted with some satisfaction, "one of the party, a mountain climber from New York, gave out entirely, but Hopkins, I and guide went on up a still increasing declivity with steep benches covered with deep, hard snow thawed into a perfect honeycomb ... breaking easily under foot, and endangering limbs in case of a fall." With faces blackened to prevent blistering and wearing "colored spectacles" to protect their eyes from the glare, they reached the summit (14,440 feet above sea level) at midday and recorded their names in the book at the monument there. Volney was enchanted to find "millions of butterflies going south over the summit" but had difficulty breathing in the thin air.

The descent was marred for Volney by a recurrence of "dysentery," and he nearly fainted at one point from weakness; "but I buckled to[,] determined not to be whipped even by Old Shasta." For several miles, the snow was smooth enough to allow them to slide down on sacks. "We would have shot down like a meteor had we not guided ourselves with our feet and Alpine pikes," Volney chuckled.

"And now I feel fine," he assured his family, "and I am proud of the greatest feat of my life." An interesting statement, that. Helping to overcome a plant disease that had nearly bankrupted a nation, receiving the Legion of Honor, changing the attitude of a nation toward grapes ... these were not the greatest feats of his life. Conquering a mountain was.

Heading Toward Home

Volney and Hopkins next traversed Klamath Valley and crossed into Oregon's Rogue River Valley, where "the terrible gulches" of the latter "almost made me sick to look down." They collected specimens at Grant's Pass and along the Rogue River, arriving in Portland on September 2 and visiting a large fruit farm. A steamboat carried them up the Columbia River to Dratilla, where they caught a train. "I feel that I must be hurrying on Eastward and homeward," Volney wrote his family with obvious homesickness.

The exact route of his return trip is unclear because of missing records. It appears that Volney went to the East Coast first to visit herbaria at Harvard and the Academy of Sciences in Philadelphia since the Pomology Division was picking up the tab for the travel. In Washington, he corrected the identification of *vitis* specimens in the National Herbarium and visited with Van Deman before turning west to St. Louis. Volney had planned to see William Trelease and George Engelmann and to spend about a week studying the latter's rich collection. But Shaw's Garden (now the Missouri Botanical Gardens) was so crowded with visitors to the city fair that he was unable to work, and Trelease was too busy to see him. Moreover, he wrote Van Deman, "my time pushing me homeward [has been] so hard," that he decided to move on after only one day and return to St. Louis at a later date.[29]

These herbaria visits were vital, in his opinion, to round out his knowledge of the American grape. "I am ambitious," he wrote, "to make the treatise and illustrations correct and thorough to the fullest degree possible, chock-full and running over with facts ... it will become the standard of reference. I am fully convinced that we have the material in our country for making the greatest viticultural country on earth." [30]

He also took time to collect specimens in southwestern Missouri and southeastern Arkansas on his way home. Volney arrived back in Denison on October 11 and found a mountain of correspondence awaiting him. "Hope to dig out some days hence," he wrote Van Deman wryly.

But the trip was to ultimately cost him much more than a few months away from his family; there would be a financial loss as well as a health toll. On October 28 he wrote Van Deman a letter which concluded with this postscript: "I find my summer's trip has by neglect of my business cost me more than $1000 loss in the Nursery, and now running behind with correspondence and fall preparations. Nearly worked down and suffering with blinding headache most of the time." [31] Will was only sixteen and still in school at this time, so Volney would have had to leave the business in much less skilled hands than his own during his absence and try to catch up on his return.

Nor did the situation with his eyes improve anytime soon. Ten months later he wrote his friend that "the oculists inform me that I have chronic *Glaucoma* in both eyes[.] They have pained me much at times during the past year[.] Blindness may result from the affection [affliction.] It is incurable and was brought on by too excessive use and straining of the eyes." [32] However, the situation must have cleared up afterwards, since he did not become blind as the doctors feared.

The final happy occasion in this very full year was learning in November 1889 that he had won a gold medal at the Exposition Universelle de Paris for his display of American grapevines. The medal and accompanying Diplome Commemoratif are today in the collection of the Munson Viticulture Center.

Munson's grape exhibit won this gold medal (actually bronze in color) at the 1889 Exposition Universelle held in Paris. (T.V. Munson Viticulture and Enology Center)

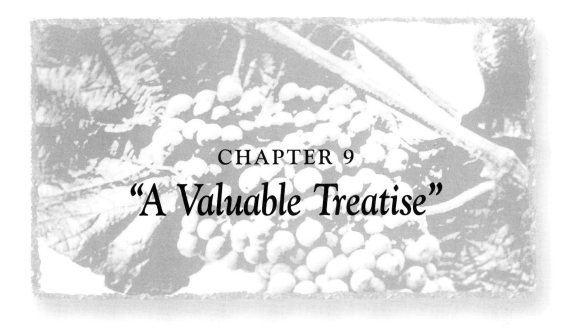

CHAPTER 9

"A Valuable Treatise"

*V*olney had now spent several years laboring on the USDA grape monograph and book, and by 1889 he was hopeful that one of the two would finally see the light of publication soon. As might be expected with a government project, however, there were still a few more problems to overcome.

Finding the Key

During the fall of 1889, Volney continued work on the manuscripts, fine-tuning his classification as a result of the western trip and of his own Denison experiments. His previously named *Vitis Floridiana,* for example, he felt should be changed to the more familiar *V. Simpsoniana* (*V. Simpsonii*) to prevent confusion between the two. "Thus we arrive at the truth slowly," he admitted to Van Deman, "by the study of the <u>growing vines</u>, while the herbarium specimens without them has us into many errors." The research had already consumed "an immense amount of time," and from the beginning of the project, he simply hoped modestly "to get through alive and give you something worth publishing."[1] Clarity and accuracy were his essentials for the book, particularly for those "who <u>run</u> as they read."[2]

The classification was a thorny problem as there was much disagreement among botanists and horticulturists about the details and the correct placement of similar species.[3] Van Deman readily acknowledged the issues but reminded Volney of the audience for whom the monograph was intended. "The construction of a satisfactory key of analysis of the genus is perhaps almost an impossibility, but surely whatever is done in this matter ought to be for the aid of the intelligent and practical vineyardist rather than for the strict scientist." The Department would back Volney's final classification all the way, he assured the Texan, as "it seems to me that it is about as near perfect as can be made, and I think the scientific world as well as the horticulturists will gladly accept it."[4]

This "key of analysis" was the heart of the problem, establishing a set of common characteristics that could be used to easily identify and classify any vine. Volney admitted that it was a task even beyond him. "The genus will not admit of the simple analysis ordinarily employed in sectionizing genera[.] Planchon acknowledges the great difficulty, Viala's work exhibits the same trouble, Millardet did not attempt a generic key. I have tried about half a dozen different sets of characteristics to see if any could be found that would well contain the species without confusion, and maintain the natural gradation of relationship to the fullest extent, but none so far satisfy in every particular."[5] The problem vexed him, for he felt that such an analysis should be part of a book designed for viticulturists, as his was. He had recently outlined still another analysis, he told Van Deman, but grudgingly conceded that it was "not radically different from others I have made out."

As can be seen by his mention of their names, Volney did not hesitate to use French help since authorities in that country had spent a great deal of time studying American grapes in the course of their phylloxera research. "I am fortunate in having the works of Prof. Millardet, and such correspondence with him and other French students of our grapes as stock and direct producers[.] They are the best judges in the world of the true value of a species in viticulture."[6] Moreover, Viala had personally brought him "the proof sheets of Prof. Planchon's great work on universal ampelideat[?], or grape family, so I can have the judgment of the ablest scholars as to classification." Then there were the American experts such as Jaeger and George W. Campbell whose "kind assistance, confidence and large correspondence" Volney relied on.[7]

A Picture Is Worth . . .

Volney had returned from the western trip to find Pierre Viala's just published report on his 1887 trip, *Une Mission Viticole en Amerique,* waiting for him. "I am shocked," he forthrightly wrote Van Deman, "at the poor character of the colored plates. If my work can have no better, I hope it may never be published, and I am sure you would not wish to see it."[8] He admitted that the work otherwise was very creditable, given Viala's "facilities for acquiring data, and his living in another country." But he disagreed strongly with much of his French colleague's classification.

The illustrations for his own book had been a major concern of Volney's throughout the project. For comparison's sake, he wanted at least one plate of each of the twenty-odd grape species (the number had risen during the course of the work from his earlier sixteen), a goal with which Van Deman concurred. "Such plates have never yet been published in this country," Volney wrote him in 1887, "though France has published such of most of our species years ago for general information." And if France could do it, he implied, the United States government certainly could, too—and should. However, Volney was a practical man and knew the size of the Department's budget. "If the Dept. cannot afford to make plates of all our species, then those of chief importance, and the new one[s] ought by all means to be plated [published in plate form]."[9] He also offered a purely practical reason for making these plates by reminding Van Deman that they could be used again in future publications Volney might write for the USDA.

A Lost Manuscript

Volney finally completed his first, short work on the American grape sometime in the spring of 1890 and shipped it to Van Deman. But in mid-August came a disturbing note from C. L. Hopkins, writing in Van Deman's absence from Washington. "I have to men-tion the most disagreeable thing," he wrote Volney with some understatement, "which has occurred in connection with the grape synopsis."[10]

Assistant USDA Secretary Willits had approved the man-uscript for publication and assed it along to the chief of the Editorial Department. That person should have sent it to the Government Printing Office. When weeks passed with no word, Hopkins had begun tracing the manuscript, only to discover the editorial chief didn't remember ever see-ing it come in and couldn't find it. Would Volney send another copy, asked Hopkins? It would then be placed on the special rush list.

Volney's response was im-mediate and to be expected; he would, of course, send another copy but he was not a happy writer. "Such loss and the hold-ing back [of] my discoveries for years," he raged, "while others get them and come forward with the credit is somewhat try-ing to a zealous worker. I pre-sume the Dep. [Department] cannot help the delay, but it will compel me to anticipate its

Swartz Studio in Denison, which did this portrait of Volney sometime around 1890, also photographed many of his grapes for use in his nursery catalog, educational sets, and exhibits such as the one he did for the Chicago Exposition of 1893. (T.V. Munson Viticulture and Enology Center)

work by a publication of my own, if procrastination continues. It is hard for me to write in this spirit," he continued apologetically to Van Deman, "but the case deserves it." He then asked the new projected date of publication, assuming that "it will be gotten out at all."[11]

Volney knew Van Deman wasn't at fault in the matter, but as the project had been long and trying, he proceeded to vent his feelings about government-funded work. "If it had been a great book 2 by 3 feet with full sized paintings of <u>game-birds</u> and <u>fishes,</u> for the accommodation of a few dozen sportsmen, the copies to go chiefly into homes of Congressmen, Congress would have rushed it right through, as we have <u>just such a work</u> already out by the Government, but a modest work that has cost me the spare hours of some 15 years of the best of my life to investigate, and thousands of dollars of my own money and considerable of the Government's to prepare, a work that will ameliorate tens of thousands of homes, and go on with increased development for ages, is beneath the notice of our Washington servants who have the position to attend to our public needs. But such is man, and his governments. As philosophers we must recognize the facts as they are and help ourselves accordingly."

An equally unhappy Van Deman understood his friend's anger. Moreover, he wanted to do a little venting himself. "I had not the slightest idea," he assured Volney in an undated letter attached to the one above, "that the manuscript was not in the hands of the Public Printer and have been daily expecting for months past to receive proofs. . . . I had expected that before this time it would have been distributed. The fault lies wholly with the Editor of this Department who has undoubtedly lost the manuscript by some unaccountable means." Such a thing had never happened before, but he felt sure the synopsis would now be rushed through because of its brevity.

The High Cost of Publishing

The surviving correspondence is scarce and woefully unhelpful as to the exact original parameters of the grape project; only a brief sentence written by Volney himself helps shed any light on the mystery. An "outline skeleton" of his classification, as he described it, was to be an "announcement" of the larger book—"an exhaustive monograph"—with the life-size illustrations William Prestele had been working on for several years.[12] But Van Deman now continued his letter above with a decided piece of bad news. "In regard to the complete manuscript [the full-length book] I fear now that it will have to be published by a special act of Congress as it will take something like $15,000 and that would take nearly half of the printing fund of this Department."

Prestele was still working on the paintings, Van Deman assured Volney, and he himself planned to make just such a request to Congress. But he had serious reservations about doing so. "This is what I had hoped not to be obliged to do as it will necessarily throw a great many of the copies in the hands of the members of Congress to distribute."[13] In the meantime he had "three of the best lithographers in the United States" preparing bids and samples on the steel engraving plates of the illustrations.

Volney, his initial anger spent, replied only that he hoped the new manuscript would "meet a better fate" than the first one. He also added a complaint that he had not been reimbursed for the cost of collecting and mailing any specimens to Prestele since the work began. Van Deman hastened to assure him on September 23 that the first proof of the synopsis was already back, then addressed the pay issue. "As to compensation for the specimens . . . I feel that we owe you a great debt. I only wish that we had an abundant supply of money out of which to pay you, but as we have very little, I will ask you to make your bill as light as possible."[14]

Personal Sorrows

Even as problems continued with the grape book, Volney was also dealing with a number of staggering personal blows—the first one literally that. On September 25 one of his horses kicked him in the left leg, breaking the bone in two places between the ankle and knee. A month later Volney, though hobbling about on crutches, was working again on a limited basis. But by mid-November two "malarial attacks" had delayed his recovery. The broken bones were knitting very slowly and still did not permit him to walk unaided. "It has been a long, painful, dreary siege," he wrote Van Deman, one that showed no sign of ending soon.

That same fall, Volney's mother, Maria Linley Munson, was stricken with a "painfull [sic] illness"; she died late in November at the age of eighty. Husband William followed her a little over a week later, leaving Volney and his siblings orphaned.[15] Volney and Maria had been very close and shared a love of horticulture; her death must have devastated him. The one unalloyed joy of the winter was the birth of Marguerite, Volney and Nellie's fifth and last daughter, the day after Christmas. (Marguerite is the French form of Margaret, and the family referred to her by both names.)

Published at Last

As Van Deman had promised, the grape synopsis, or "outline skeleton," quickly and at last went through production. *Bulletin No. 3: Classification and Generic Synopsis of the Wild Grapes of North America* has been called "one of the most painstaking pieces of botanical work ever done in this country." (*National Cyclopaedia of American Biography*) The Government Printing Office published 6,000 copies, but that was insufficient. By early November of 1890, nearly all had been distributed. Volney was gratified at the response, both personally and because he felt all the media attention it was receiving would help push the larger book through Congress.

But that money was never appropriated. Two years after the publication of *Bulletin No. 3*, Volney wrote Van Deman hopefully that he continued to correct the typewritten manuscript, adding new facts as he had them. In the meantime, however, William Prestele's health had begun to fail; the drawings were not yet complete, and he was having trouble finishing them with the press of other work. Volney, anxious to see the project finally published, suggested that photographs be used instead of colored plates to reduce costs and make the book more economically attractive to print. He even had a set of photos made and sent to Washington, but to no avail. By July of 1893, his queries regarding the book were becoming almost pathetic: "Can you tell me if I can expect my work on the native grapes ever to be published by the Department?"[16]

W. A. Taylor, who had succeeded Van Deman as chief pomologist[17] by this time, was not encouraging. Assistant Secretary Willits, he wrote Volney, believed the book would be published, but "he is not of the opinion that there is any immediate prospect. ... I think myself that carefully drawn illustrations in black which would be much less expensive would be preferable to a long delay in the appearance of the work with colored plates. ... It is certain that the work is of great value and should be placed before the public as soon as possible."[18] The public agreed. Dr. E. A. Bocking of the Scientific Society of San

Antonio, Texas, was one of those who wrote to ask the status of the long-anticipated book; Taylor admitted it had been ready for production for nearly two years.

By the fall of 1893, even Volney had given up. "The long delay in the publication of my work has caused me to regard the work as prenatally dead. Yet I hope it may somehow find resuscitation, as it certainly contains an immense amount of new and valuable material for the progress of Viticulture. Few people know the amount of labor and careful research it involves."[19]

The final Pomology Division annual report to mention the project was issued in 1892, when Van Deman included several references to the completed but still unpublished manuscript. Efforts to see it through to completion simply faded away after he left the Department in mid-1893. As late as 1899, none of the material had ever been returned to Volney but was, as far as he knew, still "shelved" somewhere within the Division of Pomology.[20]

So what did happen to the physical manuscript and illustrations?

The Prestele drawings for the book are now in the collection of the National Agricultural Library in Beltsville, Maryland. They have never been published as a whole

Painting of Vitis Monticola *by William Prestele.* (National Agricultural Library, Beltsville, Maryland)

112

William Prestele painting of Vitis Champini. (National Agricultural Library, Beltsville, Maryland)

because of the cost, though renowned French ampelographer Pierre Galet included sixteen of them in miniature in his 1979 book.

Sadly, Volney's original manuscript of the book is gone; only a few typewritten and corrected pages on *Vitis Californica* survive in the Beltsville archives. Presumably he had a personal copy, lost in the breakup of his library in the 1930s; but the USDA copy, in Volney's bitter estimate, "lies sleeping among the archives of the Department of Agriculture."[21] If so, it is truly gone. All the USDA material was transferred to National Archives II in the early 1990s, and the number of items now missing—either from poor security in the original building or from the move itself—is shocking. Volney's manuscript is among them; it is not listed in the collection index, and the archivist most familiar with the USDA documents has never seen it.

Eventually, with the USDA's permission, Volney used much of the research material in his own 1909 book, *Foundations of American Grape Culture.* He never again wrote for the Department of Agriculture.

CHAPTER 10

The World Goes to Chicago

*A*fter the disappointing denouement of the grape book, Volney began to turn his attention to a much more exciting and promising event—the 1893 World's Columbian Exposition to be held in Chicago.

The Expo

Commemorating Christopher Columbus' first landing in the New World, the Exposition was truly a world's fair, with participants from many countries. Its buildings were designed in a classical Greek and Roman style so wildly popular that the "White City," as the Expo complex came to be known, launched a new architectural style in America, the Classical Revival.

One of Expo's goals was to showcase recent phenomenal advances in industry and technology, such as the world's largest electrical generator. The scientific revolution, in less than two decades, had changed people's lives forever with the development of many modern "necessities," including the electric light bulb, the automobile, typewriter, phonograph, steam heat, elevators, and a host of other products we take for granted today.

Horticulturists wanted to show that they, too, were part of this brave new world of science. A two-day Congress of Horticulture was to be part of Expo, hosted by the Society of American Florists, American Seed-Trade Association, American Pomological Society, and American Association of Nurserymen; the latter two counted Volney Munson as a member and officer. It was, as Liberty Hyde Bailey described it, "the first national occasion, in America, which designed to call together the various horticultural trades and professions."[1] Sub rosa, "a few prominent horticulturists" hoped that the Congress would unite all the major horticultural societies in the United States into a single confederation much like the American Association for the Advancement of Science. Such a union had long

Gold medals awarded at the Chicago Exposition, such as this one won by Volney, featured Christopher Columbus on the front and one of his ships on the reverse. (T.V. Munson Viticulture and Enology Center)

been Volney's dream; he'd touted it at every possible occasion and was now taking an active role in promoting it and the Congress.[2]

Planning

As early as November of 1890, Volney had begun planning the work he would exhibit himself. William Prestele's drawings for the grape monograph would, he knew, be a real attention-grabber, and he wrote Van Deman for permission to use them: "Does it meet your approval to have prepared and framed a set of the paintings of the species according to Bulletin arrangement ... to be displayed at the Columbian Exposition?"[3] With ever the eye for publicity, and hoping to encourage publication of the grape book (which was still very viable at that time), Volney enlarged on his ideas: "I desire on that occasion to show a full set of herbarium specimens under glass, including roots, young seedling plants, &c. prepared and mounted in the most approved manner[.] To have the paintings also, alongside[,] would eclipse any show of the kind ever made, and better educate thousands of people in a short time to distinguish our species, and compel the recognition of our work, even among the stiff-necked botanists, who must <u>see</u> to <u>believe</u>[,] than anything else we could do."

Van Deman was happy to agree, although the paintings weren't yet finished. He and USDA Assistant Secretary Willits had already considered such a display and discussed the possibility; they were pleased that Volney, too, saw the potential value of exhibiting the art. Van Deman also felt that it would be helpful to have a set of botanical specimens and perhaps a map showing the geographic distribution of the various species.

Connections

Volney had several connections to the Chicago Exposition in addition to his work on the Congress and the exhibit he planned to enter himself.

Early in July of 1891, the Texas State Horticultural Society gathered for its annual meeting at Lampasas, in the Hill Country. Volney was appointed to several standing committees for the ensuing year and named chair of the Columbian Exposition Committee. In that capacity, he presented a report recommending that the Society have its own exhibit in Chicago and that it be placed in the projected Texas Building adjacent to the various county displays. The motion carried, despite some "warm" but unreported discussion.[4]

The following year his personal entry received a major boost when the Expo's Board of Managers officially requested him to mount a "complete exhibit" on the American grape. As *Texas Farm and Ranch* kindly noted (November 26, 1892), "None more worthy the honor than he, none more competent."

Other Business

Of course, the two years between publication of USDA *Bulletin No. 3* and the Chicago Exposition were not focused solely on that event. Volney stayed quite busy with personal and professional matters.

He continued to sell lots in his addition to the City of Denison as well as a large parcel to the Denison Iron Rolling Mill Company. He also became interested in the Brazos River port of Velasco near the Gulf of Mexico. Founded in 1831, the town had declined considerably since the Civil War. Now, in 1891, a new town was surveyed and a million dollars worth of lots sold the first year. Volney apparently invested and urged sisters Trite and Jennie to do the same.

In the fall of 1889, Volney and other supporters had organized the National Commercial College of Denison, which billed itself as the "largest business college in America." That may have been the literal truth, for the imposing four-story building at the corner of Main and Fannin contained 86,000 square feet of space and could accommodate 1,500 students. Perhaps reflecting his own experience with such a college as a young man, Volney was elected first president of the board of trustees. Early in 1892, the superintendent resigned with no one available to immediately fill the slot, so the board asked Volney to take on yet another position, this one as temporary head of the school. The *Gazetteer* reported that he would be spending every weekday morning at the college office until a replacement could be hired. It is not clear exactly how long he did so.

Volney also served on the city's public school board for many years, but education was not his only civic involvement. He continued on the board of the National Bank of Denison (founded by brother Ben), and he was president of the board of the Denison Canning Company, with his friend James Nimon as manager. It was planned to be the largest in the South, with 60,000 square feet.

The Land Man

Volney maintained a busy, though less hectic, pace of travel after 1889, attending horticultural meetings around the state, searching for more specimens, and conducting surveys. In the early fall of 1892, for example, he traveled to Aransas Pass on the upper Texas Gulf Coast "to investigate the natural advantages of that vicinity for fruit and vegetable culture, and especially to ascertain its adaptability to the grape and the possibilities of grape growing in that immediate locality."[5]

The three Munson brothers, pictured here in 1895, had a major impact on Denison and Texas. From left: Volney, Theo, and Ben. (T.V. Munson Viticulture and Enology Center)

Why Volney? Obviously, he was the leading authority on the American grape: who better to undertake such a task? As no less an authority than L. H. Bailey wrote that very year, his "experiments in American grapes exceed those of any other investigator."[6]

But he was also in demand as a general land appraiser. Apparently the public considered him an authority on Texas agricultural land no matter where it was because they knew his name and reputation from his many publications and papers.[7] The W. B. Munson archives in the Panhandle-Plains Historical Museum at Canyon, Texas, contain a number of letters to Volney from people across the country, asking his opinion since he was "well versed in Texas land."[8] One priceless example comes from a somewhat later period and was written by J. Woodruff of Fincastle, Tennessee (February 7, 1899). He had received the Denison Nursery catalog and shown it to all his friends. "[T]hey all git whot we call here the texas feaver. They wanted me to write to you and see if you would give some advice about whot locality you think would sute from five to eight families. [Original spelling preserved.]"[9] In another example, an Illinois man who had simply seen Volney's name in newspaper accounts of the Chicago Exposition wrote to ask him for "a correct account of your country, as to both grain and fruit" and the price of land.

These are but a few of the letters Volney received. Most he referred to brothers Ben and Theo's Munson Realty Company, but the correspondence continued to arrive on his desk for at least another decade. One of the last letters to him in the Munson Archives is from the spring of 1904 and indicates that Volney still kept his hand in with the family's business; Joseph Kohout of Chicago wrote to ask for "your price list of farms in Eastern Texas."[10]

Meanwhile, the nursery continued to prosper. In an 1891 interview with the Denison *Sunday Gazetteer*, Volney unabashedly credited that success to "having acquired a thorough knowledge of all the natural sciences and [e]specially applying them in his calling, the cash system in business, and avoiding everything having the slightest character of misrepresentation." (October 18, 1891)

Opposition

But while most people viewed Volney Munson as an authority in several fields, he did have his detractors. Among the most vocal was Dr. William Wynne Stell of Paris, Texas.

Stell (1833–1925) was a medical doctor and Confederate veteran who took up horticulture and farming after the war and opened one of the first greenhouses in Northeast Texas. For some reason which has been lost, he developed a major antagonism toward Volney; the two tangled at meetings and in print for years. A number of their confrontations took place in the pages of *Texas Farm and Ranch*.

In the summer of 1891, for example, he took Volney to task in that newspaper over transactions which Stell charged Munson had made through Denison Nursery. Stell's original attack is missing, but it was apparently couched as an article "puporting [sic] to be a candid investigation of important principles in horticulture," but which Volney interpreted instead as "really nothing more nor less than a tirade of personalities against me." (July 1, 1891) The disagreement had to do with the Stark Brothers nursery company and their traveling agents or salesmen, as we would call them today (one of Volney's long-standing dislikes), who claimed to have enticed customers from him. Moreover, the Starks sold whole root products, which Volney considered next to worthless. (See footnote 4.) What really angered Volney, though, was that the Starks' agents had lied to the public about his products, claiming they were "worthless new varieties" sold at exorbitant prices. "It was to put the people on their guard," he wrote the editor, "against such defaming peddlers that my first article in your paper was written." Volney then professed himself at a loss to understand why Stell had at that point become embroiled in a matter that had nothing to do with him.

Whatever the reason, the doctor accused Volney at horticultural gatherings and through the media of attacking the Starks and of entering into a "combine," something like a monopoly, to raise the price of one of his strawberries. Not true, Volney replied. What was more, he added testily, "I think I have a right to offer my own goods at my own prices—Dr. Stell is not compelled to buy them."

The controversy did not end there, however, as the two continued to snipe at each other. The following spring Stell went on the attack again with his article in *Texas Farm and Ranch* on "New High Priced Fruits." (April 9, 1892) His readers, Stell began, would "remember that for a number of years I have been condemning the practice of disseminating new high priced, and not sufficiently well tested fruits. ... Also in entering into trusts, combines, etc."[11] This was a twofold slap at Volney, who prided himself on the thoroughness of his experiments before commercially releasing any new plant. Stell then published allegedly genuine correspondence from several out-of-state nurserymen to Volney asking him to distribute their new berries.

The entire article is ludicrous, since even Stell couldn't produce a subsequent letter from Volney agreeing to either request. Nor would it have been unethical for him to have carried the products. What point Stell was trying to make is hard to determine, but the whole affair is typical of the acrimony between them. Southern Methodist University professor Samuel Wood Geiser noted that Stell also attacked another well-known Texas horticulturist, Henry Martyn Stringfellow of Galveston County (who planted the first Satsuma orange trees in the state and the first pear orchard in its coastal region), without any seeming reason. Geiser wrote indignantly that Stell "even invoked the *odium theologicum* against the gentle Munson" before finally pleading for mercy and giving up his tirades in 1895.[12]

But Volney was so incensed over this and other problems with the traveling agents he despised that, not long afterwards, and with the advent of son Will's coming to work for him full-time in the nursery, he changed the name of his business. Denison Nursery became Munson Nurseries "on account of unprincipled persons in many regions where we had acquired a good reputation for reliability and excellence of varieties, [those persons] claiming to be agents for and doing business in the name of 'Denison Nursery.'"[13]

Luther Burbank no doubt agreed with his stance. In an article in *Texas Farm and Ranch* (June 4, 1898), the Californian wrote, "It would be startling to the horticultural public if they knew the amount of pirating, thieving and wholesale robbery which the [plant] originator is obliged to submit to without redress."

New Products and Articles

Through the nursery, Volney released several new plants prior to the Chicago Exposition.

In 1891 came one of his most popular grapes, Carman, named for his good friend E. S. Carman (1836–1901), longtime editor of the *Rural New Yorker*, an important American

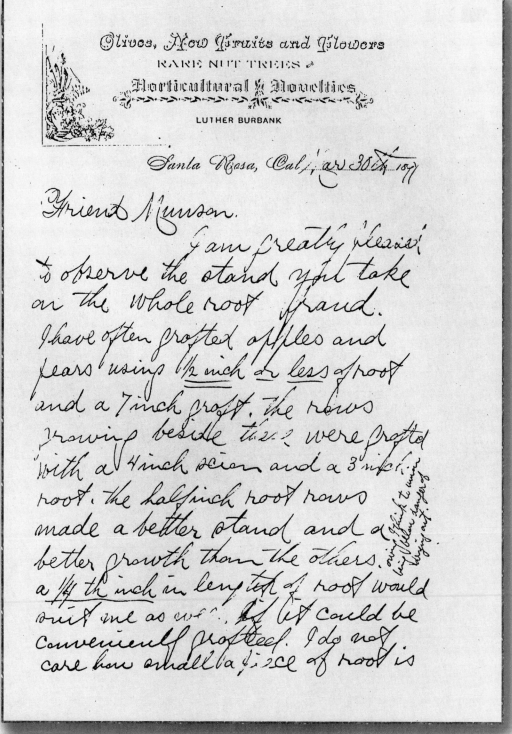

Santa Rosa, Cal, Mar 30th 1897

Friend Munson,

I am greatly pleased to observe the stand you take on the whole root fraud. I have often grafted apples and pears using ½ inch or less of root and a 7 inch graft. The rows growing beside these were grafted with a 4 inch scion and a 3 inch root. The half inch root rows made a better stand and a better growth than the others. a ¼ th inch in length of root would suit me as well, if it could be conveniently grafted. I do not care how small a piece of root is

and I think to using less & below ground trying out grafts.

Volney Munson was sometimes called "the Luther Burbank of the South." In this letter, the California horticulturist supports Munson's views of the whole root question. (T.V. Munson Viticulture and Enology Center)

used. as the results depends wholly on the quality not the size of root. I would greatly prefer an orchard grafted on smallest pieces of roots of a vigorous healthy seedling than to have "whole roots" using seedlings as they come. I do not see how any one who has had any experience in propagating plants or trees can come to any other conclusion.

Sincerely yours

Luther Burbank.

These young girls pick Carman grapes in the vineyard of W.F.D. Batjer near Abilene, Texas.
(From *Foundations of American Grape Culture*)

horticultural journal. While it ripened a bit too late for northern grape growers, Carman, "a notable triumph of a Texas grape," proved popular in the South.

A second release that year was Bailey, a black grape which honored another friend, Liberty Hyde Bailey. Representing Volney at the best of his cross-breeding work, Bailey combined a European *vinifera*, a Texas Post Oak, and a *labrusca* from Massachusetts.[14] A third hybrid, Husmann, named for Missouri (later California) viticulturist George Husmann, was also released in '91. The *Rural New Yorker* published an article on all three which was reprinted in *Le Progres Agricole et Viticole*.[15]

Volney promoted grapes such as these through his catalog, in papers he gave, and in articles he wrote for various gardening and scientific journals. For example, he cannily took advantage of the publicity for USDA *Bulletin No. 3* and quickly penned several articles reusing that material.

One of them, "The Nomenclature of American Grape," appeared in *Garden and Forest* in November 1890 as a response to a mistake the journal had made a month earlier (October 5) in printing his first article on classification, and it demonstrated several points about Volney. The first was that, clearly, he read and spoke French with ease. An editor named Stiles had changed *Vitis rubra* to *Vitis palmata* in Volney's original article. To clear up this point of nomenclature, Volney quoted from both an 1884 article by Jules-Emile Planchon and a recent letter from Pierre Viala. He presented the material to *Garden and Forest* readers in the original French, without translation, in the apparent belief that any educated person could read it for himself, too. Whether that's arrogance on Volney's part, or simply a blissful assumption that everyone was as he was, is hard to say. The second point of the article was that Volney asserted his classification and nomenclature were the correct ones by emphasizing that other important horticultural names—and from another country, too—agreed with him.

And while the tone of the rebuttal was calm and scientific, privately he was enraged over the change. Stiles printed a retraction noting that it had been made without Munson's authorization, but Volney considered that not enough. "His correction will do but should have been more apologetically expressed," he wrote H. E. Van Deman.[16] Volney also believed that Charles S. Sargent, director of the Arnold Arboretum at Harvard and overall editor of *Garden and Forest,* had really been the person behind the mistake. Van Deman soothingly agreed that Sargent was to blame and Stiles' apology insufficient.

On to Chicago

The following year of 1892 passed fairly quietly, with Volney busy in the nursery and in preparations for Chicago. Early in March of 1893 he traveled to the Windy City to install his exhibit.[17]

The Horticultural Building which housed it was on the west side of the Exposition grounds, a gigantic glass greenhouse 1,000 feet long and 250 feet wide, topped by a glass dome 187 feet in diameter.[18] Surrounding it were several acres where plants, machinery, and greenhouses were displayed. A second plot of two and a half acres was set aside for nursery exhibits from France, Mexico, and a few American firms—including Munson Nurseries, the only business in that section representing Texas or the Southwest.

Volney's exhibit, which Liberty Hyde Bailey described as "the most exact and scientific pomological" display in the building, attracted no little attention. It included "all American and most Asiatic species of grapes, represented by twenty-six [pots of] growing plants, by twenty-three [jars of] roots preserved in natural appearance in liquid, by twenty-five sections of wood from aged and young growth, by pressed leaves, upper and lower surfaces, and in all stages of development by flowers, by clusters of ripe fruit in plates, and preserved in liquid in jars, by seeds, and by fifty-five life-size photographs of woods, leaves, flowers, fruit and seeds, all labeled with their common and technical names, and presented in classified arrangement. ... In addition to this, over 150 old and new varieties, representing all manner of crossed and hybrid combinations, were shown in ripe clusters on plates for three months in succession."[19] Among the latter, the *Lincecumii* (Post Oak) crosses caused the most stir.

It was the largest and most complete display ever mounted on American *vitis* and remains so to this day. Justifiably, it received a gold award for specific merit from the Columbian Exposition judges in viticulture. The accompanying certificate, which now hangs in the Munson Viticulture Center, reads in part: "Award for a highly meritorious display, illustrating the vine in all its parts. The work represents years of labor, study, and experiment, and is of incalculable benefit to the viticultural interests of the nation, as it will probably result in the establishment of many new and valuable varieties. ... The entire Exhibit is of great scientific interest and affords superior facilities for study and comparison."[20]

After the fair closed, Volney donated the whole display to the U. S. Department of

This handsome, cardboard-mounted photograph from Volney's exhibit at the 1893 Chicago Exposition is now in the National Herbarium. Vitis cinerea, var. baileyana. (Collection of the National Herbarium, Washington, D.C.)

Agriculture. Until at least 1909, it was mounted for public viewing near the entry of the "National Museum in Washington," i.e., the Castle of the modern Smithsonian Institution. In 1934 the USDA transferred the materials permanently to the Natural History Museum to be incorporated into the National Herbarium. Today it is no longer a discrete collection. The forty-two photographs and thirty herbarium specimens—which are all that are known to have survived—are broken down by species and variety in the Herbarium's *vitis* section.

The exhibit was not the only contribution Volney made to the fair. The American Association of Nurserymen met in Chicago early in June in conjunction with the event, and he presented a paper entitled "The Nurseryman's Position towards National, State and Local Horticultural Associations." (Unfortunately, no copy has yet been located.) On August 17, Volney was also scheduled to speak to the Congress on Horticulture on "Forecast of Better Things Amongst Grapes"; for some reason, however, son Will actually presented that paper.

Disappointed in Texas

But for all the good that came to Volney from the fair, he was very disappointed at his home state's level of participation. Texas had almost no representation at the Columbian Exposition since the Texas Legislature claimed the state constitution forbade expenditures of that sort and refused to vote funds for a building. Two volunteer organizations, the Gentlemen's World's Fair Association of Texas and the Texas Women's World's Fair Association, took up the task instead. The bulk of that work fell to the ladies; headed by Austin artist Bernadette Tobin, their committee raised more than $30,000 to build and furnish the Texas Pavilion. However, it was a slow process, and the structure suffered accordingly.

Only a day or so after returning home from installing his grape exhibit, Volney wrote a blistering letter to the *Dallas Morning News* about the condition of the as yet incomplete Texas Pavilion. "Texas is getting a 'black eye' in the exposition by allowing that hull of a building to stand there as a monument to the enterprise of the biggest state in the Union ... great Texas, as large as Germany [which had a building at the fair] and needing advertising much more, can't give a cent." [21] In a tirade reminiscent of the one he'd written about Congress a few years earlier, Volney continued. "The legislature can spend great sums for newspapers for its own reading and political gambling purposes, but it can only pass laws to depress general business and keep back the general progress of the state. ... Little do I feel disposed to make long expensive journeys over the state, attending horticultural meetings, collecting and publishing valuable information for all the people, while our law-makers ... will not provide a cent for such purposes. It is a burden to live in such a state, when one gets abroad and sees what is being done in other states, making their citizens wealthy and supplying our markets with the products that such increased information brings forth." The Pavilion was an "abortion," he concluded, and gave him "the worst kind of an attack of Texas blues."

Not content with that masterful demolition of the "good old boy" network, Volney renewed his attack in a report to *Texas Farm and Ranch* on "Horticulture at the World's Fair." (July 8, 1893) His visit to the Columbian Exposition, he wrote, was a grand experience, where one could learn more than in an entire lifetime. But "Great Texas, that should have spread her effulgent rays to the farthest limits of her five-limbed star in every department of the Fair ... shows nothing anywhere in its own name. ... Texas is a stench and a byword everywhere about the Fair." On the train ride home, Volney was mortified when a businessman entered the Texas car and exclaimed at its poor condition, only to be informed by the conductor that it was "good enough for Texas." The man then disappeared, and "it is supposed," Volney commented sarcastically, "he went and drowned himself in the lake, for the circumstances had a mighty suggestion in that direction to ourselves."

He then turned his attention back to the Expo and remarked that the temperance people shunned the viticultural division "as though it were a great saloon." Oddly, and perhaps arrogantly, he described his own exhibit in the third person, noting only that it was done by "an individual from Texas" and was "the only display of its kind in the Exposition." He spent several paragraphs extolling the virtues of the grape and grape juice, which the temperance folks insisted on referring to as "unfermented wine." To the contrary, Volney wrote, the juice "is a superb beverage, a reviving, luscious food for the children, the ladies,

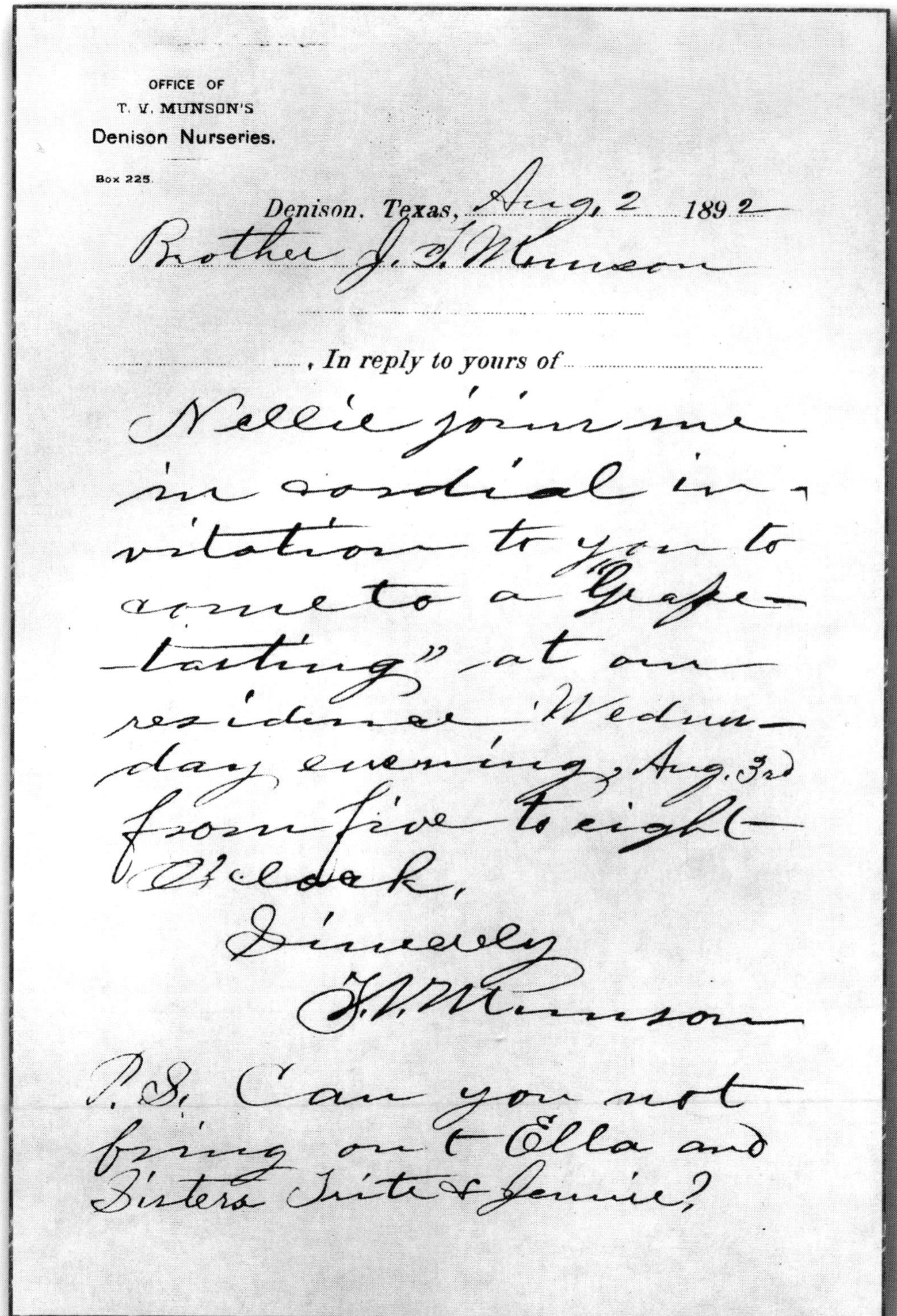

OFFICE OF

T. V. MUNSON'S

Denison Nurseries.

Box 225.

Denison, Texas, *Aug. 2* 189*2*

Brother J. T. Munson

, *In reply to yours of*

Nellie join me in cordial invitation to you to come to a "Grape-tasting", at our residence Wednesday evening, Aug. 3rd from five to eight O'clock,

Sincerely

T. V. Munson

P.S. Can you not bring out Ella and Sisters Tutie & Jennie?

Volney and Nellie frequently invited friends and family to Vinita Home to taste fruit, juice, and wines made from the grapes in his vineyards. (T.V. Munson Viticulture and Enology Center)

the weak and the strong."[22] He continued by declaring moreover that the use of dry white wine "at the table and in the sick room—no where else . . . would add to millions of homes a condiment far more wholesome and palatable than the peppers and pickles so common and dyspepsia-producing. The marvelous developments in every line in France, a great wine using nation, but non-whisky using, is a strong argument in favor of pure, dry wines, used chiefly as a condiment, as by the French."

Some have questioned whether Volney was a teetotaler who abstained from all alcohol. Those last two statements quoted above would seem to answer that. He actually made wine in the basement of Vinita Home to test the enological possibilities of his grapes, and he and Nellie served wine with meals on special occasions such as holidays. What his fastidious nature did object to was overindulgence such as was all too common in the saloons of the day.

Joy Becomes Tragedy

The year 1893, which had begun so auspiciously for Volney, closed on another happy familial note, for Rupert Scott Munson, Volney and Nellie's last child, was born in Vinita Home on December 7 and christened with his mother's middle name as his own.

Unfortunately, he was destined to share the fate of three of his brothers. On October 18, 1894, at a little over ten months of age, Rupert died of diphtheria. After the funeral services, which were held at home, the baby was buried in Denison's Fairview Cemetery next to brother Volney Earle.

CHAPTER 11
The Waning Century

As the last years of the nineteenth century ticked away, Volney devoted much time to continuing and refining his experiments, all with the idea of eventually producing his own book on American grapes. This period also marked a number of important personal and family events for Volney.

Loss of Friends

The first was a sad one, for his old mentor, Dr. Robert Peter, died in Kentucky on April 26, 1894. Peter had retired from Kentucky A&M seven years earlier and devoted himself to further research and his Winton vineyards. He had had a profound impact on Volney Munson, directing his choice of career, instilling in him a desire for pure research and knowledge, and leading him to believe in the evils of whiskey and the relative purity of wine, a stance which made neither popular in a hard-drinking era.

A year later, Volney lost yet another close friend when Hermann Jaeger of Missouri simply disappeared. On a May morning in 1895, Jaeger left his home in Joplin on business and never returned; his wife had him declared legally dead seven years later. To this day, no one knows whether he died in an accident, committed suicide, was murdered, or assumed another identity elsewhere. Not only was this a personal tragedy to Volney, but "his sad disappearance was a great loss to the development of American vine culture."[1]

Transitions

Meanwhile, many changes were taking place within Volney's family.

In 1894, oldest daughter Fern graduated from Denison High School and enrolled in the University of Texas at Austin for the coming fall. Her older brother, Will, had been intermittently taking some courses there, too, since his own DHS graduation in 1891 but had now scheduled a full year of studies for the 1894–95 term in accordance with Volney and Nellie's promise that they would give each child one year's tuition. At the university, Will planned to take as many biology and chemistry classes as possible.[2] His and Fern's departure left six children at home (until baby Rupert's death shortly afterwards), with Warder the eldest at sixteen.

Will anticipated staying at least two years at UT but was forced to leave at the end of the first term because of a severe illness that struck Volney.[3] The surviving records do not make clear what that illness was, perhaps a recurrence of the malarial attacks that had slowed his recovery from a broken leg several years earlier or the glaucoma which had troubled him after the 1889 USDA trip. Another possibility lies in a letter to Theo Munson, in which their sister Jennie remarked that she'd heard Volney "had been quite ill but was convalescing." She went on to ask Theo if the unusually heavy rains had contributed to the rash of illnesses in Denison that summer, intimating that Volney suffered from the same affliction.[4]

Whatever the cause of Volney's incapacity, Will did not return to the university, and his father's often precarious health would disrupt his plans once more in 1898. Late that year he was scheduled to sail to Havana with the First Texas Volunteer Infantry Regiment to participate in the Spanish-American War.[5] But an accident of some sort—again, unknown—befell Volney and kept Will in Denison to run the nursery, which became Munson & Son in 1895. With that change, Will managed the daily operation, leaving Volney free to concentrate on research and the introduction of new varieties of grapes and other fruits.[6]

During the last years of the decade, Warder, Neva, and Olita also graduated from Denison High School, and two continued to college. Warder attended the commercial college in town for a time, studying bookkeeping and working at several local jobs (including the nursery and the Denison Cotton Mill) before moving into banking. Neva selected the University of Texas and, later, Texas State Teachers College (now Sam Houston State University); as a teenager, she had sold her fast-growing, luxuriant hair to a Chinese wigmaker and banked the money for college. Olita didn't pursue college but lived at home, helping her parents, until her marriage.

Solving the World's Problems

Neva's high school graduation on June 7, 1900, was a real family affair since her father gave the address, one in which he expounded on his ideas about education and some of the issues of the day.[7] "The object of a good education," he told the seventeen graduates, "is not to enable the learned to outrun the weak and ignorant ... not to help one to climb into affluence and power, no matter who goes down to make room for him. The politician seeks to do that. It is not to enable one to more successfully engage in trusts, strikes, and gambling market schemes in order to grow fat on the labor of others, or pull down the man of wealth who has acquired riches justly and honorably. The greedy monopolists do that.[8] It is not to teach people to believe a mere theory, and call it truth. The man with a hobby and an ism does that."

Having explained what a good education was not, Volney said that, instead, it was one which enabled a person to "be just and generous, to live well[,] and aid others to live well."

And how did a horticulturist suggest one accomplish this? By harnessing nature, of course, in its infinite forms and energies. Moreover, he asserted, the capacity to convert raw nature into "useful articles" and to *equally* distribute that bounty among the populace marked the chief difference between a civilized culture and a savage one. "Perfect enlightenment means the cessation of all robbery, of all crimes, and the securing of a comfortable living to all." A utopian, perhaps even naive, view of the world—but wholly Munson.

In the spirit of every baccalaureate address, particularly in a millennium year, Volney then told the graduates that their generation, more than any before them, would be called upon to solve these world problems. However, if that proved too much a task, there was still plenty the young people could do. "We need many more good teachers, more scientific investigators, more inventions of useful contrivances, more originators of fine varieties of vegetables, grasses, grains, fibre plants, fruits, flowers and domestic stock; more skilled mechanics and laborers, both men *and women,* who delight in doing good work, not grudges who growl always and never care to do anything but faultfinding well." (Authors' italics)

One's first thought on reading this passage is that Volney had done many of these things and was using himself as an example. But there's more to it than simply that. He was speaking to a millennium class, one which literally would go forth and build the twentieth century. Volney had high hopes that the scientific revolution of the late nineteenth century which he'd helped lead would really blossom forth in the new and cure all the evils of poverty and sexism and cruel misuse of power which he'd described earlier. For to his last breath, Volney Munson sincerely believed that science, with God's help, could make the world the utopia he knew it could be.

His optimism shone through in his concluding paragraphs: "We live in the grandest age the world has ever known [because of the availability of public education]. . . . We live in the greatest nation on earth, greatest because possessed of natural wealth in superabundance and the intelligence to use it." Volney challenged the Denison graduates to "go forward with brave[,] cheerful hearts and constantly strive to be useful, noble, pure men and women."

Some may see in this address a similarity to John F. Kennedy's "ask not what you can do for your country" speech. Indeed, it is very much in that vein and quite different from the rambling, embittered oration Volney had given at Texas A&M twelve years earlier. This theme of service to country and neighbors recurred throughout Volney's writings and reflected his very real conviction that man was put on earth to make a success of himself and to help others do the same. Only with such a "level playing field," where all are equal, Volney believed, could true harmony and prosperity result.

Sisterly Problems

Harmony, however, was not always present within his own family.

Volney's two younger sisters, Trite and Jennie, had never been happy in Denison since moving there in the mid-1870s. After the deaths of William and Maria Munson, they used their inheritance and the money from their own real estate speculations to begin traveling. By 1895, they'd made their way to southern California and pronounced themselves satisfied. "One month here is worth a year in Texas," Jennie complacently wrote Theo. "It is a shame to spend one[']s time in that undeveloped part of the world, if one can get away and live in such a paradise as this."[9]

A few years later, they followed Katharine Tingley, head of the American Theosophical Society, to Point Loma, today part of San Diego; the sisters had first heard her speak back in Texas. Trite and Jennie shared the disdain of much of the family for organized religion. Their choice, theosophy, claimed to be based on mystical insight into nature and God, and it borrowed heavily from Buddhism and other Eastern philosophies.

As the years passed and they grew more enmeshed in the Point Loma commune, the sisters' differences with their brothers, especially Theo and Ben, became more pronounced. Trite and Jennie expected their Texas investments, managed by all three brothers in Denison, to produce money swiftly and in a satisfactory manner, in order to fund their California lifestyle. This required Ben, Volney, and Theo to be constantly dealing with their father's estate, of which the two spinster sisters had inherited the bulk.[10] Eventually their demands for money and the brothers' feelings about theosophy would cause a rift in the family.

The remaining Munson sister, Louisa, and her husband continued to eke out a living in Nebraska. Cyrus' once successful farm and orchard were slow to recover from the economic troubles that followed the 1890 drought, and Louisa regularly asked her brothers for money. In one such letter to Theo, she wrote, "You may ask yourself why does not your husband pay off debts, and keep things in shape. Simply because the man does not know how."[11]

Since almost none of Volney's personal correspondence has survived, it's difficult to say just what his response to these familial problems were. He did visit Trite and Jennie when his own business travel coincided with wherever they happened to be. He advised them on real estate dealings and sent them books and articles he thought they would enjoy reading. Other family letters frequently refer to the contents of those he wrote them. One can extrapolate from Jennie's letters that, perhaps, Volney tolerated their theosophical venture better than Ben and Theo, who, it must be said, seem to have been more incensed that the sisters donated money to Katharine Tingley and her cause. After all, Volney was a man of such liberal notions himself that he addressed the Freethinkers Association in Dallas late in 1896.[12] Some of his later writings contain a tinge of theosophy, so it may be that he was interested in that philosophy, as were many of his contemporaries.

But whether he sent money to Louisa, as Ben and Theo did, is debatable, since Volney was fond of complaining about how financially strapped he was.

Business Matters

The years 1894 and 1895 were particularly strong ones for Volney's work. He introduced no less than thirty-six new varieties of grapes in '94, some of his most important creations, including America, Beacon, Brilliant, Hermann Jaeger, and R. W. Munson. Almost two-thirds of them claimed *Vitis Lincecumii*, the Texas Post Oak grape Volney admired so much, as the mother; still others had a *Lincecumii* cross as a parent.[13] These thirty-six were "the cream from about 40,000 [seedlings and hybrids] grown. If as many as a dozen are permanently retained, I shall feel that my work has not been in vain," Volney modestly wrote in the sales catalog. Yet he later offered this declaration: "I am well aware that I am offering the most valuable grapes ever introduced for general culture east of the Rocky Mountains, many being capable of competing successfully against the leading foreign grapes grown in California, and will surely be keenly sought, even in California and France, when known."

An unidentified grower harvests an abundant field of Volney's Carman grapes.
(From *Foundations of American Grape Culture*)

He continued by unabashedly quoting from friend Hermann Jaeger.[14] "You may ask how it is possible that one man should have done more in this branch of horticulture than was accomplished in eighty years by the grape growers of the whole country," Jaeger wrote. "My answer is, that it is the difference between hap-hazard growing of chance seedlings and the methodical work of a practical man, guided by thorough scientific training."

Munson's ego aside, these *were* some of his highest quality hybrids. The famous Bushberg Nursery catalog of Missouri carried accolades on the 1894 releases from several well-known horticulturists including George W. Campbell, a Munson friend: "I believe he is and has been for some years, doing more extensive work in striving to improve our native grapes than has ever been done or attempted by any other person."[15]

This included his continuing work to develop "a unique and desirable family of grapes" specifically for the southern United States. By the mid-1890s, the result was several dozen Munson hybrids which used Scuppernong and Tomas as the mother stock and the "finer, large-bunched Herbemont" as the other parent. In his 1895 catalog (page 29), Isidor Bush commented on these grapes: "We confidently hope that friend Munson will be successful, and [we] wish that he may obtain for it such price as his great fellow hybridizer, Luther Burbank, gets for some of his wonderful 'new creations.'"

That same edition of the Bushberg catalog marked another important step for Volney. In two earlier editions (1875 and 1883), Dr. George Engelmann of St. Louis wrote the botanical classification of the American grape. But with his death, the task of providing the most accurate information possible for one of the most famous and widely used catalogs in the international world of viticulture fell to Volney. Isidor Bush wrote a glowing preface for "our friend Munson" and declared that "no other man is as well qualified for the task" as he. Bush also put in a plea for Congress to appropriate the funds needed to publish Volney's great book on grapes. It is "a work most worthy of the early attention of the Agricultural Department of our Government, for whose treasury the cost would be a mere trifle; whilst the value, the benefit to be derived from this work would be incalculable." Until that happened, Bush had been able to obtain reduced-sized "photo-engravings" of several Prestele paintings for the catalog.[16]

As to the catalog itself, Volney grumbled that his classification had been "somewhat revised and also somewhat mutilated."[17]

Notes from Around the World

Volney's work took another strong international turn during this period as well. Early in 1895 he shipped a number of grape cuttings, mostly his own hybrids, to the Imperial College agricultural station at Yokohama, Japan, and the Denison newspaper report noted that he already did a great deal of business in England and France. Three years later Volney journeyed to Parras, Mexico, on behalf of an unnamed French firm (probably the Richter family) to investigate the native grapes there.[18] The French paid the expenses for the two-week trip and a "liberal fee for his professional services."

Also in 1898, Volney was elected a foreign corresponding member of the Societe Nationale d'Agriculture de France, an honor of which he was very proud. He was later named an honorary member of the Societe des Viticulteurs de France.

Another international note also came from France. A memorial to Jules-Emile Planchon, in appreciation for his work with grapes and phylloxera, had been discussed for several years. Now complete, the columnar statue, sculpted by A. Baussan, was erected in Montpellier under the initiative of the Societe Centrale d'Agriculture de l'Herault and dedicated on December 9, 1894. Still in place today, the statue is capped with a bust of Planchon, and the column ornately carved. Wild grapevines twined around the base commemorate the American grapes which revived the French industry after the horror of phylloxera. Holding up a bunch of grapes to Planchon is the full-size depiction of a stalwart, handsome man who looks remarkably like Volney Munson. Dressed in working clothes of open-throated shirt, leather gaiters, and hat, he carries a hoe for digging vines over one shoulder; his leather collecting bag rests on the other side of the column amongst the grapes. According to an account of the dedication, that man is "*un vigneron, vigoureux gras de nos belles plaines du Languedoc*" ("a winegrower, a robust lad from our beautiful plains of Languedoc").[19] However, the statue matches the description of one depicting Volney which the Munson family has long asserted is somewhere in France.[20] Two further factors—the physical resemblance of the vigneron to photographs of Volney in the mid-1880s (the time of Viala's visit to America), plus the fact that the column is also dedicated to the American grape—lead the authors to wonder if he may have been the model.[21]

During this period, Volney published several articles in the Parisian journal edited by Pierre Viala, *Revue de Viticulture*. "Explorations Viticoles dans le Texas" (1894) detailed his grape findings in the Hill Country counties of Bell, Burnet, Blanco, and Williamson, while "Les Porte-Greffes des Terrains Crayeux Secs" (1895) discussed vines such as *monticola* and *berlandieri* for hybridizing. Then there was a remarkable five-part series (1894–96) on

The "robust vinegrower" holding grapes up to Jules-Emile Planchon may have been modeled on Munson. (Photo by Roy E. Renfro)

"Les vignes Americaines en Amerique" which featured line drawings of leaves and woods that he likely drew himself. (This series was translated by someone in France, probably because of the time involved; but all the others appear to be Volney's own work in French.) Volney also co-authored a piece with B. T. Galloway, then chief of the Section of Vegetable Pathology of the U.S. Department of Agriculture, on "Le Traitement du Black Rot en Amerique" (1896) for the same publication.

Revue des Hybrides Franco-Americains was another journal to which Volney regularly contributed letters and brief articles. So familiar was his name in France by this time that the *Revue* often referred to him simply as "Munson" and described him as "*l'eminent hybrideur et ampelographe des Etats-Unis.*"

American Publications

Volney published several of his most important English-language works during the last two years of the nineteenth century. First was another landmark series for *American Gardening* entitled "Fifty Years Improvement in American Grapes," which traced hybridization efforts in the U.S. back to the 1840s. Volney divided the series into three parts: northern varieties, southern grapes, and a summary of his own work.

He began Part Three by apologizing for what might appear to some as egotism, but he felt he should share his work "so that others can utilize it with that of my co-workers to carry forward a broad and complete development of American grapes." The fifty-six-year-old Munson then added, "Like Rogers and Bull and Jaeger, I, too, will soon be numbered among those who have ceased their labors."[22] There followed a brief history of his introduction to grape research which he would repeat, almost word for word, in two later publications including *Foundations of American Grape Culture,* his *opus major.* And like those later works, "Improvement" modestly makes no mention of his work on behalf of the French but notes only that Viala visited him and took notes.

The second publication was released in November 1899. *Investigation and Improvement of American Grapes at the Munson experimental grounds, near Denison, Texas, from 1870 to 1900* was the first of two bulletins he would write for the Texas Agricultural Experiment Station based at Texas A&M College.[23] It is a lengthy work of more than 200 pages which sets forth the history of his interest in grapes in much the same manner and wording he used in *American Gardening.* It is memorable for Volney's oft-quoted description of the grape as "this most beautiful, most wholesome and nutritious, most certain and profitable fruit."[24]

Bulletin 56 is also the first time Volney wrote of being called "the vine crank," a nickname bestowed for his near obsession with the field. "This, to me, was good evidence that my object was not comprehended," he noted, "that the special knowledge necessary to comprehend it, and become interested in it, was lacking generally among grape growers, and even among the majority of writers upon grapes. It was clear to me that in order to make the work generally beneficial, and regarded with any degree of favor by the public, that, THE PEOPLE MUST BE EDUCATED."[25] Hence, he continued, he had spent many years in teaching Americans about their native fruits.

Volney would incorporate much of the structure and wording of this bulletin into his own book several years later.

Munson & Son

Volney and Will continued to expand the nursery during this period, and in the spring of 1899 they built a new greenhouse on the north end of the property. Munson & Son had an international clientele though Texas remained the single biggest market; orders for the 1893–94 season alone came from forty-two states and territories of the United States and seven foreign countries. Every year, Volney diligently produced a new catalog of offerings addressed to "My Numerous Intelligent, Discriminating Customers" and featuring on the cover his most famous medals: the Wilder Award from the American Horticultural Society (1885), the gold medal from the Exposition Universelle de Paris (1889), and the French Legion of Honor (1889).

Grapes continued to be the business' strong point, sold to private individuals, vineyards, and other botanical collections. Half the varieties (88 of 169) planted at the Texas A&M Experiment Station, for example, were purchased from Munson & Son. *Texas Farm and Ranch* (July 30, 1898) was quick to praise his work with the school at College Station. "The splendid results of Prof. T. V. Munson's long years of experimental work have here a signal demonstration, for no man can go over this experimental vineyard and not be strongly impressed by the wonderful vigor of many of these new hybrids."

Volney's expertise above all others was now generally acknowledged. "Mr. Munson, of Texas, is doubtless the largest experimenter in this line in the Union, and many of his new creations as grown in Texas, are certainly very fine," asserted George W. Campbell. "The variety which I have tested most successfully, named Brilliant, is very satisfactory, and promises to be among the best red grapes I have grown for general use, and worthy of extensive trial."[26] Andrew S. Fuller, in his 1899 edition of *The Grape Culturist*, credited Volney for his labors in gathering "together a mass of material heretofore scattered through the works of various botanists, and [arranging] it in a convenient order for study. This grouping of the species, or varieties, is probably as near correct as the present state of botanical investigation will permit."[27] And Emmett Stull Goff, writing in *Garden and Forest* in 1895, didn't mince words: "It has been said that Mr. Luther Burbank, of California; Mr. E. S. Carman, of New York; and Mr. T. V. Munson, of Texas, have each of them done more for horticulture in the way of improving varieties than all of the [agricultural] experiment stations combined." Perhaps old friend Liberty Hyde Bailey expressed it best when he wrote that "no other single individual is making a stronger impression upon American horticulture."

Even into the twentieth century, botanical writers would continue to praise him: "But of the many hybridizers of American species ... [Munson] leads all. ... During his mature years, say from 1880 to 1910, he was recognized in America and Europe as the leading breeder of American grapes and the chief authority on their botany."[28]

Moreover, this accredited authority had begun lending his name and endorsement to others. Luther Burbank did not disdain using Volney's complimentary words about his new plum in his 1898 catalog.[29] And T. V. Munson & Son sponsored an advertisement for the *Western Fruit Grower* of Missouri, "the best horticultural paper published," they promised.

Following his success at the Columbian Exposition of 1893, Volney continued to be sought after for similar events. In the summer of 1895, Texas governor Charles Allen Culberson appointed him representative from North Texas to the Cotton States and

International Exposition in Atlanta, Georgia, which was intended to showcase the "New South" and to rival Chicago's fair. Unfortunately, information on attendees outside of Georgia is scarce, and the extent of Volney's participation could not be determined.

Yet he did not allow such honors to "go to his head." He was as willing to speak to the horticultural society in Perry, Oklahoma, as to stand before the American Association of Nurserymen, where he was elected vice-president for Texas in July of 1895. Any group willing to listen was, in Volney's estimation, a chance to educate yet more people; and he took advantage of every opportunity to do so.

Leading the Aggies?

One of the most important acknowledgments of his international status came in 1898. Lawrence Sullivan "Sul" Ross, former Indian fighter, Texas Ranger, governor of Texas, and at that time president of Texas A&M College, died abruptly the first of the year. The school's directors began searching for a new president and soon settled on seven possible candidates—among them Volney Munson. And it was an impressive list, too. Joseph D. Sayers (1841–1929) was a lawyer, former Texas legislator, and U.S. congressman who successfully won the governorship later that year. Fred W. Mally (1868–1939) had been with the U.S. Department of Agriculture briefly before becoming a nursery owner near Dickinson, Texas. An expert on the boll weevil, he later became A&M's first professor of entomology and, eventually, state entomologist. Col. Woodford Mabry (1856–1899) was adjutant general of Texas from 1891 to 1898; Camp Mabry in Austin was named for him after he died in Cuba. Frank P. Holland (1852–1928) was easily among the most qualified for the position. Founder of *Texas Farm and Ranch Magazine* and later of *Holland's Magazine* for women, he promoted the Texas State Fair, sponsored the first Farmers Institutes (later the Texas Agricultural Extension Service) and agricultural clubs for farm children (later Texas 4-H), and was a champion of agricultural schools. L. L. Foster (1851–1901) was a state legislator who became the youngest Speaker of the Texas House, was commissioner of agriculture under Governor Ross, and served on the Texas Railroad Commission. Finally, there was Henry Harrington, a professor at A&M and, coincidentally, Sul Ross' son-in-law. Munson appeared on the short list because of his stature in the horticultural world and his long connection with both the U. S. Agriculture Department and Texas A&M. But on June 7, the directors selected Foster, who, ironically, would serve only three years before dying in office as his predecessor had. Whether Volney would have accepted the presidency is another matter. Presumably, at this stage in his life, he was at least willing to consider it since he allowed his name to remain in nomination.

The Munson System

In the late 1880s, perhaps even earlier, Volney turned his knowledge of grapes into another direction and developed a new training system which he introduced in the *American Agriculturist* in 1890.

Commercial grapevines must be trained onto some kind of support or they will simply flop all over the ground, damaging the fruit. These supports also help protect the grapes from too much sunlight and allow air to circulate freely, preventing rot. In Munson's day,

Beacon, shown here growing on a trellis system of Volney's design, was one of his most popular hybrids.
(From *Foundations of American Grape Culture*)

support systems generally fell into three categories—upright, drooping, and horizontal—all referring to the way in which the canes grew on the supports. William Kniffin of New York developed one of the more popular drooping systems, which used two support wires rather than the three needed for an upright design.

Volney's initial plan[30] was a variation on Kniffin's, but he set two posts in the same hole and flared them out at the top to create a V. This formed an open platform, or trellis, on which the vines could spread. Volney believed his system more closely followed that of nature's since a canopy of leaves formed along the top to shade the roots and fruit, and air could circulate freely. He later refined this "earlier cruder form" and considered it both economic and efficient since it saved time in pruning and tying. In the "Munson Three-Wire Trough Trellis," posts of a strong wood such as bois d'arc, oak, or cedar were placed at the ends of each row with intermediate posts of a lighter wood. Cross-arms were attached to each post with wire, which made them more flexible than nailing or bolting. A center wire was all that was required until the vine grew large enough to need the support of the two additional, lateral wires. Volney also addressed a natural hazard resulting from so much metal and advised running a wire from the earth up the post to serve as a lightning rod. "A vineyard thus provided with lightning arresters," he wrote in *Foundations* (page 225), "will take off, noiselessly and harmlessly, the heaviest of charges from a cloud. I have had rows of vines, not provided with ground wires, almost destroyed by lightning."

Such a trellis, he assured his readers, was cheap and simple to build, made it possible for the grower to stand erect while working on the vines, distributed heat, light, and air equally, and allowed "ready passage from row to row, without going around, thus getting larger and better crops at less expense and increasing length of life of vineyard and the pleasure of taking care of it." (page 227)

To advertise it, Volney and Will frequently set up an exhibit of the "Model Three-Wire Grape Trellis" at national nursery and horticultural conferences. Reaction was generally favorable, especially in the hot and humid South, though Andrew Fuller chastised Volney for making extraordinary physiological claims to make his technique sound better. Texas A&M College tested the Munson System for five years before declaring it "the best system we know of for such vigorous growing varieties as Herbemont and Lenoir."[31] George Husmann, Jr. agreed, writing in a 1902 article that "of late years many use the Munson trellis or a modification of it." Such was its popularity in the South that several variations eventually appeared. In 1945 it was still being listed as one of the eight most popular training styles, and it continues to be listed sporadically in today's literature.

Phylloxera Again

Volney's old nemesis, the phylloxera louse, came back into his life during the 1890s, this time in the United States. First discovered in the *vinifera* vineyards of California in 1873, the louse spread much more slowly there than it had in France because of the absence of the winged female. However, California growers were equally slow to take action. They were not enduring the rapid devastation that had struck Europe and thought they had plenty of time to address the problem. Meanwhile, the experts disagreed over which rootstock would work best. And eventually the state's wine industry, already suffering from overproduction and low prices, was affected by the national economic slump of the mid-1890s. For that reason, some thought phylloxera might actually be a blessing by lowering production and consequently raising prices.

Like the French, Californians first tried any eastern rootstock they could obtain, Lenoir being a particular favorite. Others advocated *Vitis californica* but soon discovered what Viala, Munson, and others already knew, that it was worthless as grafting stock. Amazingly, despite the wide dissemination of Volney Munson's new classification in 1885 and 1886, many growers continued to believe that all varieties within a species were the same. As viticultural historian Charles L. Sullivan later wrote, "it was as if a person with Cabernet Sauvignon wondered at why his neighbor's Zinfandel was more susceptible to rain damage than his grapes. After all, they were both vinifera."[32]

Finally, Arthur Hayne of the University of California agricultural school went to Montpellier himself to see what the French had found, so as not to duplicate efforts. A year after returning, in 1897, he published his recommendation of several *rupestris* varieties, including St. George and Gloire de Montpellier.

But Volney had already beaten him to the punch.

Not only had he traveled himself in California to study its grapes, but he was in regular communication with Luther Burbank in Santa Rosa, old friend George Husmann (who had moved to Napa), and others such as William Pfeffer, editor of *Pacific Tree and Vine*. Pfeffer, in fact, on hearing of Hayne's upcoming trip to France, had written Volney in the summer of 1896 and asked his opinion. Volney believed that *rupestris* would prove best for their needs but disagreed with Hayne's heavy reliance on the French experience, declaring that "the climate and soils of California are very different from those of France, and California must do much original experimentation for herself." In a separate letter to *Pacific Wine and Spirit Review* (August 22, 1896), Volney expressed his preference in these circumstances for St. George, Mission, and Martin.

He continued to stay current with happenings in California for many more years and to offer his advice and experience when needed. St. George, his recommendation, remains in use in California today as a rootstock and has recently enjoyed a resurgence of popularity because of its resistance to new strains of phylloxera.

Volney received little credit in California for his work and assistance. In Arthur Hayne's 1897 book, *Resistant Vines: Their Selection, Adaptation, and Grafting,* for example, phylloxera-resistant American vines were duly noted as "the only practical defense" against the aphid. But Hayne declared that this discovery was "made by Europeans" and does not mention any American involvement. Decades later, Vincent P. Carosso wrote a history of *The California Wine Industry, 1830–1895: A Study of the Formative Years* (1951) in which he quoted Hayne and enlarged on his findings. "The discovery of the resistant vine," wrote Carosso, "was the most important contribution made toward checking the ravages of the phylloxera. It was upon this vine that the entire wine industry—both Californian and European—was saved." But nowhere did he mention Volney Munson. Moreover, Carosso continued, "the classification of native species according to their adaptability to soil and climate and their compatibility to different European *vinifera* grafts inaugurated a new era in viticulture."[33] Again, no credit was given to the man who had actually formulated that classification nearly seventy years earlier.

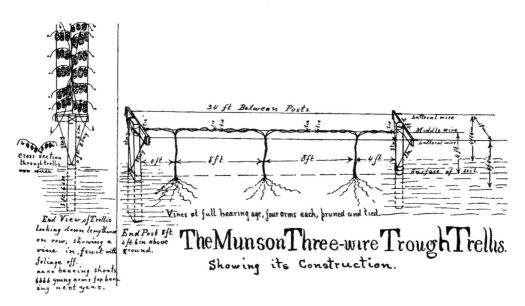

Volney never patented his grape canopy design but exhibited it at horticultural shows. It was a popular design in the humid South for decades. (From *Foundations of American Grape Culture*)

A plum tree originated by T.V. Munson. (Photo by Roy E. Renfro)

Other Fruits

Volney Munson is best known for his work with grapes. But they were neither his only interest nor his only area of experimentation. From about 1890 to 1910, Volney originated and/or introduced more than two dozen other fruits, most of them in the *prunus* family. (Of course, he carried both his own hybrids and the standard varieties for sale through the nursery and its catalog.)

Prunus encompasses several popular fruit and ornamental trees: cherry, plum, peach, apricot, and almond. Most are original to Europe, although the ornamentals tend to be Asiatic. Both Henry Van Deman and Cornell professor Liberty Hyde Bailey were particularly interested in *prunus* and urged Volney to devote time to their study. (In an 1890 letter to the former, Volney even mentioned one specific variety, *P. australis*, which, he wrote, he and C. L. Hopkins named on one of their research trips together.) According to Samuel Wood Geiser, Volney originated and/or introduced eleven new varieties of peaches and seven plums, as well as the Munson

The only surviving illustration of the Munson Cling Peach appeared in the Denison Nursery's 1917-18 catalog. (T.V. Munson Viticulture and Enology Center)

141

The cover of the first sales catalog printed after Volney's death featured the persimmon he originated. (T.V. Munson Viticulture and Enology Center)

apricot, two persimmons (of the Ebony family), and the Munson mulberry (Mulberry family).[34]

In addition to the Parker Earle strawberry, he also introduced the Goree in 1907, "the best strawberry in cultivation." His apples included Bradford's Best, Shirley (which he had found and introduced in 1888), and Rutledge (introduced by Volney and grown from the original tree in Travis County, Texas). The Lexington pear, which he introduced, had actually been found in Kentucky by his father-in-law, C. S. Bell, in 1882 and sent to Volney for grafting.

Some of his research was done at the request of the USDA's Pomology Division; he regularly sent samples to Van Deman for research and publication or to Dr. Vasey to be deposited in the National Herbarium. He also helped them identify new varieties of fruit, especially grapes, which were in question.[35] In turn, Volney was on the list of select nurserymen to whom the department routinely entrusted new seeds and cuttings for testing. Between 1887 and 1892, his name appeared at least eleven times in the "Journal of Receipt and Distribution of New Fruit Specimens." Volney tested apples, nuts originated by Luther Burbank, Persian peaches, figs, chestnuts, persimmons, and others. These usually were

This vineyard at Denison Nurseries was interplanted with rows of peach trees. In the background is the Denison Cotton Mill, started by Volney's brother Ben. (From *Foundations of American Grape Culture*)

plants which the department believed had potential commercial value in the agricultural world; they relied on trusted growers around the country, such as Volney, to try them out and report back.

He had a special interest in Asiatic varieties and regularly experimented with them. The *Report of the Pomologist for 1890,* for example, included an illustration of an ornamental shrub from Japan called *goumi* (*Eleagnus Longipes*), drawn from a specimen sent them by Volney. He was also intrigued by the *kaki,* or Japanese persimmon, which he grew from seeds sent him by Van Deman and from grafted stock he obtained from Japan.

A 1910 price list for the nursery also listed a variety of crepe myrtle known as Munson's Pink, presumably a flowering tree originated by Volney.

Sweet Nuts

Beyond *prunus* and persimmons, Volney experimented a great deal with pecan trees, one of Texas' great natural treasures. The same Hill Country that was full of grapes was also home to some of the best pecans, and Volney looked for both on his travels. Back in Denison, he grew the original trees and also grafted the different varieties to see what new ones he might develop. One of his favorite hybrids he named "Gonzales" for the county in which he found the original specimen; it was, he wrote, "the finest nut I have yet secured in Texas." The Pomology Division agreed, calling Gonzales a promising new variety. (*Report for 1893*)

As with grapes, he collected examples of pecans everywhere and provided the USDA with several lists of native Texas pecans.[36] Apparently, he expected to do a monograph bulletin for the department on this subject, too, for in a letter to Van Deman, he wrote, "I thought you would like to illustrate these nuts and they will come in as part of my work on 'Wild Fruits and Nuts.'"[37]

That Volney maintained his interest in this tree for many years is evident in his 1907 article for *Texas Fruits, Nuts, Berries and Flowers* in which he posed the question, "Can a Race of Commercial Seedling Pecans be Evolved?" He wrote that he planted his first grove in 1885 after buying "a pound of very large, fine, thin-shell pecans" while attending the Cotton Exposition in New Orleans. At the time, pecans were not grown commercially in Texas to the extent they are today, and Volney believed that was a loss. He also knew, however, that building a premium grove was a lifetime's work. "Start some of your boys at this work as a side line," he adjured his readers. "It will be splendid recreation and a constant source of pleasure and possibly a fortune in their old age. Horticulture cannot stand still. It must ever move forward, and breeding is the lever that pushes it more than all else into new fields of excellence."[38]

"Pushing the envelope" was a phrase unknown in Volney Munson's day, but he would certainly have understood the concept.

CHAPTER 12
Meet Me in St. Louis

*T*he first years of the twentieth century would bring Volney significant new international respect and another treasured gold medal, as well as the natural diminution of his family as the children began to marry and move away. Regrettably, he also received a painful slap in the face from Texas' vocal prohibitionists.

Recognition

Several distinguishing international accolades soon came Volney's way. In 1900, Liberty Hyde Bailey wrote a history of American hybridization efforts for the journal of the Royal Horticultural Society of England. Within the grape field, he noted, "[Jacob] Rommel and [Hermann] Jaeger and [T. V.] Munson" were among the western hybridists who "have worked with the species of the mid-continental region; and Munson, in particular, has bred with the excellent and variable *Vitis Linsecomii* of the post-Oak regions of the south-west." Bailey later discussed the great work being done in the relatively new American agricultural experiment stations and added that their research was paralleled by private experimenters. "Various private individuals, [such] as Munson and [Luther] Burbank, are making useful experiments along similar lines and on a commercial basis."

In 1901, Volney was elected a member of the International Conference of Hybridizers, a society of horticulturists and botanists from the United States, Canada, the West Indies, and Europe who were devoted to the latest in plant science. The conference met the following fall in New York City under the auspices of the Horticultural Society of New York, and Volney presented a paper on "Advantages of Conjoint Selection and Hybridization, and Limits of Usefulness in Hybridization Among Grapes." To illustrate his talk, he exhibited photographs of different species and varieties of grapes, a collection he afterward donated to the New York Botanical Garden.[1]

Portrait of Thomas Volney Munson by Pat Pierce. (Photo by Roy E. Renfro)

The nursery's 1926-27 catalog honored its fiftieth year in business.
(T.V. Munson Viticulture and Enology Center)

V. BERLANDIERI

NCECUMII

V. CANDICANS

Three paintings by William Prestele for Volney's never published book on the American grape for the U.S. Department of Agriculture. (Left) Vitis Berlandieri, *(above)* Vitis Lincecumii, *(above right)* Vitis Candicans. (National Agricultural Library, Beltsville, Maryland)

The medal awarded to a Chevalier du Merite Agricole in the French Legion of Honor, 1888. (T.V. Munson Viticulture and Enology Center)

Volney created this hybrid, "Last Rose," in 1902. (Photo by Roy E. Renfro)

Nymph.

Originated by T.V. Munson 1872

Drawn & Painted July 18 1872

Pink verbena from Volney's 1872 Sketch Book. (T.V. Munson Viticulture and Enology Center)

Natural selection alone, Volney told his audience, was too slow; but indiscriminate crossing could prove injurious. His theory of combining varieties to produce the best hybrids of plants, he continued, was the opposite of the quack doctor who thinks that if he combines all medicines into one pill, it will cure every ill. Not so, of course. "The true principal is to produce special varieties for special soils, climates and uses." In this same way, he had crossed *berlandieri* with *vinifera* to achieve a grape well-suited to the hot, dry climate of Southwestern Texas with its limy soils and phylloxera. "It is quite clear, from all our experience, that little can be gained by combining more than three or four species for any particular region, and those species should be, as far as possible, the best selected natives of the region to be supplied."[2]

This sounds elementary to us but was still widely debated in Munson's day, and the paper well illustrated both his grasp of the "new" science and his familiarity with the latest publications on those subjects. At its conclusion, the presiding chair congratulated Volney and pointed out that it "comes from one who has followed the work in this line for a long time."

Volney continued his commitment to expansion of those ideas when he helped organize the Society for (the Promotion of) Horticultural Science in 1903, following the meeting of the American Pomological Society. "The object of the Society is the strengthening of horticultural investigation and teaching on its scientific side[,] and the aiding in the development of horticulture as a science. ... The field of the Society ... lies between that of the popular societies on the one hand and that of the societies for general science on the other and connects them."[3] These were certainly goals the sixty-year-old Volney had advocated for decades. He supported the new organization, which attracted many of America's most forward-looking figures in that field, by serving as an honorary vice-president from 1903 to 1907.[4] At the 1905 conference in New Orleans, he presented a paper on "Improvement of Quality in Grapes" in which he spoke of his own work. "Here we are on the threshold of the far reaching and intricate subject of breeding grapes for various purposes, seasons, soils, climates, varieties of color, flavor, etc. ... It is in this field that the writer has worked diligently during most of his spare moments from business for thirty years[,] producing many thousands of hybrids and including combinations of nearly every one of the 24 American species and the Asiatic kinds of grapes." To illustrate the positive results of the methods he advocated, Volney declared: "Thus in two generations I have raised the sugar content of Scuppernong [an important grape for the South] 40 to 50 per cent, greatly thinned the skin, reduced the seeds, increased the clusters and productiveness, without getting any susceptibility to disease."[5] Yet there was still, he acknowledged, work to be done to make Scuppernong even better.

His support of new organizations such as the SFHS led him to also join the American Plant and Stock Breeders' Association, which organized in St. Louis in 1904 following the Louisiana Purchase Exposition. (The name was later shortened to the American Breeders' Association, and in 1913 it became the American Genetics Association.) One of its goals was to boost understanding and knowledge of "artificial evolution" in breeding.

His expertise in the viticultural field was recognized by many of his peers, such as fellow Texan Gilbert Onderdonk, a renowned authority himself on peaches as well as grapes, especially for the coastal South. In several manuscripts and articles written early in the century, Onderdonk commented on his own positive experiences in raising Munson hybrids and referred his readers directly to their source. "But if the writer were transformed into a young man again [Onderdonk was then in his seventies] and wanted to plant a vineyard in North or Central Texas he then would apply to Prof. Munson for advice." Recommending Perry, Carman, Muench, and Sweetey for the coast, Onderdonk added that "if my labors had not been cut short by my attack of paralysis I believe I should have found others of Munson's new creations that would be valuable in the coast country."[6]

Another compliment of a more unusual sort came Volney's way in 1906, when horticulturist Patrick O'Mara penned an acrimonious letter to the *Florists' Exchange* regarding the Carnegie grant of $100,000 recently awarded to Luther Burbank of California. Why should Burbank receive that, O'Mara wanted to know, when there were many others just as, if not more, deserving? Among those he listed to debunk the notion that Burbank was a "solitary pioneer" was Volney Munson.[7]

Growing Up

Meanwhile, the Munson family circle in Denison was changing. The first of Volney and Nellie's children to marry was Fern, in 1901. She and Archibald Acheson, an employee of the Denison Post Office, were married at the Grayson County Courthouse in Sherman on October 30.[8] Will followed suit a few months later when he took Nina Cummings of Austin as his wife and built a house on a section of nursery land not far from Vinita Home. Warder and Neva, the third and fourth surviving children, remained single. She taught school in Denison and lived at home, as did Warder, a bookkeeper at several local businesses. But Olita married N. C. Calvert on April 23, 1903, in a very elegant wedding with the reception afterwards in the Munsons' "beautiful suburban home."[9]

Volney and Nellie began giving the children pieces of his extensive real estate holdings, mostly in the Vinita Home vicinity, as they became adults. Will and Fern had each received such a gift in 1899; three years later, so did Warder, Neva, and Olita. In 1910, Marguerite and Viala would also share in this distribution, while the older siblings received additional lots in T. V. Munson's addition.

In other family matters, Nellie's father, Charles Stewart Bell of Kentucky, became blind in 1900 as a result of cataracts and his general health began to fail; he died in 1905, widely praised for his lifetime of horticultural work with the Lexington Cemetery. His passing must have been a great personal loss for Volney, too, since Bell had been as much of a mentor to him as Robert Peter.

Within Volney's own family, a rift grew up in this period between Trite and Jennie in California and the brothers in Denison. Volney's views aren't clear, but Ben and Theo[10] disagreed vehemently with their sisters' theosophical lifestyle. As early as 1900, Trite was imploring Ben to understand "the great and good work that is going on here for humanity."[11] By 1904, the argument became more heated, particularly over money and the brothers' management of the sisters' funds. In one complaint about a delay in receiving their statements, Trite and Jennie wrote Munson Realty's office manager, "We certainly deserve simple business courtesies and promptness of attention in matters of business, even if they cannot give us *true brothers interest and affection.*" (Authors' italics)[12] This would seem to refer to all three brothers since Volney continued his part-time involvement with the business and is mentioned in many of Jennie's business-related letters. However, the siblings eventually settled their differences or at least learned to live with them.

A Booming Business

The Munson Nursery kept Volney, Will, and the rest of the family quite busy during these years.[13] In the first decade of the century, sales of fruit trees and grapevines alone more than tripled from their previous level. A 1906 biographical sketch of Volney noted that he had shipped products to every state in the Union and abroad to France, Italy, Australia, Spain, Germany, East Africa, Natal (South Africa), Brazil, Japan, and Mexico.[14] These were the results of years of work which he described in a letter January 9, 1909, to N. M. McGinnis of the New York Agricultural Experiment Station in Ithaca. In addition to the several hundred hybrids he'd introduced commercially, Volney wrote, "I originated a vast host—75,000 or more—that died nameless as they either showed serious defects in vine or fruit. Most of them were not even carried to fruiting age." He added that, in his opinion, "it requires 6 to 8 years of fruiting to fully bring out the true permanent character of a variety." In a postscript to the letter, Volney commented on this percentage of return. Of 75,000 hybrid seedlings, he considered less than a hundred to be worthwhile; "that is one of value in every 1,250 hybrids." But, he continued cheerfully, "with present knowledge and material, I think I can produce about one of value in every 100 hybrids."

His success extended to other products, too. In 1911, Volney reported to the Agriculture Department that he had sold over 5,000 grafted or budded pecan trees in the last five years and at least that many more seedlings. His decision to locate in the railroad town of Denison had been good for business since its rail connections made it possible for him to reach a much wider market.

He continued to test new products for the USDA. In 1899, for example, staff pomologist George C. Husmann, Jr. (son of Volney's old friend) had sent him some eighty vines, mostly French, for field experiments. Six years later, Volney reported back in his meticulous manner, noting those he considered had done well, those too weak to be of value, and

In this view, the Munson & Son Nursery office building stands to the right of Vinita Home, which had been painted white. (T.V. Munson Viticulture and Enology Center)

varieties he had long ago discarded after his own extensive testing. In the summer of 1905 he attempted to grow orange trees for the Department but all four died soon after planting, the victims, Volney believed, of freeze damage suffered before he and Will received them. Department staff members such as William A. Taylor traveled to Denison periodically to see these and other experiments in progress. Nor were they Volney's only visitors. Munson Nursery was a popular destination for many individuals and institutions such as the Biltmore Herbarium in North Carolina, which sent a collecting team there in the summer of 1903.

Of his own grape hybrids, Volney released several new varieties. The most important was Headlight, created in 1895 and disseminated commercially in 1901. U. P. Hedrick (*The Grapes of New York*) considered it "one of the most promising of Munson's many valuable grapes," as worthy of trial in the North as in the South.[15] Volney also sold a number of his hybrid cuttings to the Agriculture Department for a grape experimental station just opened in California.[16]

Volney believed firmly in the importance of such stations to further horticultural research. In 1901 he and two other Denison businessmen traveled to Fort Worth to lobby the state to establish its new station in their town. More than two dozen towns and cities made presentations to the Texas A&M Board of Regents at that meeting. However, the station ultimately went to Troup in East Texas, and Denison never was selected.

Running the Good Race

Other than his service on the Denison School Board, Volney had long eschewed any role in local government. However, in June of 1904 he yielded to the "importunities" of many people in Grayson County, "especially ... farmers who have had dealings with him for years," and allowed his name to be placed in nomination for the Democratic primary for the state legislature.[17] The *Denison Sunday Gazetteer* supported his campaign enthusiastically: "He is a very conscientious man, scrupulously so in all business transactions, and can always be depended upon to do what he believes to be right, regardless of consequences ... we believe he would prove a valuable member of the lower house. It would be a credit to Grayson county [*sic*] to elect a man who has earned such an enviable reputation both in this country and abroad for his researches. ... Every agriculturist and horticulturist should feel proud of an opportunity to cast a ballot for Prof. T. V. Munson."

Because the announcement of his candidacy came just two weeks before the primary, with, consequently, no time for many personal appearances, Volney chose to reach the public with an "Open Letter ... to the Democrats of Grayson County." He had no ambition for public office, he wrote, but could not reject a call from his fellow citizens to do his duty. He pledged to support agricultural interests as well as those of the public school system and closed by signing himself "Yours for wise, pure and beneficial legislation."

Unfortunately for Volney, his past very much came back to haunt him at this point. The very research and work the *Gazetteer* lauded made him extremely unpopular with Texas' Prohibition Party.

Since 1887, prohibitionists had striven to save Texans from "rum tyranny." As a result, well over half the state's counties were dry, and many churches had even given up fermented wine for communion. In 1903 Grayson County had finally adopted local option, making it the most populous dry county in Texas.[18] Local liquor proponents took those election results all the way to the U.S. Supreme Court but lost the fight just shortly before Volney announced his candidacy.

Several factors worked against Volney's campaign. First, H. A. Ivy, secretary of the Texas Local Option Association and a longtime spokesman for the Prohibition Party, lived just ten miles away in Sherman; his proximity allowed Ivy to easily influence voters.

Second, Volney himself had been very outspoken with his opinions on prohibition and wine drinking. In the 1904 *Texas Almanac and State Industrial Guide,* released early that election year, he declared that "the prohibition laws have practically destroyed the industry of winemaking in Texas." Volney also believed that prohibition was costing Texas a profitable industry and potential wealth that could be even greater than that of France. He extolled the benefits of growing grapes for table use but enraged the temperance folks by concluding that "a pretty bottle of sparkling wine adds much to the beauty of the table and greatly to the condiments and healthful appetizers," a sentiment he had expressed in print several times earlier.[19]

Here, then, was a man who not only encouraged people to drink wine but had helped resuscitate the struggling grape and wine industries of Europe. To the prohibitionists, Volney Munson was practically the poster boy for wine, and they set out to defeat him soundly. H. A. Ivy even issued a circular targeting prohibition opponents such as Volney around the state. In addition, the party enlisted the labor vote on its side.

Volney's brother Theo and Sherman attorney Don Bliss tried to salvage the campaign by hiring M. W. Bowles to "put one of your brother's [Volney's] cards into the hands of every working man in the town as well as many farmers and others as he could, but especially the working men, he [Volney] being one of them and a member of their orders."[20] But Bliss was unable to raise the funds needed; apparently the prohibitionists had scared off contributions. The race was really lost, however, when "that fellow Gafford got up that talk that your brother [Volney] was the enemy of organized labor."

The result was foregone, though it took several days and the official canvass of the vote to settle the question of whether Volney or a Mr. Witcher had won the third seat: Volney came in fourth, losing by 40 or 45 votes.[21] The *Gazetteer* (July 17, 1904) reported that he "made a splendid race," considering that the prohibitionists and labor voted "almost solidly against him."

Meet Me in St. Louis

Volney never ran for public office again but instead turned his attention after the election to the next big event in his life, the Louisiana Purchase Exposition in St. Louis, Missouri.

The Palace of Horticulture there was the largest building erected up to that time to display horticultural products; covering six acres, it cost nearly a quarter of a million dollars to build. Designed much like its predecessor at the 1893 Chicago Expo, it had a large and lofty central hall for fruits and nuts with two wings devoted to flowers. There was also a library and a gallery for meetings of various related societies. Inside the Palace were exhibits from twenty-six U.S. states and territories and seventeen foreign countries.

Volney's association with the fair, which commemorated President Thomas Jefferson's

The Festival Hall and Grand Basin at the St. Louis Exposition, 1904.
(T.V. Munson Viticulture and Enology Center)

The magnificent Palace of Horticulture in St. Louis. (T.V. Munson Viticulture and Enology Center)

purchase of the Louisiana Territory a century earlier, had begun in 1903 when he was named to the Texas World's Fair Association. Soon after, he was selected a member of the International Jury of Awards for horticulture in group 107 (pomology) and certain classes of 110 (seeds and plants for gardens and nurseries). "For the most part the members of ... [the horticulture juries] were professors of Horticulture in the various agricultural colleges, horticulturists of the state experiment stations, or fruit growers and florists of national reputation."[22] The selection process for jurors was rigorous. Every department chief submitted a list of men he thought best qualified; the director of exhibits made the final choice, which then had to be approved by the Exposition Committee. Almost half of those selected were from other countries. Texas had two judges in pomology, Volney and E. C. Green of Texas A&M College, "in recognition of the work being accomplished by the Texas horticulturists." Only Missouri had more representatives on that jury.[23]

By the time he lost the Democratic primary on July 9, Volney had already selected a sampling of Texas pecans, all grown in Grayson County, for the U.S. Department of Agriculture exhibit at the fair and had sent up his own first specimens of grapes, peaches, plums, and prunes. With some effort, he and Will maintained a constant display throughout the event, sending their first fruits at the end of June and the last in early September. Most of the grapes were his own hybrids, and he must have been equally gratified to see them on display from other agencies and states as well. The Texas Experiment Station at

The Texas Pavilion echoed the state flag's five-pointed star design.
(T.V. Munson Viticulture and Enology Center)

Troup and the states of Illinois, Arkansas, Missouri, New York, Michigan, Texas, and Kansas all sent in Munson varieties such as Brilliant, America, and Beacon.

Nor were grapes his only entries. His Kawakami persimmon was the only Japanese-American hybrid at the fair, and his total persimmon exhibit of thirteen varieties the largest. The only Buerri pears in the Texas exhibit were from Munson Nurseries, and his Ray peach was the first one received. He and Will also had an entry in the evergreen section (Group 108).

Jurors began convening on September 1, 1904, to judge the various exhibits, and Texas and the Munson Nurseries did very well. The state was awarded grand prizes for both its exhibit in the Palace of Agriculture (which included Munson products) and for its collective exhibit of fruits in Horticulture.[24] For their grape exhibit, Volney and Will carried home one of only six gold medals given to Texans, as well as a bronze for their evergreens.[25] Volney received a certificate for his jury service and a second which accompanied his gold medal. These documents, featuring an allegorical image of a woman draped in an American flag, are now in the Munson Viticulture Center collection.

In 1905 the Texas State Horticultural Society published a review of *Texas Fruits at the World's Fair,* 1904 which included many flattering references to Volney and Munson Nurseries. In fact, the only full-page photograph in the booklet was of Volney, with a smaller view of Will.

The bronze medal for evergreens won by Munson & Son Nursery, 1904. (T.V. Munson Viticulture and Enology Center)

Making a Scientific Name

During this decade, Volney concentrated his efforts in three areas: originating and releasing more new hybrids, writing for publications, and presenting papers. Much of this may have been done with the idea of solidifying and expanding his professional reputation in advance of publishing his own book on American grapes.

At least eighteen new Munson grapes were commercially introduced in the decade 1900 to 1910, including some of his more popular hybrids: Ellen Scott (named for wife Nellie, 1902), Headlight (1901), and San Jacinto (1908). The latter proved a valuable parent stock, though not even Volney knew its exact heritage; it was the result, he later wrote, of a Scuppernong accidentally pollinated by one of several *Lincecumii* varieties nearby. Toward the end of the decade, Volney also began working on a new group of hybrids, five of which would be released posthumously in the 1917–18 catalog and named for his grandchildren.

A Munson Nurseries brochure in the Leyendecker Collection at the Center for American History (Austin, Texas) lists eleven of his varieties and can be roughly dated to

1902. Volney was not modest about his grapes. This pamphlet is forthrightly headlined "Munson's Newer Grape Creations—Still They Come ... Better Than Ever."[26] It began with an extraordinary statement. "To provide against the possible loss to the world by delayed introduction of some of the best grapes ever originated, we have concluded to offer for the first time the following selections from our creations, although a very poor method of bringing money to the originator." The price list warned customers that no more than twelve plants of any one variety would be sold in an individual order; moreover, "no cuttings of these [will be] supplied at any price." Perhaps Volney was feeling pushed by another grower threatening to release similar varieties. Or perhaps he was trying to create a demand by making them hard to obtain, since the vines sold for a pricey $1.00 or $1.50 each. The philosophy behind this particular marketing effort is difficult to understand.

Volney's publications during this period included a number of papers he gave to prominent American horticultural and scientific societies. Some of these were also reprinted by other periodicals. His 1905 paper to the American Breeders' Association, for example, appeared only a few weeks later in the *Nebraska Farmer*. In "Breeding Grapes to Produce the Highest Types," Volney detailed his efforts to create an extra early red market grape. Headlight was his best result so far but not yet his ideal, he reported.

In the early years of the twentieth century, French viticultural experts Pierre Viala and Victor Vermorel published a massive seven-volume series entitled *Traite General de Viticulture: Ampelographie,* an intensive study and classification of the world's grapes so valuable it is still in print. Volney wrote a number of articles for the series—in French— and is listed prominently on the front page as one of the contributors, some eighty names comprising a veritable "who's who" of the world's most famous viticulturists of the day.

He wrote another monograph for the Texas Agricultural Experiment Station which was released in 1906, when Volney was sixty-three. *Bulletin No. 88* was ponderously entitled *Length of Life of Vines of Various Species and Varieties of Grapes: Profitableness: and by What Diseases Seriously Affected.* In it, he reported on the eight-acre vineyard he had planted in 1886–87. Now nineteen years old, the plot was still in use, still partially producing and profitable, and that with only one fertilizer treatment of cottonseed. As he frequently did, Volney included tables and photographs to illustrate the slim volume of eight pages; the Experiment Station printed 15,000 copies of it. Thriftily, he used some of the same material, word for word, in a *Farm and Ranch* article in which he scolded Texas Gulf Coast horticulturist H. M. Stringfellow for recommending *Vitis Aestivalis* in the state. (February 17, 1906)

Though most of his writings were on grapes, he also produced a brief article for the Waco, Texas, *Searchlight* on causes of glacial and warm periods on earth, another demonstration of his far-reaching scientific interests. Volney believed that "there are perpetually in operation, motions and conditions in the solar system and universe that will produce such irregular periodical warm and cold epochs ... affecting the sun and all the planets and satellites."[27] This theory, he declared, was "entirely original with me, so far as I know, and first occurred to me about May 1, 1907."

Volney studied the work of the German monk Mendel, whose research and theories were rediscovered about 1900.[28] In the same paper for the American Breeders' Association described earlier, he declared that Mendel's law had "little significance" to fruit growers since they relied on cuttings, graftings, etc., rather than seeds for their new generations.

Volney stayed busy with the various associations to which he belonged and traveled to a number of their meetings. In 1905, at Kansas City, he was elected first vice-president of the American Pomological Society, a position he held for the ensuing four years; he hoped, he said, to "prove worthy." And he continued to consult regularly with the faculty and staff at Texas A&M. In his annual report for 1906, extension station horticulturist E. J. Kyle noted that Volney had "made suggestions ... [regarding crossing established vines with native grapes] which this Department could undoubtedly follow with profit."[29]

In 1906 his alma mater, the University of Kentucky, presented Volney with an honorary doctorate, an occasion which took him back to Lexington for a visit. After that, according to current Texas A&M Extension horticulturist Dr. George Ray McEachern, the Texas college extended him full professional credentials, which may help explain why he was considered for an Adams Fund grant there.

In Aggie Land

In 1907 Volney began discussions with E. J. Kyle, who was also professor of horticulture at Texas A&M College, and J. W. Carson, assistant director of the state experiment station there, regarding work he proposed to undertake with financial support from the Adams Fund. These were monies disbursed at the discretion of the college president and the director of the Experiment Station for strictly "investigational and research work, of a scientific and original character." In this case, some $1,500 to $2,000 would be available to Volney for equipment and his services.

In his project outline, Volney proposed to continue his work with grapes but to delve specifically into more complex questions and problems with breeding. He listed those areas he wished to explore, all of them continuations of his own thirty-odd years of research:
- The relative influence of each parent on progeny
- The best weather in which to pollinate, the length of time pollen retains its virility, and other related questions
- The most successful geographic ranges for each variety
- How soils, heat, and moisture affect varieties
- How to determine maximum sweetness and acidity, and best times to harvest
- Capabilities of each variety to endure shipment
- Obtaining descriptions and photos of less well-known varieties
- "To produce new varieties better suited for profitable and more varied grape culture ..."

Regrettably, only two letters about this project have survived, and neither provides the outcome of the funding request.

Mr. Sargent

Among Volney's many correspondents was C. S. Sargent, director of the Arnold Arboretum, Arnold Professor of Arboriculture at Harvard University, and director of *Garden and Forest* magazine (published 1888–1897). The only known surviving letters between the two are those from Sargent to Munson written between 1908 and 1912, and in the collection of the Arboretum.[30]

Of course, with Volney's expertise on the American grape, their letters most often discussed that topic. In fact, Sargent, in his very first letter, offered Volney seedlings of vines collected for Harvard in China. He also encouraged Volney to prepare a bibliography of his publications, especially on the grape, as "it is very hard to find papers scattered through the horticultural press unless there is some clew [*sic*]." (April 15, 1909)

However, as professor of arboriculture, Sargent's main interest was trees, particularly those of Texas, with which he and the rest of the country were still little familiar.[31] Over the next few years, he made several trips to Denison to visit Volney and accompany him on collecting expeditions in the area. At other times he hired Munson, at the Arboretum's expense, to collect Texas specimens.[32] He expressed his pleasure at Volney's initial research on plums and urged him to do more: "There is a great deal of work to be done in the Texas plums and I do wish you would take up a systematic study of them. No one else is so well placed and so well equipped for doing this work." (October 15, 1909) "I am counting on you," he reiterated later, "to make it possible to work up this whole Texas situation." (May 6, 1910) Indeed, Sargent seems to have thought Volney as equally devoted to the search for Texas plums as he himself was. His letters frequently repeated the mantra that "it is evident we have more work to do," while almost every one included a request that Volney send him more specimens, no matter how distant and inconvenient to obtain.

The fact that Volney was now in his late sixties and not in the best of health seems to have weighed little with Sargent (who was actually two years older than Volney himself) and his urgency regarding what he considered the importance of the work, as these excerpts demonstrate: "I am sorry to hear that you have been unwell and used up by your efforts in my behalf. I hope the cool weather will restore you to health. ... We haven't got to the end of this Texas plant business yet and I am going to write you at length about it before long. ... I will point out just what I think we ought to try to accomplish this year. ... I hope to be in Texas in March and I hope you will feel like making a trip with me to Dallas, Cherokee County, Lampasas County and San Antonio." (October 21, 1910; December 16, 1910; January 11, 1911; and February 15, 1911)[33]

Nonetheless, Sargent's admiration of Volney's body of work is plain in his letters. Whatever the Texan needed from Harvard, Sargent made clear, he was to receive. Sargent even sponsored him as a corresponding member of the Massachusetts Horticultural Society, to which Volney was elected in November of 1909. And it was Sargent who named one of the plums Volney found, *Prunus munsoniana*, the wild-goose plum native from Ohio to Texas. "The name lanata," he wrote Volney, "was given to a different plum from your Texas Big Tree species which will be called *Prunus munsoniana* if I have anything to do with it." (September 9, 1910) And indeed, the tree still bears Volney's name and has become widely cultivated.

CHAPTER 13
Munson's Magical Flying Car

$\mathcal{T}$he last few years of T. V. Munson's life are noteworthy for three creations: the book that thumbed his nose at his religious critics, the book that has earned him enduring fame, and his final, most amazing invention.

Theophilus Speaks Out

Volney decided at last to speak out in 1906 about his personal beliefs in a booklet he entitled *The New Revelation*. He wrote it under the pseudonym of Theophilus Philosophius, which, by his own translation, meant one who was both a lover of and believer in God and one who was a lover of learning the truth.[1] Thirty-three pages long, it was privately printed in Denison by B. C. Murray, who also published the Munson Nursery catalog. Even after its completion and release, however, and as late as 1912, Volney continued to refine its message. His annotated and revised copy has survived and is still in the possession of a family member who graciously shared it with the authors.

Several factors may have contributed to his decision to write *Revelation*. That Volney was still being persecuted for his religious beliefs, or lack of them, at this late time in his life is evident in a letter from old college friend John Leet to Ben, written after Volney's death: "I am pleased to see that Volney died game. It takes pluck to resist a hostile religious environment."[2] Volney must have been amazed and outraged, even embittered, that such fanaticism, as he would have termed it, could yet thrive in what he considered the enlightened and scientific twentieth century.

But more than that, it appears that he himself was still struggling with the nature of God and religion, some forty years after he first began to question them. In the whole of *Revelation*, he mentioned only two specific events, the Great Storm of 1900 in Galveston, Texas, and the San Francisco earthquake of 1906. The hurricane that struck Galveston and the Texas coast killed an estimated 10,000 people and remains the greatest natural disaster in American history. Frisco's deadly April quake, the worst to hit an American city, leveled and burned it and killed hundreds more. This latter disaster in particular must have been fresh in his mind, for he was either writing *Revelation* at the time or began it soon afterward; and he was plainly, painfully dismayed that God had not appeared to answer the pleas for rescue of all those who died. His impassioned cry of "Where then was he, who had stilled the boisterous waves of Galilee, and his promises to answer earnest prayer?" indicate that Volney still yearned for the omnipotent and perfect God he spent most of the book declaring cannot logically exist.

An important point to remember in reading *The New Revelation* is the time period. The great scientific discoveries that began in the 1850s and 1860s and continued through the remainder of the century at breakneck speed irrevocably changed not only the physical world we lived in, but the very way in which we perceived that world. For centuries, that view had been couched in religious terms: God created the world and everything in it. Now science was presenting an entirely new view, one that did not seem to allow for God. Volney was not the only person frightened and confused. Church membership in America dropped sharply in these decades, and numerous new churches and cults—some respectable, others frankly off-the-wall—sprang up to provide alternatives.

Revelation presents Volney's own dilemma. He believed in a God-like being whom he chose to call Nature, but not in the Bible and certainly not in the religious hierarchy that had translated it to the public for centuries. His book expended much effort and ink to explain why no thinking person should believe in the God presented by the Bible, then closed by declaring that "if you must call nature God, then I am a son of God." In the end, perhaps, he wrote it to convince himself as much as others.

Making His Case

Volney divided the booklet into four sections: (1) Why Theophilus Became an Infidel, by Himself, (2) Dualistic and Monistic Doctrines Compared, by Philosophius, (3) The New Revelation, by Philosophius, and (4) Philosophius' Funeral Oration, by Himself. His intent appears to have been to use Theophilus, the lover of God, to refute churchly doctrine and Philosophius, the lover of truth, to present the opposing, scientific viewpoint.

He began by stating that those who refused to profess the tenets of Christianity and join a church were considered "unbelievers" interested only in giving rein to their evil natures and shirking their moral responsibility. Just so had he been labeled. Therefore he deemed it proper "to refute once [and] for all such unkind and unjust and untrue charges."

He recalled his decision to join the Disciples of Christ Church (Campbellite) at age seventeen, where he had prayed earnestly, taught Sunday school, and condemned "unbelievers, and other Christian sects than my own, as all good orthodox Christians have generally done." But his scientific studies at Kentucky A&M "completely wrecked my faith," and he left the church on graduation. He then listed and discussed five reasons (six in the annotated version) why he had given up on Christian doctrine:

• He loved the truth and saw the only means of securing it as "unprejudiced, fact-enlightened reason."

• Christianity is founded on the myth of the creation and fall of Adam and Eve and on

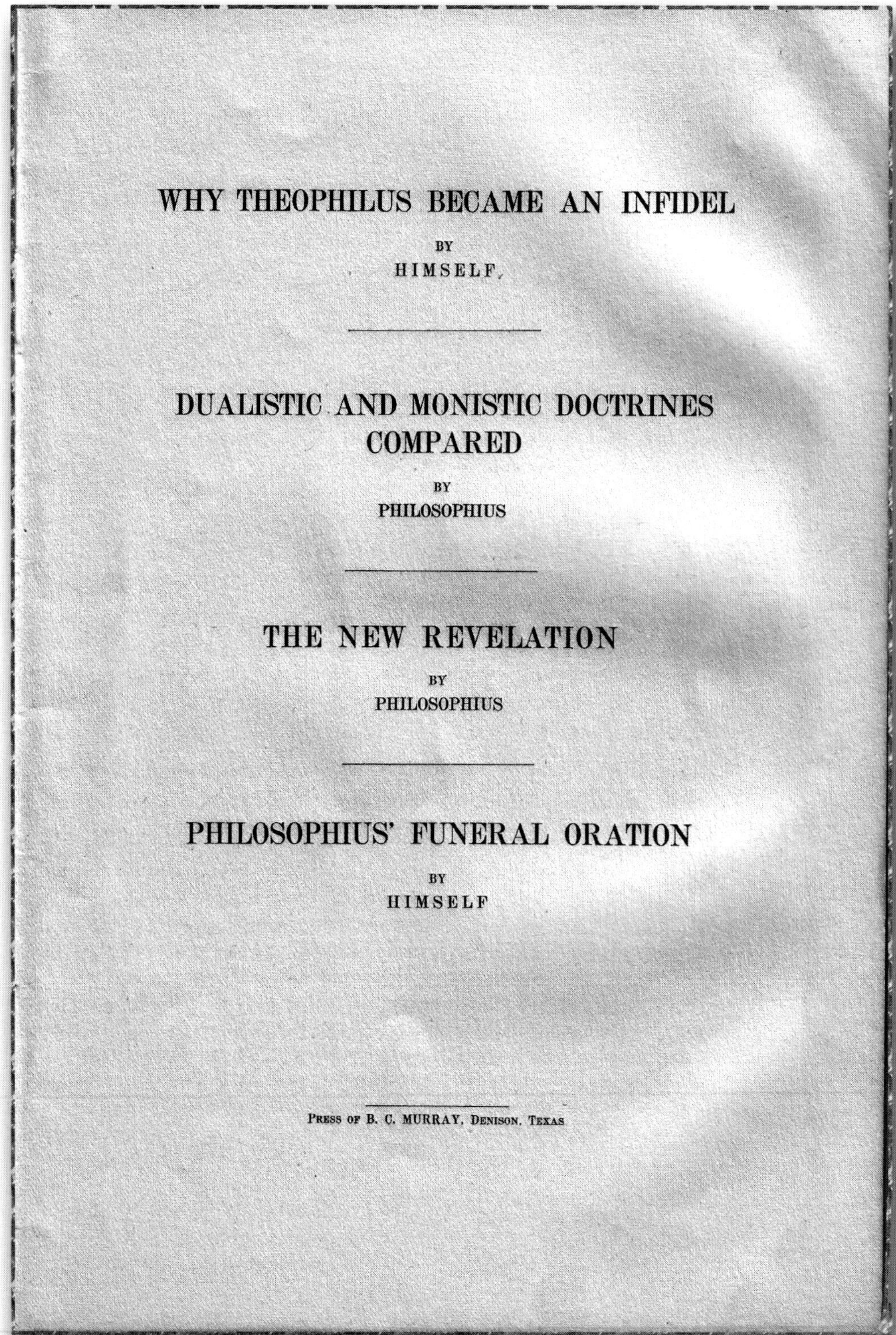

WHY THEOPHILUS BECAME AN INFIDEL

BY
HIMSELF.

DUALISTIC AND MONISTIC DOCTRINES COMPARED

BY
PHILOSOPHIUS

THE NEW REVELATION

BY
PHILOSOPHIUS

PHILOSOPHIUS' FUNERAL ORATION

BY
HIMSELF

PRESS OF B. C. MURRAY, DENISON, TEXAS

The New Revelation set forth Volney's intense personal beliefs and views on organized religion.
(T.V. Munson Viticulture and Enology Center)

the claim that Jesus of Nazareth was the begotten son of God. (This is a lengthy section which includes his querulous statement that Jesus' complicated genealogy "requires some amplification." The basic tenet here is that a perfect God would have created a human being with perfect moral character, unable to err and, hence, to fall. At one point, he asked if the God who drove out Adam and Eve could possibly be the loving Father of Christianity. "I say no, a thousand times No!! And all true history and science and plain common sense say no! No! NO!!! ..." This portion of *Revelation* is replete with chapter and verse references to points in the Bible he argued against, indicating his thorough reading and study of that book.)

• There are many contradictions, both theological and historical, between the various chapters of the Bible, such as who was the father of Jesus; Volney listed five different attributions.

• God does not answer prayer, as evidenced by the huge loss of life in Galveston and San Francisco. God as portrayed in the Scriptures is not infinite, but a finite, living being. All man-invented gods are non-entities and "spooks." God would be better called "The Infinite Phenomenifera," or "Phenomenon Maker."

• Christianity has a bad record of persecution and misery in the name of the Lord: crucifixions, martyrdom, the Crusades, the Spanish Inquisition, etc. But now the sun of science is reaching its zenith, teaching that knowledge and labor are not curses but blessings "and also teaching that woman is man's mother and *rather more than his equal*, and not his slave." (Authors' italics)

• Church attendance had declined from near 100% to 40% in the United States and Europe by 1912. Christianity was only a western sect of Buddhism, "adulterated with more or less of Mazdaism, Mithraism, and Egyptian, Grecian and Roman mythology."[3] (In this late, handwritten addendum, Volney stated that he did not regard all religion as an unmixed evil; indeed, he saw it as a necessary part of man's evolution.) He believed that the basic doctrines of all religions were descended from the pronouncements of the wise men (shamans) of ancient times which became accepted as dogma over the centuries. "So far as effecting the organization and cooperation of society, and a brotherly feeling, they were good. But the dogmatizing and tyrannizing priestcraft was evil, and a vast clog to progress. *It is this part of religion that I eschew.*" (Authors' italics)

The second section of *The New Revelation* is "Dualism and Monism: Or Religion and Science in a Nutshell." The former, Volney declared, is based on the belief that there are two substances, matter and spirit, and that matter was created by spirit out of nothing, as taught in the Christian church. Volney called it the philosophy of conflict, from its thousands of warring sects. Monism, on the other hand, the belief in one element, embraced peace and progress for it was based on "the testimony of the senses and logical reasoning therefrom." His handwritten footnote here adjures the reader to "read this carefully, studiously, and you will get an idea of the philosophy now entertained by the majority of the great thinkers of the world," among whom he clearly included himself, for monism was based on the proven fact, on science. "This scientific Monism has but one faith," he explained, "that is the faith or conviction that this infinite being—nature—is uniform in all its parts in conforming to the great law of action and reaction in all its varieties and ramifications of form and motion." He concluded with another handwritten addendum: "Faith glories in its invented spooks, yet it is frightened to death (in reason) by them. Reason grows fat and happy on fact and truth, and hence has no ghosts to fear."

"The New Revelation," part three of the publication, is a three-page poem to which Volney kindly added an "interpretation." It is basically a history of religion and science through the ages but is written so fancifully that the reader would come away with quite a different meaning without the key provided.

Part Four is the most fascinating, for here Volney wrote his own funeral oration, which he adapted from the writings and style of Lafcadio Hearn.[4] The sea is used throughout as its analogy, one he'd used before and an interesting choice for a man who'd spent his life roaming woods and prairies. Its very stylized prose suggests that Volney believed in a form of reincarnation, for the mother sea promises the wave that "thou wilt break and sink, only to rise in new forms." Death, the sea said, was "only the completion of the rhythm of life ... Suffice thee to learn, that because thou wast, thou art, and because thou art thou wilt formulate again." Volney likened himself to one of the waves, declaring that he came "to the under-tow of death, after enjoying well—because I have also suffered well—the fitful yet glorious form and voyage of human life." If death is a happy sphere, fine; but if not, "we shall ever try, as in the life just left, to make it better." He will not be truly gone, he wrote, for "nothing that is or has been can ever cease to be in essence and potentiality. My substance will take part in other forms and give rise to other phenomena—to other life, as happy in their spheres as in this just ended."

Volney reminisced about the way he'd tried to spend his life. "It has ever been a delight to me to aid in ameliorating the conditions of life about me. Especially have I tried to leave a good impress on my children and neighbors." Then comes perhaps the most telling sentences in the whole of *Revelation*: "In matters of belief in ancient religious dogmas, *my reason outran my faith and made me happier.* But where I may have made mistakes, only better reason, not faith, can correct. Herein, have those who know little and think less, condemned me the most." (Authors' italics)

Death, he concluded, was not destruction but the "eddy line of change among forms ... It is life condition that is painful, not death. Death is the door to other life." Therefore, Volney asked his friends and family, "return to your pursuits as joyfully as though you had planted a seed that will perennially spring up in beautiful trees and flowers. ... Farewell. Plant a vine on my grave, and see it clasp its hands with joy."

In addition, Volney penned a song late in 1910 (when he was in poor health) to open the funeral service. The first stanza read:

> One idle day, my wits astray,
> 'Gan looking round, for firmer ground
> On which to build, a better guild,
> That will not pray to gods of clay
> But make a sound far more profound,
> With truth instilled, the mind more filled.

A second song, entitled "On Knowledge, Not Faith, Stands the True Religion," was to be used after the oration. It began with a simple plea: "The world needs one religion true/That every child can understand." Volney also included a page of instructions to be carried out after his death. "I want no cold rock set in vanity at my grave. Man creates for himself his own proper monuments in his works and children. Print 10,000 copies of this little book, and mail one to as many preachers that they may learn to teach truth rather than fable. Also send a copy to each of the leading magazines and journals. The religious theories must be reformed in order for man to make better progress."[5]

Only a few months after Volney made his last changes to this testament, "Philosophius' Funeral Oration" would indeed be read over his grave by a longtime friend.

A Firm Foundation

The New Revelation, enlightening, mystifying, and sad by turns, was not the only major publication Volney wrote during this period. And while it may have been the more self-satisfying to him in some ways, it was not the book that would live on after him.

Volney had dreamed for years of writing the definitive book on the American grape. He had put aside his own plans for publishing to accept just such a commission from the U. S. Department of Agriculture, but that massive, illustrated tome never came to fruition, due to lack of funding. (See Chapter 9.) Now, in the waning years of his life, Volney once more turned his attention to the book that would incorporate his research, experience, and knowledge.

In 1907 he began to write, pulling together decades of field notes, correspondence with other horticulturists and viticulturists, and, probably, his unpublished draft manuscript for the USDA. He also selected almost ninety life-size half-tone engravings of grapes and their components taken by Denison photographer G. W. Moore. Daughter Viala, then nineteen and home from college for the summer, painstakingly typed all the material into book form, 252 pages in length.

In October of that same year, Volney traveled to New York City with two purposes in mind: to present a paper on grapes to the International Conference on Plant Hardiness and Acclimatization and to seek a publisher for his book manuscript, *Foundations of American Grape Culture.*[6] But while the paper was a success, the search was not. In an article written a few years later, Volney recalled that "the 'panic' was coming on and I found no publisher willing to undertake the job."[7] This referred to the Panic of 1907, when many large New York banks and several railroads failed, causing a ripple effect through the economy. However, in the 1908–09 Munson Nursery catalog, he declared that "the author has failed to get publishers of ordinary agricultural works to publish for him, fearing there is not enough demand for such a fine work to justify the undertaking." This frustrating setback may have affected his health, too, for his handwriting shows a definite deterioration in the summer and fall of 1908.

With "the assistance of my able son, Will B. Munson," Volney finally decided to publish the book himself "if 1,000 subscribers can be obtained" to offset the cost of printing. In *The National Nurseryman,* he described the planned work: "The first edition will be an autograph edition, with good recent photo-engraving of the author. The printing will be on 100 pound coated book paper, bound in buckram [a stiffly finished cotton or linen], gold lettering on back and cover, costing delivered about what we ask for it, $3. It is aimed to be and remain a classic on the subject." He added that it would be "printed and bound by a first class [publishing] house, well equipped for such work." Eventually he hired Orange Judd Company, a New York firm that specialized in agricultural and horticultural titles, to publish it.

The Layout

The title page of the book carried those awards and titles of which Volney was most proud: vice-president of the American Pomological Society, honorary member of the American Wine Growers Association, honorary member of the Societe des Viticulteurs de France, foreign corresponding member of the Societe Nationale D'Agriculture de France, Chevalier du Merite Agricole in the Legion of Honor, and "practical viticulturist and nurseryman."

Volney then acknowledged the assistance of his family, especially "my noble son William Bell Munson" and "my attentive and talented daughter, Viala Lausell" (Green), who proofread and typed the manuscript, respectively. Volney's "devoted wife" and other children he thanked for helping him test varieties, and he declared the book his memorial to all of them.

A poem of dedication followed, which described how the ancient Sumerians brought the first "wild sweet grapes" from the hills into town and began cultivating them. Volney wrote of those early viticultural efforts and briefly paid tribute to America's role in preserving the Old World grape:

> The seeds were scattered round the huts of mud;
> Some grew and clambered up the walls, and bloomed all sweet,
> At length bore fruit, and cooled the huts with shade;
> Some few bore better grapes than from the wilds he brought;
> Such vines he loved and saved and kindly trained, betimes.
> He always gathered from the new and better vines,
> And planted vacant places with their seeds, select;
> He gave to kith and kin, who likewise grew and gave.
> Thus on and on, through old, ten thousand years,
> Have come adown to all mankind the twining vines
> Of Ararat, in Muscats, Flame Tokays and Cornichons.
> The sons of men still hand them on with loving care,
> Well mingled with those from our free American hills.

The preface briefly and modestly reviewed Volney's career; except for his listing the Legion of Honor on the title page, for example, there was no description or even mention within the text of his work with Viala, Planchon, et al. Chapter I presented the botany of the American grape and Volney's own classification of twenty-six American and two foreign species, while directions for breeding and hybridizing them were covered in Chapter II. Next came a listing of varieties, followed by Volney's assessment of the adaptability of each to various regions, which he candidly described as "one of the most valuable pieces of grape literature ever presented." Chapters V through VIII gave a succinct presentation on growing and marketing grapes as well as their value for adornment around the home. The book concluded with several indices to photographs, varieties, topics, and tables.

Thoughts on Creating

Buried amid the factual material in Chapter II of *Foundations* is a fascinating paragraph which tells us much about Volney's character. Entitled "Personal Qualifications Necessary

Studio portrait of Volney made for the frontispiece of Foundations of American Grape Culture.
(T.V. Munson Viticulture and Enology Center)

in the Originator," it details mental and emotional qualities he felt necessary to someone who wished to follow in his line of research.

"This work," he wrote, "requires not only theoretical knowledge, but also direct personal knowledge, experience, skill, and much of the intentive faculty, with great patience and perseverance, without the stimulus of money-making in it, for there is little to the originator. There is no law providing protection to the inventions (varieties) of an originator, as there is to the less meritorious mechanical inventor. The originator must have a great fund of enthusiasm, and an ambition to add something to the general fund of human development for the benefit of the world at large, and, that he may reap some personal compensation, or enjoyment, he must have an intense love of close communion with nature, causing him to admire the infinite correlated life movements; to study the loves and hates prevailing in all organic life and growth, discovering the great fundamental truth in ethics, as well as in the development of organic beings, that *love breeds life, hate breeds death.*" (Munson's italics) (pages 129–130)

Good News

Foundations was released in the fall of 1909 to positive reviews.[8] *Science*, for example, published a four-column review by Los Angeles engineer Theo. B. Comstock, who called the book a "veritable boon" for grape-growers, "so filled with meat, so well and completely arranged and thoroughly indexed, so copiously illustrated with most excellent reproductions from life, and so thoroughly digested, that it is impossible to characterize its contents in a sentence."[9] Comstock particularly praised the photographs, "executed with consummate skill … under the jealous scrutiny" of Munson, as "far beyond anything heretofore brought out in black and white in this line." It was his further opinion that *Foundations* "must long be regarded as a model of its kind," advancing grape experimentation in the United States by a quarter of a century.

Other horticulturists and vineyardists throughout the country hastened to add their testimonials, a number of which Volney reprinted in his February-March 1910 nursery price list. William Pfeffer, a California grower and writer, called it a "charming book … invaluable to all viticulturists no matter where they live." Volney's old friend H. E. Van Deman declared the pomological world "deeply indebted" to him. Texas horticultural legend Gilbert Onderdonk termed it "a very valuable book, such as never before appeared

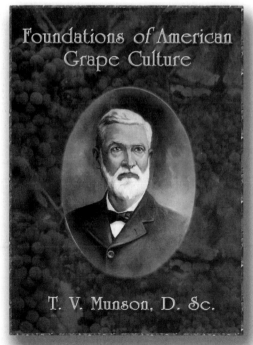

Foundations of American Grape Culture *was most recently reprinted in 2001 by Eakin Press because of the continued demand for the book, nearly a century after it was first published.* (Courtesy of Eakin Press)

in our literature." And another old friend, Liberty Hyde Bailey of Cornell, later characterized it as the "faithful work of a lifetime."

Volney sold the book through the Munson Nursery catalog for $3.00. After his death, it would be reprinted five more times. In 1934 daughter-in-law Minnie Secoy Munson, who managed the nursery after Will's death, reissued *Foundations*, adding a foreword to it describing Volney as "one of the world's foremost authorities on grape culture" and his book "the standard guide and textbook of thousands of grape growers."

In 1966 the Denison Public Library, through the generosity of the W. B. Munson Foundation, reprinted the book once more. However, this edition was found to have errors, so it was corrected and reissued by the Library and the Foundation in 1974 and again in 1985. The most recent reprint was done by the Munson Viticulture Center and the Munson Foundation in 2001. Because of its classic nature and value to grape growers, *Foundations of American Grape Culture* remains in print and in demand today.[10]

The cross-like shape of the Munson trellis was developed to prevent the support wires from twisting and breaking in high winds. (From *Foundations of American Grape Culture*)

The Safety Car

Volney had tinkered and invented all his life, mostly with agricultural tools and equipment, but his most unusual creation came about in 1911.[11] That fall, he suffered a long bout of ill health which likely necessitated a good deal of bed rest and, consequently, long hours in which to think. Looking back over his lifetime of work and study, Volney must have remembered the near 80,000 miles he'd traveled on foot, horseback, and by rail-

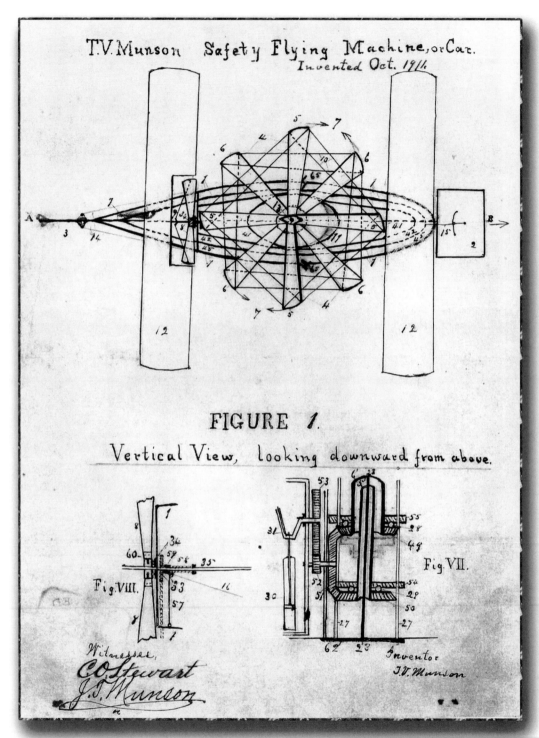

"Munson's Safety Flying Machine." (Courtesy Mr. and Mrs. John Maki)

road—all hot, dusty, dirty miles, many of them painfully slow. And one can conjecture, knowing Volney, that his next thought must then have been "how to do it better."

Over and over in his horticultural research, Volney demonstrated that he stayed abreast of the latest in science and technology. So he was undoubtedly familiar with attempts to build various types of aircraft; just a few years earlier, in 1908, Orville and Wilbur Wright had dazzled Europe while on tour with their fixed-wing biplane. But Volney was interested in a machine that could take off and be set down again in a limited space. Such a craft would have been invaluable in his own grape hunts, which were often in far-flung and remote locations.

A helicopter (a Greek word, meaning "spiral wing") was not unheard of, even in 1910. Some scholars claim to have found a drawing of one in fourth century China, Leonardo da Vinci sketched his own design in 1483, and a British scientist of the 1840s built an unpiloted steam-powered copter that flew ninety feet high. *Texas Farm and Ranch* even carried a cartoon on September 4, 1897, depicting a man of the near future "fishing for birds from an airship." The cigar-shaped vessel was powered by a rear rotor that looked suspiciously like a Texas windmill. But it was not until 1907 that a Frenchman, Paul Cornu, piloted a machine vertically into the air and hovered a few feet above the ground. Two years later, Igor Sikorsky, a Ukrainian who eventually fled revolution for the United States, built his first prototype but couldn't get it to fly.[12] All of these inventors faced the same problem: building an engine that could provide the vertical thrust needed and rotors that would whirl in such a way as to cancel out torque.[13]

Volney's "Stable, Safety Flying Car, or Machine" was a "new and useful" design that would, he promised, "provide a stable and safe flying-car in which to navigate the air."[14] The fuselage was streamlined, or "fish-form," to lower air resistance and, from the rear, resembled a lopped-off triangle set on wheels, which allowed it to be rolled about on the ground. It was to be constructed of strong, light metal with a wood floor and sides of leather or cloth that could be rolled up or even left off completely in good weather to increase speed. At the pilot's eye level were mica or glass windows. The bottom was watertight so that the car could also be propelled through water by means of the vertical propeller on the upper back. A rudder in the front, controlled by the pilot with wires or cords, raised or lowered the car when flying. Above the car were two galvanized sheet iron or brass "turbines"—the rotors—powered by gasoline and rotating in opposite directions; they were strengthened with wire braces to prevent the metal from flattening or breaking upward. Thus Volney's design foreshadowed, in its use of three rotors, the modern helicopter in which the opposing rotation of the two large blades along with the forward thrust of the third on the back of the ship overcomes torque.

One of Volney's two major improvements to the machine was a parachute above the rotors which remained folded and closed until time to descend. The simple motion of falling would open it, and springs closed it automatically on coming to rest. The other improvement was two detachable gliding planes, like wings, fore and aft, to help steady the vessel in any weather. He thought of comfort, too, and put springs on the seats of both pilot and passenger to prevent rebound "should [a] hard fall occur."

To fly the car, the pilot entered and adjusted his seat so his weight balanced the opposite end. He fired up the engine for the horizontal, lifting rotors and began to rise; at the desired height, he turned on the third, vertical, propelling rotor—located on the back—and set forth.

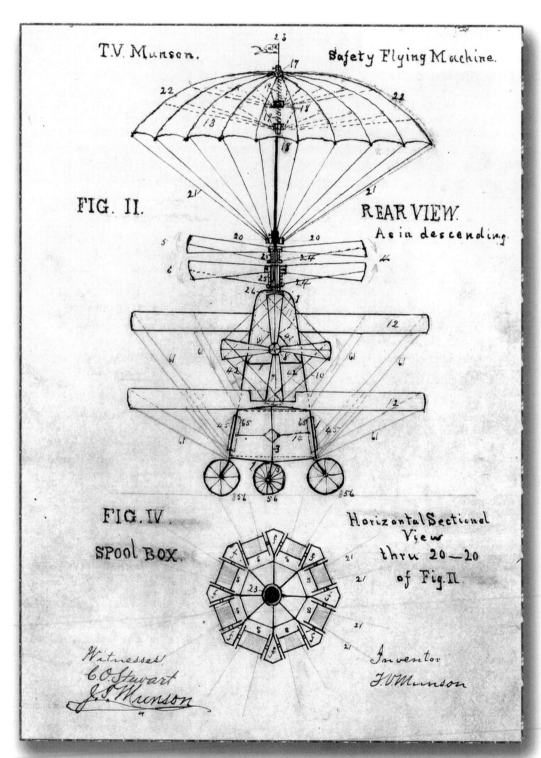

"Munson's Safety Flying Machine." (Courtesy Mr. and Mrs. John Maki)

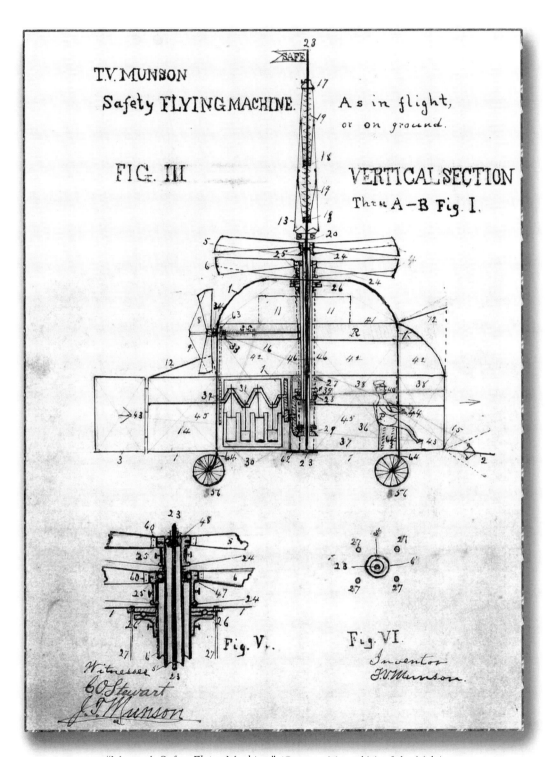

"Munson's Safety Flying Machine." (Courtesy Mr. and Mrs. John Maki)

But Does It Fly?

Before submitting his invention to the U.S. Patent Office, Volney sought the opinion of an outside expert, Theo. B. Comstock, a mining and metallurgical engineer in Los Angeles who had written a glowing review of Volney's grape book for *Science*. Comstock examined the drawings and narrative and wrote Volney with his conclusions on September 20, 1912. He felt that Volney had "two very good novel ideas in the elevator machinery and the automatic parachute, and that your detachable appliances are decided advantages. I am just a little uncertain whether your horizontally rotating engines, in opposite directions, constitute a gyroscopic adjunct in some degree, or if they are intended to react, as it were, upon a spirally enchained air column."

Assuming that Volney had allowed for the correct proportioning of weights within the car, Comstock declared that "there does not appear to me any reason for failure in action. On the contrary, my own judgment is favorable and I do not discern any cause militating against stability and other most desirable qualities of a safe-working flyer."

But the world never had the chance to try the Munson Safety Flying Car.

To properly submit the invention for a patent, Volney had to build a working model, and he apparently became sick before he could do so. Four months after hearing from Theo Comstock, Volney would be dead.

CHAPTER 14
"A Great and Good Man"

ith so much of his great work now accomplished, one might think that Volney Munson would have settled quietly into retirement. Not so. He continued to be active until a few months before his death.

Family Doings

Misfortune struck the Munsons in 1906 when the marriage of Nina and Will was annulled, and Nina was institutionalized in Austin, Texas. She died there in 1910 and was buried beside her father.

A happier occasion in 1908 was the marriage of Viala and W. C. "Billy" Green on September 10. Possessed of a magnificent singing voice, Vee had attended the College of Music in Cincinnati, Ohio, then did further training in Chicago and New York before returning home to teach private classes in the Denison area and to supervise the music program in the public schools there.[1] She sang professionally and would, over the next few years, reject lucrative contracts to stay in Texas. Husband Billy was a partner in several business ventures with Volney's brother Ben Munson and one of his favorite hunting and fishing companions; at the time of his marriage to Vee, Billy worked for the electric company in Denison, Texas Power and Light.

The Munson siblings—Louisa, Ben, Volney, Theo, Jennie, and Trite—and many of their children gathered in Denison for Christmas of 1911, a family gathering such as hadn't been enjoyed in four decades and the last time they would all be together. Trite and Jennie came in from Point Loma and Louisa, now widowed, from Nebraska. Volney and Nellie kicked off the celebration with a dinner at Vinita Home on December 21. Then followed a week of driving parties, a visit to the Denison Opera House for a play, the Elks Club

The six Munson siblings gathered in Denison to celebrate Christmas 1911 together. From left to right: Louisa, Theo, Volney, Ben, Trite, and Jennie. (Courtesy of T.V. Munson Viticulture and Enology Center)

Christmas ball, dinners at the homes of local family members, and a six-course Christmas feast at the Katy Hotel hosted by Ben and Ella.

On Boxing Day, Volney arranged for group and individual photographs to be made of the siblings as souvenirs of the occasion. These show him without the beard and mustache he'd worn for so long; family members recall that he shaved them (sometime after 1903) because his hair had turned snow white and he felt so much white hair made him look old.

Professional and Civic Activities

At the age of sixty-seven, Volney might have been forgiven for decreasing the amount of time he spent on civic work. But in 1910 a new organization caught his eye because it had an agenda very similar to his own. A group of state businessmen formed the Texas Industrial Congress in San Antonio that April and invited all "patriotic, public-spirited Texans ... who want their state to grow and prosper and who are ready and willing to do everything in their power, in every reasonable and right way, to promote that end."[2] The Congress had an impressively optimistic set of goals which included the united and harmonious development of the state's resources, the development of Texas in every respect,

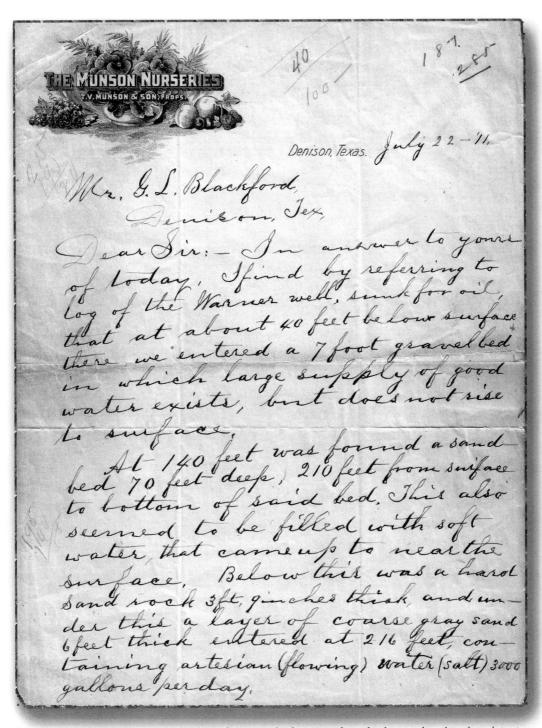

THE MUNSON NURSERIES
T.V. MUNSON & SON, PROPS.

Denison, Texas. July 22 – '11.

Mr. G. L. Blackford,
Denison, Tex,

Dear Sir:— In answer to yours of today, I find by referring to log of the Warner well, sunk for oil, that at about 40 feet below surface there we entered a 7 foot gravel bed in which large supply of good water exists, but does not rise to surface.

At 140 feet was found a sand bed 70 feet deep, 210 feet from surface to bottom of said bed. This also seemed to be filled with soft water, that came up to near the surface. Below this was a hard sand rock 3ft, 9 inches thick, and under this a layer of coarse gray sand 6 feet thick entered at 216 feet, containing artesian (flowing) water (salt) 3000 gallons per day.

Munson continued to be consulted on a wide range of subjects, such as this letter asking him for advice in locating a water well. (T.V. Munson Viticulture and Enology Center)

2

Below 212 feet all waters were arterian and salt, — 13 layers passed thru from 212 to 812 feet, — 600 feet of salt water.

Think it would not be safe to go deeper than 175 feet, for fear of saltness, unless such is desired. Still safer to stop at about 44 feet, in the fresh water gravel bed, if supply is sufficient.

It is in this bed that the Katy's second well at Warner got inexhaustible water, under my suggestions, I understand they pump over a million gallons a day, yet make no impression on the supply.

Of course, whatever depth of strata at your place, overlie the forma-

174

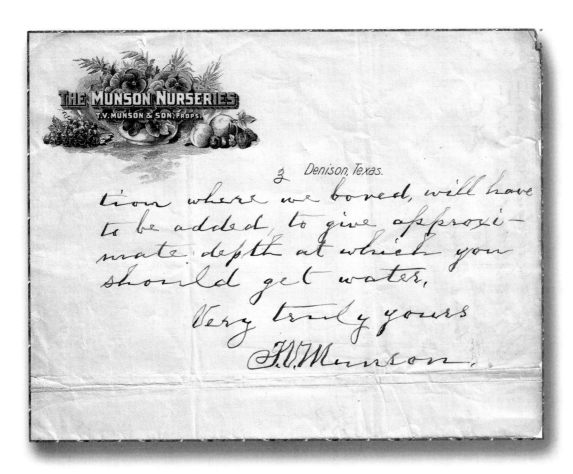

tion where we bored, will have to be added, to give approximate depth at which you should get water,

Very truly yours

T. V. Munson.

and equal rights for all. Of course, Volney had been advocating some of these same issues for many years, so he joined readily. However, very little information has survived on the TIC, and the extent of his involvement is unknown.

Despite his age, Volney continued to take an active role in the nursery, developing a new group of grape hybrids (released after his death) and planting a new apple test orchard in 1911 and 1912. In that latter year, he set out the last of the grape seedlings he'd been working on. Business was so good that he and Will even considered expanding the nursery lands by renting part of Trite and Jennie's east Denison farm.

Volney published several more scientific articles on breeding and hybridity as well, primarily for the journal of the American Breeders' Association. In 1911 he penned the entry on "Texas" for friend Liberty Hyde Bailey's *Standard Cyclopaedia of Horticulture.* His own book, *Foundations of American Grape Culture,* was in such demand that it was translated into several languages. Volney also maintained a role in the various organizations to which he belonged, serving as vice-president of the Society for Horticultural Science in 1911 and on the International Commission of Viticulture. Within the American Pomological Society, Volney was a member of the committee on new fruits of American origin, a position which helped him to stay current with the latest developments.

His last great professional honor came in 1912, when he was elected a life member of the American Breeders' Association. The letter from secretary W. M. Hays noted that this

Undated portrait of Volney made in Little Rock, Arkansas.
(T.V. Munson Viticulture and Enology Center)

honor was generally "reserved for special services" and was specifically bestowed in this case "in recognition of your public service in the breeding of grapes and other agricultural work."[3] Hays added that "only a very limited number of elections have been made." That select list was composed of:

- Hugo de Vries, director of Amsterdam's Botanical Garden, a botanist and evolutionist
- Sir Francis Galton, the English founder of the sciences of biostatistics and eugenics
- William Bateson of Cambridge, England, a Darwinist known as the father of genetics (which word he coined)
- Dr. William Saunders, who founded the Canadian Experimental Farms Service and directed it 1886–1911
- and colleague Luther Burbank, the only other American elected to that point.

No mean company. And no doubt a gratifying acknowledgment of his lifetime of work by his scientific peers.

The Transition of Death

Volney's lifelong history of poor health now began to take its final toll. At the time of the 1911 family reunion, sisters Trite and Jennie had noticed that Volney "looked so thin and white and seemed to suffer so with the least bit of cold" that they feared he lacked the vitality to fight a serious illness.[4] Late in December of 1912, he did become sick again, this time with *la grippe* (influenza), and his condition worsened considerably as pneumonia set in. Around the first of January he rallied, and the family felt that he'd passed the crisis. But he failed to regain his strength over the next few weeks, and additional complications finally brought his life to a close, though not before time to call Warder, Fern, and Marguerite home.[5]

Thomas Volney Munson died at 8:00 on the evening of January 21, 1913, at the age of sixty-nine and a half.[6] Daughter Fern later recalled that he spoke his last words calmly and with "unperturbed spirit" to the family gathered around his bed: "Death is merely a transition—I am ready for it. I have led a full and eventful life; without any regrets, I leave you in the fullness of my powers, thus will you always remember me; and I would have it so."[7]

The funeral took place two days later "in the presence of a vast concourse of friends and sorrowing relatives."[8] Trite and Jennie remained in

Volney's tombstone was exuberantly carved with the wild grapes he loved. (T.V. Munson Viticulture and Enology Center)

177

California since both were ill themselves with *la grippe*; Louisa, too, was sick. So the task of burying the first of the siblings to die was left to brothers Ben and Theo. From Vinita Home, Volney's body was carried by horse-drawn hearse to the auditorium of the XXI Club in Denison, which Theo had donated decades earlier.[9] That space was so completely filled that some mourners were turned away for lack of even standing room. Dr. John Ellis, a member of the Denison School Board, spoke first, likening Volney to Longfellow's poem:

> *Lives of great men all remind us*
> *We can make our lives sublime*
> *And departing leave behind us*
> *Footprints in the sand of time.*

After that, R. S. Legate, a longtime friend of Volney's and president of the School Board, read the funeral oration which Volney had written in 1906. Among the actual and honorary pallbearers were some of North Texas and Denison's most important citizens, including fellow school and bank board members, doctors, lawyers, and nurserymen.[10]

From the auditorium, the funeral procession moved a few blocks to Fairview Cemetery, where Volney was buried. Ben delivered the eulogy: "Dearest brother, we now surrender you back to the bosom of the great universe into whose mysteries during life you so loved to delve. . . . Your heart was loving and tender as the flowers you grew, your resolution in the discharge of duty as strong as the oak and firm as steel: your energy was tireless, your patience most wonderful: your character and conduct spotless and clean: your love of nature was only surpassed by your love of man. . . . Your life will live in the moral fibre of your posterity, in the impress your life has made upon those who knew you, in the finer fruit of the vines you created, and in the wider intelligence your writings have wrought. The world will be better and happier for your having lived. Rest in peace."

Tributes

Telegrams and letters of tribute flowed into Vinita Home from around the world. Henry Van Deman wrote that Volney's memory "has the perpetual fragrance that accompanies a noble life" while horticulturist E. J. Kyle, with whom Volney had worked at Texas A&M, declared he had "established a name that will live as long as people study and persue [*sic*] horticulture."[11] Nurseryman F. T. Ramsey of Austin mourned that "we feel that the greatest has gone and there is no one to take his place."

E. J. Krause of the Waco *Searchlight* referred to Volney's years of struggle with established religion in his letter. "Your dear father," he wrote the family, "was a rare example of living the higher life of the spirit (ahead of the time); the unselfish life of love, of the brotherhood of humanity, which is not understood nor comprehended by the average majority of the present day humanity."

Many writers referred to Volney's modesty regarding his professional achievements. P. C. Edmondson of the United States War Department was one of them. Volney, he declared, "permitted this quality of character to carry him to the extremes of modesty. For, most men with his record of accomplishments, his mental equipment and with the meritorious things he had for sale, would have advertized [*sic*] them more lavishly and would have personally profited more from them. One result was that, though doing a work in plant breeding probably more important than any one else in his day, he was much less widely known among the masses than he deserved to be—enjoying a reputation mainly among scientists and progressive horticulturists. ... It always seemed to me that he viewed his business more as a labor of love than one of commerce." Frank A. Waugh, head of the Division of Horticulture at Massachusetts Agricultural College, was far more blunt about Volney's reticence: "It seems rather strange that a man like your father who has back of him a hundred times the substantial accomplishments of a Luther Burbank, should have received less public attention."

"Such men never die," W. A. Harshbarger of Kansas' Washburn College wrote Nellie. "It is given to only the few, nature's noblemen, to render such great service to so many. ... I can but rejoice with you that he so clearly saw the light, and so steadfastly followed where it led." Employing the same theme, the American Pomological Society declared forthrightly that "a great horticultural light has gone out."

Agricultural, horticultural, and scientific organizations hastened to also add their praise. The Massachusetts Horticultural Society pronounced *Foundations of American Grape Culture* "one of the most important contributions [to the field] in recent years." The American Pomological Society, to which Volney had belonged for twenty-seven years, said this of his many publications: "As a writer in the scientific and rural press of the country he occupied a prominent place, and the productions of his pen will be missed with regret, for his words were always those of wisdom."[12] Wilber Dubois, editor of the *Florists' Exchange,* called the loss irreparable; "he wrought with a master's skill, his years of patient labor resulting in the creation of a new race of Grapes of inestimable value." (February 8, 1913) The *Texas Horticulturist,* journal of the state society, placed Volney's obituary and photo on the front page of the February issue, concluding sadly that "truly a great man, and a great horticulturist has passed away."

The *Bulletin of the Texas Department of Agriculture* (September–October 1913) included a sketch of "The Life and Labors of the Late T. V. Munson" presented at the Texas Farmers' Congress meeting by friend and fellow nurseryman John S. Kerr of Sherman. "The impression he made," Kerr wrote, "was of one thoroughly at home among his new creations, one easily the complete master of the laws of plant selection and reproduction. Not only was he a master in his line, but to his great abilities were added the highest qualities of a courteous and cultured gentleman, kind and wise husband and father, warm and loyal friend, and useful and broadminded citizen." Munson went even further than other renowned plant scientists of the day, Kerr continued, for he "utilized his knowledge of the deep things of nature in the production of new and rare plant creations. ... His great mind was a rare storeroom of knowledge." Kerr called for the organizations to which Volney had belonged to join with the family in writing a biography of him.

Legacies

Volney's will had been written in 1889, before Margaret and Rupert were born and just before he left for the U.S. Department of Agriculture trip to California, and was never revised. At that time, he wished to leave his affairs so that his "wife and children shall derive the greatest benefit from my estate." He appointed wife Nellie his executrix, with Ben to succeed her; brother-in-law John M. Bell of Kentucky was a second backup.

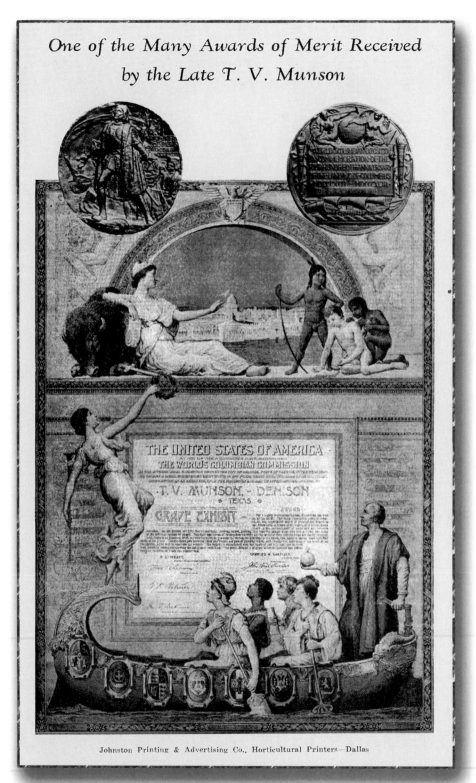

Volney's medals and certificate from the World's Columbian Exposition were pictured in a Munson Nursery Catalog after his death. (Courtesy of T.V. Munson Viticulture and Enology Center)

Nellie was to receive the entire estate. But if she remarried, then 75% (his own half plus his half of the community property) would go to the children instead. In a series of very complicated instructions, Volney left directions for the estate to go to his sisters, then to his brothers, and finally to Nellie's relatives, in the event that all the children died, an understandable precaution for one who had already lost three.

The inventory filed with the probate records[13] lists almost $20,000 in real estate holdings: unsold lots in Volney's First Addition and about sixty-five acres, most of that in one parcel. There were also seventy shares of stocks in several of Ben's companies, worth less than $2,000. What's missing are the value of Vinita Home and its furnishings, the value of the nursery stock and business, and some land.

Margaret, born after the will was written, had to transfer her legal share in the estate to Nellie. Later that fall, the children also sold to their mother for a token amount their interest in Lots 10 to 18 of Block 5 in the addition so that she would have full title.

What Happened to the Family

Life resumed after Volney's death, as it always does.

Will took over the nursery and moved into Vinita Home to be closer to the business after he remarried in 1916. Minerva Secoy, known as Minnie, was sixteen years younger than Will; she'd grown up in Denison and was living there when they married. The couple later built their own home on Hanna Street. (She and Will had a daughter in 1922 and adopted a son in 1929.) Will tried to maintain his father's grape varieties, but he did not continue Volney's research; indeed, much of that leadership shifted to the New York State Agricultural Experiment Station in Geneva, which had been established in the early 1880s. Will's own interest was in shrubbery, and he was a well-known writer on the subject. He served as president of several state and regional horticultural societies.

Fern, who became a widow herself just seventeen months after her father died, raised her sons in Denison and traveled with her mother and sisters on vacations from her job at the Denison Post Office and after her retirement in 1941. She rented part of her home on North Scullin Avenue to boarders to supplement her income. Eventually she and Neva both retired to a Denison nursing home, where Fern died in 1974, the last of Volney and Nellie's children. She bequeathed her interest in the family's history to her granddaughter, Joyce Acheson Maki, whose help in writing this book was invaluable to the authors.

At the time Volney died, Warder was living in Houston, where he was a bookkeeper at a bank. He stayed in Denison briefly and later moved to Ardmore, Oklahoma. A diabetic like his brother Will, Warder died at fifty-one when he slipped on an icy sidewalk and sustained a severe head injury.

Neva remained at home with her mother, teaching school and helping with the nursery. When Will remarried and moved into Vinita Home, she and Nellie relocated to a house at 1020 Sears Street, near Vee and Billy. She taught in the Denison public schools until 1920 and then served as assistant librarian at the public library (built on the site of her Uncle Ben's first home) for several years. After Margaret died suddenly, Neva moved in and helped raise her children. Like Fern, Neva traveled extensively and also devoted much time to her painting, holding several shows in the 1930s and 1940s. She died in a Denison nursing home in 1973.

Will, the elder Munson son, took over complete management of the nursery after his father's death. (T.V. Munson Viticulture and Enology Center)

Olita and Col Calvert ran a boardinghouse in Denison; her Uncle Theo lived there for a time, too, until his death. The Calverts retired to Ruidoso, New Mexico, in 1947 and Olita moved on to California after Col's death; she died there in 1967. Olita had retained a number of items and documents pertaining to her father, including photographs, nursery catalogs, his manuscript journal from college and Nebraska, and many of his medals. In 1945 she donated them to the Dallas (Texas) Historical Society.[14]

Viala died tragically young, just a few days short of her thirty-seventh birthday, after emergency surgery. She and Billy had no children.

Margaret married her longtime fiancée J. W. Thompson, an employee of the Missouri Kansas & Texas "Katy" Railroad, in 1914; they'd been engaged since before she left for college. She, too, died in her late thirties, and sister Neva moved in to help care for her two teenaged daughters.

Ellen Scott, Volney's beloved Nellie, lived on in Vinita Home for several years, helping with the nursery and tending her extensive rose garden. Neva was with her, as were Will, Margaret (until her 1914 marriage), and Warder (for a short time). Nellie did some traveling, including a 1920 summer in California with Neva, Olita, Margaret, and their three girls, but generally lived quietly. In January of 1923 she became ill with influenza. To help her regain her strength, Will took her to Corpus Christi that spring, but she soon suffered a relapse due to a heart condition she didn't know she had. At her request, Will carried her home to Vinita, where she died on May 29 surrounded by her children. They buried her next to Volney, a white rose from her garden in her hand.[15]

Volney and Nellie's grave in Denison's Fairview Cemetery.
(Photo by Roy E. Renfro)

Vinita Home still stands at the corner of Hanna and Mirick, though the acres that once surrounded it are now filled with houses. Olita inherited the property on Will's death, and her daughter Eleanor subsequently lived there with her family. When the Dwight D. Eisenhower Birthplace was opened in Denison in the 1950s, Fern and Neva donated a number of family pieces to it, some of which may have come from Vinita Home. The house was purchased by Mr. and Mrs. Sidney Johnson in 1960, and they have restored it extensively. In 1967, through the instigation of Eloise Munson, Ben's youngest child, the house was awarded two Texas Historical Markers for its association with Volney. Georges A. McClenahan, French consul in Texas, attended the ceremony to represent France's interest in Volney and his work.

Brothers and Sisters

Theo was the next of the six siblings after Volney to die, in 1919. He had been in poor health for several years and slipped into a long coma before finally dying on June 9. Theo had amassed a small fortune through his various surveying and real estate businesses and his stock in Ben's business ventures, and he had long ago set up a trust to distribute that to his nieces and nephews. Keenly intelligent, Theo had shared a fascination for geology with his brothers; his library was said to be the largest and best in town; and he had a strong philanthropic bent, which he had exercised to Denison's benefit.

Louisa and Ben died within a few weeks of each other in 1930. Ben had definitely been the entrepreneur of the family, and his vast holdings in oil, land, insurance, cotton mills, livestock, gold mines, cattle, coal, and railroads made him a wealthy man. At one time, Ben and his partners owned or controlled hundreds of thousands of acres in Texas, particularly the Panhandle area, including the T-Anchor and the 6666 ranches. The Munson Realty Company he and Theo founded—and with which Volney worked intermittently—is still in the family and in business in Denison after 130 years. Many of Ben's descendants continue to live in the town he helped build, and a number have followed his first career of law.

Trite and Jennie Munson remained in California, Trite dying in 1935 and Jennie eleven years later at the age of 93. Both left their estates to the Theosophy Center, where they'd made their home for so many years.

The Fate of the Nursery

Will continued to manage Munson Nursery after his father's death, maintaining its reputation and success; he even shipped grapes to Russia, a market Volney had never penetrated. Munson hybrids were the backbone of the flourishing grape industry in 1920s Florida, where growers were especially fond of Carman (popular all along the Gulf Coast). And Volney's bunch grapes for table use were standard throughout the South.

But two major problems badly hurt the nursery and its business after Volney's death. World War I led to a nationwide shortage of labor and a concentration on production of war-related items, both of which caused a drop in planting. The second factor was the general paranoia that overtook the country during the national prohibition (1920–1933), when many people stopped growing grapes, even for table use. In 1900 there were more

The Munson Nurseries

NURSERIES

BAILEY GRAPE
See Page Five

CATALOGUE
NUMBER 60

DENISON, TEXAS

The Golden Anniversary catalog of 1936-37 was the last one published before Munson Nurseries was sold.
(T.V. Munson Viticulture and Enology Center)

than 1.3 million bearing grapevines in Texas, most of them in the northern part of the state; that number sank to 712,000 a decade later as prohibition fever grew, and then to a low of 404,000. And though it rebounded to close to a million by 1930, the damage had been done to nurseries such as Munson. Moreover, Munson hybrids were "tainted" by their association with Volney, a man who had advocated wine drinking and had helped save the industry in Europe.

Will had held the business together but died in a Sherman hospital on September 13, 1931, after a lingering illness that was probably complicated, if not caused, by his diabetes.[16] His widow, Minnie, then took over management of the business and ran it for about seven years but was unable to reverse the financial woes of the past two decades.[17] Both she and Will had allowed the number of Munson hybrid grapes in the inventory to decrease until only a few dozen varieties were left for sale.[18] With this depleted supply and a lessened demand due to prohibition, Volney's grapes began to disappear from America's vineyards.

In 1938 Minnie sold the nursery stock, including the grapes and all the remaining reprints of *Foundations,* to Ross R. Wolfe of Wolfe Nursery in Stephenville, Texas.[19] Thus Munson Nursery closed after more than sixty years. Wolfe himself died in 1947, and the subsequent management failed to promote Munson's varieties, until only five or six were left in stock. Today Wolfe Nursery is no longer an independent chain of nurseries, and the present management knows nothing of the final disposition of the Munson grapes; nor do they carry Munson grapes any longer.

More Losses

This was also the period, in the 1930s after Will's death, when Minnie apparently disposed of much of Volney's equipment and books and sold his Legion of Honor medal. Grandson Marcus Acheson recalled visiting (probably after Will's death) and seeing one of Volney's gold medals being used to crack hickory nuts and to hold open the dining room door.

As well as Volney's grapes and his library, his archives of personal and professional papers seem to have disappeared during this time, too. Another grandson, Volney Acheson, remembered that Volney's correspondence was stored in the attic when he was a young boy. At least half of that cross-shaped room was lined with shelves which were filled with letters, most from Europe and the remainder from U.S. correspondents, especially Luther Burbank. Though Minnie later told a family member that she had given them to the Dallas Historical Society, no trace has ever been found of them.

One of the biggest mysteries surrounding their fate involves T. C. Richardson (1878?–1956), an editor with the *Farmer-Stockman Magazine* in Dallas. In the 1970s a rumor arose that Minnie had given the papers to Richardson for a biography he planned to write, but that he died at home in Palestine, Texas, before completing it; the papers, so said the rumor, were then in the trunk of his car but subsequently vanished.

A more careful analysis of sources as well as new information now indicates that Richardson had the papers as early as 1934, for he and they are mentioned in a brief sketch of Volney written at that time by Liberty Hyde Bailey. (The latter described Richardson as chairman of the Texas Agricultural History Committee.) Since Richardson did not die

An Ideal That Protects You

Reproduced above is the decoration presented to the late T. V. Munson, founder of the Munson Nurseries, by the Republic of France in recognition of his services in saving the Grape Industry of France.

The late T. V. Munson, who founded the Munson Nurseries more than fifty-six years ago, was a great scientist and one of the world's greatest authorities on grape culture and growth.

But he was more than that. He was a great business man. He understood that business is not a one-sided thing. Both the seller and the buyer had to profit in every transaction. The product had to be not only of the highest quality, but also at the right price.

That ideal has dominated every transaction we have made from that day to the present time. In order to make clear to our customers and to keep it always before our minds, we have put that ideal into a sort of business philosophy such as this:

(1) Every plant shipped from the Munson Nurseries must be grown under the most favorable conditions.

(2) Each plant must be true in name and color.

(3) Each plant must grade number one quality and be priced as such.

(4) Each plant must, to the best of our ability and knowledge to make it so, be free from disease.

If an ever growing list of customers and friends is any indication, then the Munson Nurseries have never deviated in any manner from any one of those commands.

Today as in the past we say that any plant you buy from us must measure up to those promises. If it does not, we will make it good.

Such a pledge is worth money to you, much more money than the cost of any single item. For if the plants you buy are not of the best grade and stock, the success you have in growing them is limited or eliminated altogether.

The 1935-36 catalog carried a drawing of one of Volney's prized French medals and declared Minnie Munson's resolve to continue his philosophy of business. She sold the nursery the following year. (T.V. Munson Viticulture and Enology Center)

until 1956, while coincidentally on business in Palestine, it seems more likely that he had long since returned them to Minnie. The authors have failed to find an extant collection of Volney's papers other than those given by Olita Munson Calvert to the Dallas Historical Society.

By this time, too, only a few of his contemporaries remained. Among them was U. P. Hedrick, who tried to keep Volney's reputation alive in his 1945 work, *Grapes and Wines from Home Vineyards*. "But of the many hybridizers of American species," he wrote, "T. V. Munson, [of] Denison, Texas leads all. He made every combination of American species possible, fruited thousands of seedlings, from which came many good varieties, especially for the South and Southwest. . . . During his mature years, say from 1880 to 1910, he was recognized in America and Europe as the leading breeder of American grapes and the chief authority on their botany." (page 151)

Nevertheless, by the 1950s, almost all trace of T. V. Munson's lifetime of work had vanished. Only a handful of nurseries still carried a few of his hybrids, his personal library and archives were destroyed, even his beloved vineyards in Denison had disappeared beneath the onslaught of post-World War II housing. His very name, once a household word in much of the country and the world, was in danger of being forgotten.

And then came resurrection.

CHAPTER 15
A New Beginning

*T*he fate of T. V. Munson and his grapes seemed bleak by the 1950s. Wine drinking was out of style, his hybrids had largely disappeared, and many Texas growers had given up on grapes entirely. But the popularity of foods, like fashions, tends to be cyclical, and the pendulum was about to swing again.

Fading Away in Texas

In 1900, Texas had a million bearing-age grapevines and about twenty-five wineries, the beginnings of the profitable industry that Volney Munson had envisioned. In fact, Texans exported more than 300,000 gallons of wine *to* California that year. This growth was due to several factors:

- the heavy influx of wine-drinking European immigrants into the state in the latter part of the nineteenth century and their consequent encouragement and development of the industry,
- the presence within Texas of so many native grapes,
- the development of new hybrids specifically for the Southwest, and
- the influence of Volney and other Texas horticulturists.

But prohibition—that "moment of sublime aberration," as author Leo Loubere described it—swept away much of this progress. In Texas, only Val Verde Winery, founded in 1883 by an Italian immigrant, managed to stay open by producing sacramental wines and by switching much of its production to table grapes. (Val Verde remained in operation and today is the oldest winery in the state.) Commercial ranchers, oil men, and bankers came to the economic forefront in the next few decades, pushing aside the farmers and growers who had helped build the state, as Texas' economy began to change from rural and agrarian to urban and industrialized.

Some grape cultivation and research continued but on a very limited basis, particularly after freezes and Pierce's Disease[1] devastated the table grape crops of the Lower Rio Grande Valley in the late 1950s. Professional and extension horticulturists, for all intents and purposes, stopped recommending that Texas growers plant grapes. This was despite A&M's establishment of a Fruit Investigation Lab in Montague County on the Red River in 1937 to focus on grape research and despite its then-presence at the Winter Garden Station in South Texas.

And so matters continued into the 1960s. A&M actually closed both its facilities in 1962 because no one wanted to know about grapes anymore. It seemed that the industry in Texas was pretty much dead.

Changing Times

But in the late 1960s and the 1970s, several changes began to occur, and not just in Texas. California vintners began to aggressively market their wines for, in the free and easy new American lifestyle, wine drinking had suddenly become popular again. Meanwhile, in Texas, the price of oil and cattle started fluctuating much more sharply than oilmen and ranchers liked, so they began seeking alternative revenue sources for their land. A few growers braved the high possibility of natural disasters and diseases and slowly, tentatively, reexamined grapes as a commercial crop.

Texas A&M encouraged this process greatly when it released research that touted West Texas as *the* new hot spot for grapes, thanks to irrigation, though nearly all the state, it declared, had potential. Several professors at Texas Tech University in Lubbock also planted trial vineyards after noticing how well their grape trellises, planted for shade, grew; Bob Reed and Clinton McPherson went on to plant their first vineyard as a science class project at Tech. The eventual result was the Llano Estacado Winery (1976), now the second largest wine producer in the state and winner of many national and international awards.

Grapes began to seem the up and coming new crop, but people weren't sure which varieties to plant: *vinifera,* French-American hybrids, or native Texas stocks. In 1973, Texas A&M planted demonstration plots across the state to try all three. Based on those findings, most growers settled on the hybrids, though that, too, would change in a few years. At the same time, Val Verde, then the only winery still open, re-entered the commercial wine market. Its success convinced others that winemaking and grape-growing could indeed be profitable. By 1977, groups and individuals throughout the state had organized into the Texas Grape Growers Association to enhance communications and education between growers, professional horticulturists, and botanists, and to lobby for favorable legislation.[2]

Here Come the French

Texas is unusual in that, beginning in 1876, it set aside 50 million acres of public land, mostly in the western part of the state, to benefit its public schools and colleges.[3] That vision was fulfilled in the 1920s when oil was discovered on university lands and made the state's school fund one of the richest in America. But in addition to managing the wealth that comes out of the ground, the fund also must care for and conserve surface lands and make them productive through livestock and crops.

The Board of Regents of the University of Texas, one of the biggest recipients of school land funds, became intrigued by the potentially high economic return of grapes. In the mid-1970s, they voted to invest a portion of the money in an experimental vineyard and enology (winemaking) program. By 1984, the university had about a thousand acres of vineyards in a very unlikely spot, the high desert of West Texas, one hundred miles south of Midland in the mountains of Pecos County. Then an alliance developed to manage the properties: the historic French firm of Richter (which had imported Volney Munson's vines a hundred years earlier), the Domain Cordier group of Bordeaux (another leading exporter), and Richardson Gill (who was later chairman of the board of the Llano Estacado Winery). The result was Ste. Genevieve Vineyards, which immediately became the "big cheese" in Texas with a first-year production of 275,000 gallons and a winery capacity of 1.5 million gallons. (Today it produces 400,000 cases annually, is the largest winery in Texas, and the thirty-first largest in the United States.)

Ste. Genevieve put Texas back on the viticultural map in this country with its sheer size. Other wineries soon followed: Fall Creek (1975), La Buena Vida (1978), Moyer Texas Champagne (1980), Pheasant Ridge (1982), Messina Hof (1983), and Texas Vineyards at Ivanhoe (1984), among them. Moreover, passage of the Texas Farm Winery Act in 1977 allowed vintners to make wine even in "dry" jurisdictions, a real boom to the industry.

Re-enter Mr. Munson

So what does all this have to do with Volney Munson?

Quite simply, the blossoming of this new Texas industry regenerated interest in a man—a Texan—who had once been the world's expert on American grapes. It awakened the modern Munson family to a new recognition of his accomplishments. And it brought together two other men who would lead the fight to bring his work back into the light.

By 1960, four of Volney and Nellie's seven adult children had died, and Olita would die a few years later; only Fern and Neva lived into the 1970s. And just a handful of the third generation had any direct memory of their grandfather. Moreover, many of the family had begun to leave Denison and scatter across the globe. Most of Ben's descendants had stayed in North Texas, but their memories of Volney had also faded with time and the loss of the nursery.

In the early 1960s, however, some of Ben's children and grandchildren rediscovered *Foundations of American Grape Culture* and became interested in Volney. In 1963 they visited France to learn more and established the first contacts with that country's modern viticulturists. Among the latter was Dr. Max Rives, director of the Bordeaux agricultural experiment station, who was conducting genetic studies using Volney's hybrids. Thus began a series of visits between the two countries. As a result of this enthusiasm on both continents, the family began gathering information and locating copies of Volney's publications. At the same time, the University of Kentucky celebrated its centennial by establishing a Hall of Distinguished Alumna; both Volney and Ben were inducted into that inaugural class (1965).

Eloise Munson, Ben's youngest child and one of the few left who had personally known her uncle, instigated the first reprint of *Foundations* and the awarding of Texas Historical Markers for Vinita Home (see Chapters 13 and 14). "Miss Eloise," an astute businesswoman who carried on many of her father's projects, also created the entity that would play such an important role in subsequent events. When she died in 1969, her will established the W. B. Munson Foundation to continue channeling many of her considerable assets into the charitable work she'd undertaken for years.[4]

And soon the Foundation would be approached about a particular project involving grapes and Volney Munson.

Grape Warriors

While all these events were happening within the family, two other men also became interested in Volney. W. E. Dancy was born and raised in Cooke County, which adjoins Grayson on the west. After a distinguished career as a comptroller for Gulf Oil Company, Dancy had retired to the rocky hills of Eureka Springs, Arkansas, where he took up grapes as a hobby, remembering from his youth the acres and acres of grapevines planted along the Red River before prohibition uprooted them. Because of his North Texas heritage, he was familiar with Munson and now wanted to know more about his work, to the point that collecting information and rare Munson hybrids became his personal crusade.

Enter John Clift, state editor of the *Denison (Texas) Herald* and a longtime oenophile, who wrote a column for his newspaper entitled "The Wine Rack." Clift had been in the Denison Public Library one day and noticed the reprints of *Foundations of American Grape Culture* for sale. A casual inquiry elicited the response that, thanks to the burgeoning fascination with wine in America, W. E. Dancy's Munson crusade, and the family's own interest, the book was selling very well. Clift contacted Dancy and quickly became infected with the Munson bug, too.

W.E. Dancy, at his home in Eureka Springs, Arkansas. (Photo by Roy E. Renfro)

192

The two began seeking out Volney's surviving hybrids, hoping to save them. They wrote hundreds of letters to vintners, nurseries, and private growers around the country and built up a list of sources. The number of survivors was discouragingly low. Clift, reporting on their progress in "The Wine Rack," moaned that "in many cases, I was just too late, the last of the Munson vines having been plowed under" only recently.[5] Initially they located only nine private collections, but word of the project spread rapidly and in little over a year, the list grew to forty—still a small number.

Then the pair decided to expand their campaign. They approached the W. B. Munson Foundation and asked its directors if they would like to fund some kind of tribute to their namesake's brother. Yes, the directors would and they, Clift, and Dancy began to explore ideas. A memorial vineyard to showcase Munson's 300-plus hybrids seemed the perfect project, but how could two men and a Foundation board accomplish it?

Back to College

In the spring of 1974, the group carried their proposal to Grayson County College, a junior (two-year) college midway between Sherman and Denison which had a strong horticultural and agricultural program.[6] Ben Munson, III, then president of the Foundation board, explained their reasoning: "I've asked the college to assist in this project because I feel that this area has great potential for the grape industry [just as Volney Munson had always said]. Grayson County College has the potential management skill[,] more so than any other institution in the area. I feel that we need this kind of expertise to make the project work."[7]

The college was enthusiastic, and an advisory committee was formed of Clift, Dancy, representatives from the Foundation and GCC, and an elderly Denison man named Horace Foster, who had actually worked for Volney at the nursery as a young boy.[8] Roy E. Renfro, Jr., co-author of this biography and then agricultural instructor at GCC, was appointed director of the project.

The group enlisted the help of viticultural experts throughout the world who gave advice, helped find cuttings, and visited the Denison site. Among them were Dr. Max Rives from Bordeaux; his associate, Alain Bouquet of the National Institute for Agronomic Research in Bordeaux; Dr. Harold P. Olmo of the University of California /Davis; Dr. Pierre Galet, author of more than thirty books on grapes and on the staff of Pierre Viala's old school in Montpellier, now the Ecole Nationale Superieure Agronomique; and Dr. George Ray McEachern of Texas A&M University, one of the pioneers of the new Texas wine and grape industry. There were also plans to bring in U. A. Randolf, who had previously headed the Montague County grape research facility for A&M.

A Hillside of Crosses

Through the summer and fall of 1974, Renfro and the committee worked on the basics. Where should they plant the vineyard? Soil tests around the college's physical plant located the best site on a hillside on the West Campus. More than four hundred Munson canopy trellises were constructed of Red River cedar and planted on the slope, their peculiar shape causing people to remark that the place resembled a cemetery more than a vineyard. Wide concrete walkways between the rows allowed college staff access and also per-

193

Roy E. Renfro plants the first Munson hybrid, Carman, in the T.V. Munson Memorial Vineyard, spring 1975. (T.V. Munson Viticulture and Enology Center)

mitted the eventual visitors to tour the vines. By the spring of 1975, the vineyard was ready for planting.

While Renfro and his crew prepared the site, Dancy continued his hunt for hybrids. Rootstocks and cuttings began arriving from all over the United States, and Dancy excitedly discovered the major surviving Munson collections at the University of California/Davis, the University of Arkansas Fruit Experiment Station/Clarksville, and the Missouri State Fruit Experiment Station/Mountain Grove. The newly christened T. V. Munson Memorial Vineyard received its first plantings of five varieties that spring and quickly grew to more than sixty; by the summer of 1979, the Vineyard was hosting its first public tours.

With all the excitement generated by this activity, GCC decided in 1974 to also offer a viticulture and enology curriculum. One of the first of its kind in an American community college, it was one of the few degreed programs in that subject available in the United States and the only one in Texas. In just two years, it attracted over 200 students and has grown every year since.

The Munson Center

A museum to house Munson memorabilia and papers was also one of the projects discussed from the very beginning. In the mid-1980s, with the vineyard in good shape, Grayson College and the Munson Foundation began planning for this next phase. Long and low, with a steeply pitched roof and dormer windows, the 5,000-square-foot brick

Dr. Roy E. Renfro, Dr. Pierre Galet, and David M. Munson examine grapes in the new vineyard, 1980.
(T.V. Munson Viticulture and Enology Center)

structure was designed to reflect the Italianate style of Vinita Home and was poised atop the hill overlooking the new vineyard and greenhouse. It was constructed entirely with private funding from the W. B. Munson Foundation, the Clara Blackford Smith and W. Aubrey Smith Charitable Foundation, and the Oliver Dewey Mayor Foundation, all from the Sherman/Denison area.

Inside the Center are classroom and office spaces as well as facilities to process plants, juice, and wine. An elegant library and conference room housing Munson artifacts is located in the front quarter of the building, and framed reproductions of some of William Prestele's drawings for Volney's never-published United States Department of Agriculture book grace the entry area.

Cognac Partnership

The fall of 1988, which marked the centennial celebration of Volney's receipt of the Chevalier du Merite Agricole award, saw a flurry of activity. The new T. V. Munson Viticulture and Enology Center was dedicated on September 10 with representatives from the Munson

The T.V. Munson Viticulture and Enology Center was dedicated on September 10, 1988.
(T.V. Munson Viticulture and Enology Center)

W. B. Munson III, Volney's great-nephew and chairman of the W. B. Munson Foundation, cuts the cer-emonial ribbon at the dedication of the T.V. Munson Viticulture and Enology Center, 1988. (T.V. Munson Viticulture and Enology Center)

family and wineries around the world, as well as French and U.S. dignitaries, present. The festivities then moved abroad to Cognac where, in October, American and French admir-ers of Munson dedicated a plaque to him at the Cognac Research Center. There, one hun-dred years earlier, Pierre Viala had conducted his grafting experiments on the grapevines Volney Munson had helped him find in Texas.

Dr. Max Rives, who had first met and worked with the Munson family twenty-five years earlier, led a full-day symposium in Cognac on the accomplishments of Munson, Viala, and the others who led the attack against phylloxera.

Interest Mounts

The international spotlight focused on Denison and Grayson County by the Munson project inevitably spun off in other directions. As early as 1974, Denison officials began talking of a grape festival; today the Denison Wine Renaissance Festival is a major and successful March event, combining art and wine and attracting thousands to Volney's hometown.

Spurred by the success of the 1988 Denison-Cognac dedications, leaders from both cities also explored the Sister City Movement which pairs American cities with ones throughout Europe. (Coincidentally, this program was started by President Dwight D. Eisenhower, who was born in Denison in 1890.) In 1992, a four-day celebration in Cognac kicked off their participation as a Sister City. Today an important part of the

THOMAS VOLNEY MUNSON

(1843 – 1913)

THOMAS VOLNEY MUNSON –DE DENISON, TEXAS– SPECIALISTE EN VITICULTURE ET
AUTORITE INTERNATIONALE EN LA MATIERE, A RECU DU GOUVERNEMENT FRANCAIS
LE TITRE DE 'CHEVALIER DU MERITE AGRICOLE' EN 1888 POUR SES RECHERCHES
SCIENTIFIQUES SUR LES PORTE-GREFFES AMERICAINS RESISTANT AU PHYLLOXERA. PIERRE
VIALA, PROFESSEUR DE VITICULTURE A L'ECOLE NATIONALE D'AGRICULTURE DE MONTPELLIER,
A L'INITIATIVE DES VITICULTEURS CHARENTAIS, REALISA EN 1887 UNE MISSION VITICOLE
AUX USA ET. GRACE A LA COLLABORATION DE T.V. MUNSON, PUT RAMENER DES ESPECES
DE PORTE-GREFFES ORIGINAIRES DU TEXAS QUI ONT CONTRIBUE A LA RECONSTITUTION
DES VIGNOBLES FRANCAIS RAVAGES PAR LE PHYLLOXERA.
    "CETTE PLAQUE A ETE POSEE LE 6 OCTOBRE 1988 EN PRESENCE D'UNE DELEGATION
AMERICAINE DU CENTRE T.V. MUNSON, DENISON, TEXAS, ET DES AUTORITES LOCALES
ET REGIONALES FRANCAISES.
    UNE PLAQUE IDENTIQUE A ETE POSEE AU CENTRE T.V. MUNSON, DENISON, TEXAS,
                          LE 10 SEPTEMBRE 1988"

The Centennial Celebration medal was given to French dignitaries at the dedication of the Munson plaque in Cognac. (Photo by Roy E. Renfro)

Denison/Cognac Sister City program is the visit of high school students between the two cities for several weeks each summer. Exchanges of medical doctors and business leaders have also strengthened the relationship.

Dr. Roy Renfro, with the assistance of John Clift, organized the Texoma Chapter of the American Wine Society in 1989. It is now one of the largest in the United States, with over 250 members from across the country, many of whom simply want to belong to a chapter associated with Volney.

Education has always been an important element of the Munson Viticulture Center's mission, even before the physical structure was built. In 1980, with the cooperation of Grayson College, the first Thomas Volney Munson Symposium was held in Bonham, Texas, by the Sam Rayburn Foundation and East Texas State University (now Texas A&M at Commerce). Among the speakers was Dr. Pierre Galet, on his first lecture tour of the United States. He spoke on "Munson and the Use of the American Vine in France" and told his audience there were still Munson hybrids being grown in Europe; he had even encountered "Bailey" in Brazil and enjoyed a wine made from it.[9]

More recently, the Viticulture Center sponsored an International Symposium in Viticulture and Enology in June 2000 which also marked the 25th anniversary of the vineyard and program at Grayson College. The symposium combined a summit on the Denison/Cognac Sister City program with a viticultural conference featuring speakers from France, Italy, Australia, and the United States, and concluded with the Denison Wine Renaissance Festival. In the intervening years between these two events, Dr. Renfro, director of the Munson project at GCC, has given well over one hundred programs on Munson across the United States and in Europe.

Most Texas winery owners and cellarmasters have now attended classes at the Viticulture Center, which averages a total of 250 students per semester. Grayson College offers two degrees, the certificate in Viticulture/Enology and the Associate of Applied Science in Viticulture/Enology. The program was honored in 1998 as the most outstanding one in Texas community colleges by the Texas Association of Community Services and Continuing Education.

Tour groups from around the world also visit the Memorial Vineyard and the Viticulture Center every year. Though most are from Texas and the southwestern United States, others have come from France, Germany, Italy, China, Japan, Australia, and South Africa.

Vineyardist Jack Dempsey in the T.V. Munson Memorial Vineyard.
(T.V. Munson Viticulture and Enology Center)

Munson Wine

Historically, wines have been made from Munson hybrids; Volney mentioned a few from time to time in his writings. In *Foundations of American Grape Culture* (page 191), for example, he quoted William Pfeffer, editor of *Pacific Tree and Vine* in California: "[Munson's] Post-Oak family of hybrids are most vigorous, good bearers, the best resistants, and make very fine wines. In fact, there is no Zinfandel, Mataro or Carignan wines, not to name a long list of standards, that can equal a wine of Neva or of Big Extra, and no light claret as refreshing as one of Elvicand."

Despite the resurgence of interest in Volney today, however, there are still only a few wineries using his hybrids, though he considered forty-three of them suitable for that purpose. In Ontario, Canada, T. G. Bright & Company (established 1874) produced a President port from the grape of that name for many years. Mount Pleasant Vineyards in Augusta, Missouri, was founded in 1881 by Volney's mentor, Frederick Muench. From an acre of Muench grapes, which Volney named for that pioneer viticulturist, Mount Pleasant today makes about 1,000 gallons of wine each year, blending it with other sweet wines before bottling. The University of Arkansas has a test plot of Munson grapes in Altus, from which the Post Familie Vineyards produces a deep red wine that is among their most expensive brands. A few other vintners in Missouri and Arkansas also use Munson vines for blending with other varieties.

In colder regions, Vintinto V65115 is often recommended for winemaking. A cross between Volney's black hybrid Lomanto and Colobel, used to deepen the color, Vintinto produces a dark table wine. Lomanto has also won awards on its own. In the San Antonio (Texas) Regional Wine Guild Competition of 2002, for example, a Lomanto wine tied for first place with *Cabernet Sauvignon* in the dry red grape category and placed second in sweet red grape. Volney's white hybrid, Captivator, did similarly well in San Antonio's 1999 contest, capturing a first and two seconds. Delicatessen, another of his favorite hybrids, took first in Native and Red Wines at the 2003 Greater Kansas City Cellarmaster competition.

Munson hybrids are included on a number of lists which recommend grapes for winemaking. The United States Bureau of Alcohol, Tobacco, and Firearms lists six on its approved list of type names for wine designation: Beacon, Captivator, Dixie, Ellen Scott, Fern Munson, and Muench. Bailey and Champanel appear on other lists.

Volney's great work was honored in 1991 when Moyer Texas Champagne in San Marcos, near San Antonio, released a Cuvee T. V. Munson champagne. (The American debut was held at the Denison Rod and Gun Club, not far from Vinita Home.) A "Texas Grand Vin Brut," this cuvee was 85% pinot noir and 15% chardonnay—in other words, made with no Munson grapes. It was bottled in jeroboam size (4/5 of a gallon, or the equivalent of four regular bottles) and packaged in a handsome presentation box.

Other Uses

Though still somewhat scarce in nurseries, Munson hybrids have a prominent place in the grape world outside of enology. Geneticists at many modern research stations use them in cross-breeding programs since they have a high degree of disease resistance, especially to Pierce's Disease. Dr. Justin Morris of the University of Arkansas, for example, is only one of many using Munson grapes in his research. The Missouri State Fruit Experiment

Station has also developed several new hybrids using Munson grapes as the parent stock. Ozark Prize resulted from crossing Volney's Dr. Collier with Sheridan, a Concord-type grape, while Blue Eye was a true Munson offspring, using both Ellen Scott and America.

Other studies have touted them for making juice and as table grapes. In 1995 the Viticulture and Enology Research Center in California compared several Munson cultivars—Cloeta, R.W. Munson, Bailey, and Extra—with Concord and concluded that they produced a better quality of juice with higher yields. The North American Fruit Explorers recommends several Munson varieties such as Xlanta and Champanel for eating, noting that they are "softer and juicier than modern varieties." The latter is also popular for jellymaking and for landscape use.[10] New York, North Dakota, and Kentucky are among the many state agricultural extension services which endorse Munson hybrids as useful for their areas.

Of course, many also remain important as rootstocks. (See Chapter 7 for a lengthier discussion.) Volney's hybrid, Champanel (a *Champinii* cross), as well as his native finds Salt Creek and Dog Ridge (both of them *V. champinii* selections), are among the mostly widely used in the world today. To ensure their continued availability for research, many of his most important creations are preserved through the National Plant Germplasm System, more than two dozen cooperating facilities which "collect, maintain, characterize, document, and distribute plant germplasm from all over the world."

New Recognition

Volney Munson's reputation and achievements are now being acknowledged by a new generation of horticultural and viticultural authors. In 1976, Drs. D. V. Fisher and W. H. Upshall, writing a *History of Fruit Growing and Handling in United States of America and Canada, 1860–1972*, declared that, nearly six decades after his death, "Munson's work is still recognized as the most authoritative in this field." (page 183) Frank Giordano pronounced him one of "America's wine prophets" in 1984, while Texas A&M professor George Ray McEachern, in a three-part history of the Texas wine industry (1996), wrote that Munson "was 100 years ahead of his time, breeding grapes to match the variety to the climate and to naturally prevent insect and disease damage." Other authors such as James Wilson (*Terroir*) and Thomas Pinney (*A History of Wine in America*) have also devoted space to Volney's labors.

Volney's very name came to stand for excellence when the Texas Wine and Grape Growers Association established the T. V. Munson Award in 1985 to "recognize a person who has made significant contributions to the Texas wine and grape industry through their efforts in viticulture." The first of the distinguished recipients was Dr. Roy Renfro.[11] The Dallas (Texas) Opera also inaugurated an award in Munson's name in 1993 to honor an outstanding American winemaker. Dr. Renfro and Stanley Marcus, a noted wine connoisseur and head of the renowned Neiman-Marcus stores, presented the award to the first recipient, Daryl Groom, winemaster at Geyser Peak Winery.

Volney himself has been inducted into the Texas Heritage Hall of Honor, established in 1992 by the State Fair of Texas to recognize lifetime achievement in agriculture. As of this writing, less than forty Texans have been so honored.

Horace Foster, who worked for Volney as a boy, holds a copy of Munson's book.
(Photo by Roy E. Renfro)

Today's Volney Munson

Volney's book, *Foundations of American Grape Culture,* continues to receive praise nearly a century after it was first published. The Michigan State University Extension Service, for example, still recommends it, while the North Carolina Department of Agriculture and Consumer Services declares that Volney "laid the foundation of *all* modern viticultural techniques." (Authors' italics) Nor is its use confined to this country. Universidad de la Rioja, in the famous wine country of northern Spain, is among the foreign institutions which include *Foundations* in their viticultural courses.

Volney has also become the only person for whom an official Texas wine trail is named. Six of these seven winery and vineyard routes bear the designations of geographical or geological features, but the last is the Munson Trail in Northeast Texas.

In this electronic world today, Volney Munson can be found on the Internet at a number of gourmet and wine websites such as epicurious.com, cooking.com, and wein-plus, all of which note his contributions to the French wine industry. Anthony Hawkins' "Super Gigantic WWW Wine Grape Glossary" lists recommend French-American, American hybrid, and native grapes. Among the hundreds on the site are a dozen or so of Munson's— Albania, Bailey, Bell, Captivator, Champanel, Ellen Scott, Headlight, Husmann, Muench, President, as well as Dog Ridge. The website for Cognac, France, meanwhile, forthrightly proclaims that "if Cognac is still today the commercial capital of a blessed region, it is to a large extent thanks to Texas" and Volney Munson.

There are also a number of plant nurseries both in this country and abroad which sell Munson grapes and advertise over the Net. They can be located by searching for specific hybrid names such as America, Bell, Ellen Scott, or Fern Munson.

The Picture in Texas

At the start of a new millennium, the Texas wine industry is growing fast and is back on the track that it once occupied before prohibition. From a mere 14,000 gallons in 1979, the state now produces well over a million every year, a bounteous harvest with an impact of about $100 million on the economy.

Nearly fifty wineries and 150 commercial vineyards can be found in six designated wine-growing regions: High Plains, Escondido Valley, Hill Country (at 15,000 square miles, the largest official American Viticultural Area in the United States), Bell Mountain, Fredericksburg, and the Davis Mountains or Trans-Pecos. October has been officially designated "Texas Wine Month," and Grape Fest in the city of Grapevine is the largest wine festival in the southwestern United States. Texas ranks fifth among American producers, with nearly 4,000 acres of land devoted to growing wine grapes. The state even has the world's only bonded winery located inside an airport, La Bodega at Dallas-Fort Worth International Airport.

Most Texas wines are sold within the state, but a few go to such countries as Canada, France, Japan, and Russia. Val Verde, Llano Estacado, Messina Hof, Ste. Genevieve, and several others of the second generation of wineries are still in operation and flourishing. And the wine industry is the fastest growing segment of Texas' agribusiness economy.

Texas vintners say their wines are unique because of the state's own special "*terroir,*" a

French word used to describe the combination of soil, rainfall, and climate best suited to the vines. More than 80% of all Texas grapes and 40% of its wine grapes are now grown in the High Plains around Lubbock and in the Trans-Pecos Region in the Big Bend area, a fact which would have amazed Volney since these were deserts in his lifetime. Only 10% come from North Central Texas, his home and an area he once compared to the finest vine-growing regions of southern France. Among those are sixty-five of the hybrids he created which now climb the hillside in the Munson Memorial Vineyard.

Once scorned outside the state, Texas wines now hold their own against other countries. In the fall of 2000, for example, the Wine Society of Texas and LeNotre Culinary Institute staged a Texas-French Shootout, a blind taste testing in which Lone Star wines won five out of six categories. Overall winner was Blue Mountains Vineyard in the lovely West Texas mountain town of Fort Davis.

Volney would be gratified to see this rebirth, for he always knew Texas could rival any other country, including France, in grape and wine production. His monumental, lifetime pursuit was the educational advancement of the grower-producer, and the development of superior grape varieties and vineyard practices to ensure a healthy and prosperous industry.

A Thoroughly Modern Viticulture Center

W.B. Munson III and Dr. Roy E. Renfro break ground for the T.V. Munson Viticulture and Enology Center. (T.V. Munson Viticulture and Enology Center)

The T. V. Munson Viticulture and Enology Center has kept pace with these developments and continued Volney's work; in doing so, it has garnered several "firsts." In the fall of 1997, Dr. and Mrs. John Anderson donated their winery, Schoppaul Hill (formerly Texas Vineyards at Ivanhoe), to the Grayson County College Foundation as an educational site for GCC's viticulture and enology classes. The facility, rechristened the T. V. Munson Instructional Winery, is located at Ivanhoe in Fannin County, approximately thirty miles east of the Munson Viticulture Center. The Andersons' gift made Grayson the only community college in this country with a full commercial winery as part of its program. Eventually, the winery will release products made from Munson grapes as it focuses efforts over the next few years on varieties with the most potential for winemaking.

The Center has its own website (www.tvmunson.org) and plans to offer viticulture and enology courses over the World Wide Web. (For more information, see GCC's website: www.grayson.edu.)

T.V. Munson Viticulture and Enology Center, with a portion of the vineyard in the foreground.
(Photo by Roy E. Renfro)

The facility's collection now includes photostatic or original copies of virtually all of Volney's known publications and papers, and photographs of the family and business. The Dallas Historical Society has graciously placed its Munson material donated by Olita Calvert on long-term loan there, including award certificates and medals, Volney's surviving journal, family scrapbooks, and one extant sketchbook. Mrs. Calvert's descendants have since generously donated other family items to the Viticulture Center, including Nellie Munson's portrait (on page 17) and Volney's desk. In keeping with its educational mission, Center staff are also building an impressive library of both rare and recent titles on viticultural and enology subjects from around the world. It includes the books and notes of geologist James Wilson, author of *Terroir* (1998), as well as the research conducted for this biography.

Epilogue

Munson's lifetime of achievements, thanks to the efforts of people such as William Dancy, John Clift, Roy Renfro, and the W. B. Munson Foundation, is finally coming into its own once more. Never completely forgotten by the botanists and viticulturists, he is only now becoming familiar again to the general public in the United States, who know him vaguely as "the man who saved the French vineyards."

Perhaps his knowledge and experience alone did make the final, crucial difference in France. Certainly his name, of all those Americans who worked tirelessly to help find the cure to the phylloxera epidemic, remains foremost among winery owners and grape growers throughout Europe and much of the rest of the world. Volney wouldn't have cared about that particular accolade anyway; he never mentioned it in any of his writings. He was most proud of his wife and children: they were his true monuments, he felt. After that came his "other" children, the hybrids he spent his life creating. As long as even one of them remains to climb a trellis or a post high to the sky and to "clasp its hands with joy," then he lives on.

And that's a memorial not many of us can claim.

Appendices

Appendix I.

PUBLISHED WORKS OF T. V. MUNSON
JOURNAL/NEWSPAPER ARTICLES AND BOOKS
(Listed in chronological order)

No date or date is uncertain:

"The Shakespear-Bacon [*sic*] Controversy." Privately printed.

How to Organize a Horticultural Society, also the Constitution for a Local Society. With Mrs. J. R. Johnson. (Texas State Horticultural Society, circa 1888.)

"History of Parker Earle Strawberry." *Indiana Farmer,* circa 1891.

"Hybridity and Its Effects." Typescript.

"Berries in Texas." *Prairie Farmer,* circa 1894.

"Among the grape-growers some new varieties." *Prairie Farmer,* circa 1894.

Munson's Newer Grape Creations. Circa 1902.

"Le Black Rot en Amerique." *Actualities,* pages 666–669. (Incomplete reference)

1873   "Experiment with Potatoes." *Farmer's Home Journal.*

Description of Lincoln, Nebraska. *Kentucky Gazette,* March 1 and December 20.

1874   "Nebraska." *Farmer's Home Journal.* December 29.

1875   "Nebraska." *Farmer's Home Journal.* January 9.

1880   "Fruit Culture in Northern Texas." *Burke's Texas Almanac and Immigrant's Handbook for 1880,* pages 55–57. (Houston, Texas: J. Burke, Jr., 1880.)

1881   "Early Peaches." *The Gardener's Monthly and Horticulturist,* Volume 23, page 177. (Philadelphia: Charles H. Marot, Publishers, 1881.)

1883   "Forests and Forest Trees of Texas." *American Journal of Forestry,* Volume 1, pages 433–451, July 1883.

Native Trees of the Southwest. (There is a reference to this as a USDA Bulletin, but the authors could not confirm.)

"Systematic Horticultural Progress." *Transactions of the Mississippi Valley Horticultural Society for the Year 1883*, Volume 1, pages 136–139. (Indianapolis: Carlon & Hollenbeck, 1883.)

1884   "Trees Peculiar to Texas." *Transactions of the Mississippi Valley Horticultural Society*, Volume 2, pages 47–51. (Indianapolis: Carlon & Hollenbeck, 1884.)

"Origin of Herbemont." *The Wine and Fruit Grower*, September 1884. Reprinted in the same journal August 1885.

1885   "Address on Native Grapes of the United States." *Transactions* of the American Horticultural Society, Volume 3, pages 128–140. (Indianapolis: Carlon & Hollenbeck, 1885.)

"Native Grapes of the United States." *Pacific Rural Press*, Volume 30, page 2, 1885.

"Native Grapes of the United States." (Revised) *The Wine and Fruit Grower*, Volume 7, pages 83–86. August 1885.

"New Classification of the Native Grapes of the United States." *Ohio State Horticultural Society Annual Report for 1884–1885*. (Columbus: Myers Brothers, 1885.)

"Sur une nouvelle classification des vignes des Etats-Unis d'Amerique." Translation of address to American Horticultural Society, by M. G. Bourgade. *Le Progres Agricole et Viticole* (Montpellier).

1886   "American Grapes: Importance of Botanical and other Scientific Knowledge to the Progressive Horticulturist and Especially to the Viticulturist." *Proceedings of the Twentieth Session of the American Pomological Society, 1885*, Volume 20, pages 95–100. (Lansing: Trop & Godfrey, 1886.) (Also published as a separate publication by the same printer.)

"Comparative Hardihood of Grapes." *American Garden*, August 1886.

"A New Species of Grape, and the Scuppernong." *Gardener's Monthly*, Volume 28, 1886.

1887   "The relative times of germination, leafing, blooming, ripening and size of fruit of species of American grapes." (Tabular form.) *Proceedings* of the Society for the Promotion of Agricultural Science, 1887.

1888   "What Shall My Profession Be?" Address given to graduating class of Texas A&M College, June 5, 1888, and printed privately. (Denison, Texas: Murray's Print, 1888.)

"Experiments in the treatment of grape disease made in 1887 at Denison, Texas." In United States Department of Agriculture, Division of Botany, Bulletin No. 5.

1889   "Hints to Southern Fruit Growers." *Southern Horticultural Journal*, Volume 2, no. 12, September 15.

1890   *Classification and Generic Synopsis of the Wild Grapes of North America*. United States Department of Agriculture, Division of Pomology, Bulletin No. 3. (Washington: Government Printing Office, 1890.)

"Les Vignes Sauvages de l'Amerique du Nord." *Le Progres Agricole et Viticole*, Volume 14, pages 532–535. (Montpellier, 1890.)

"Vitis Baileyana Munson." (Reprinted from above.) *Garden and Forest,* October 1, 1890.

"A Classification of American Grapes." *Garden and Forest,* Volume III, pages 474–475, October 1890.

"The Nomenclature of American Grapes." *Garden and Forest,* pages 637–638, November 5, 1890.

1891 "Possibilities of our native grapes." *The American Garden,* Volume 12, pages 580–586 and 658–661. (New York: The Rural Publishing Company, 1891.)

"Grapes in North Texas in 1891." *Report* of the Texas State Horticultural Society. (Brenham, Texas: Banner Steam Book and Job Printing House, 1891.)

1892 "Grape explorations across Texas." *Dallas Morning News,* April 22, 1892.

"Structural Botany." *Dallas Morning News,* July 22, 1892. (Reprinted in *Texas Farm and Ranch,* August 6, 1892.)

1893 "Progressive Grape Culture for the North." *Report* of the Michigan Horticultural Society.

1894 "Les Vignes Americaines en Amerique." Part I. *Revue de Viticulture,* Volume 1, pages 81–84, January-June 1894.

"Explorations Viticoles dans le Texas." *Revue de Viticulture,* Volume 2, pages 369–372, July-December 1894.

1895 "Classification of the Native Grapes of North America." AND "Viticultural Observations Upon the Native Species of American Grapes." *Bush & Son & Meissner Illustrated Descriptive Catalogue of American Grape Vines: A Grape Grower's Manual.* 4th edition. (St. Louis: R. P. Studley & Co., 1895.)

"Les Porte-Greffes des Terrains Crayeux Secs." *Revue de Viticulture,* Volume 3, pages 81–84, January 26, 1895.

"Les Vignes Americaines en Amerique." *Revue de Viticulture* (translated by M. Pierre Viala):
- Part 2: Volume 3, pages 157–161, February 16, 1895.
- Part 3: Volume 4, pages 245–251, September 14, 1895.

1896 "Les vignes Americaines en Amerique." *Revue de Viticulture* (translated by G. Gouirand):
- Part 4: Volume 5, pages 157–167, February 15, 1896.
- Part 5: Volume 6, pages 421–427, October 31, 1896.

"Le Traitement du Black Rot en Amerique." With B. T. Galloway. *Revue de Viticulture,* Volume 6, pages 545–549, December 5, 1896.

"Resistant Vines." *Pacific Tree and Vine,* August 29 or September 9, 1896.

1897 "The Unattained Ideals in the Grape." *Garden and Forest,* Volume 10, 1897. (New York: The Garden and Forest Publishing Co., 1897.)

1899 "Fifty Years Improvement in American Grapes." *American Gardening:*
- Part I: "Development of Northern Varieties." September 9.
- Part II: "Development of Southern Grapes." October 7.

- Part III: "Brief Summary of T. V. Munson's Work with American Grapes." November 4.

Investigation and Improvement of American grapes at the Munson experimental grounds near Denison, Texas, from 1870 to 1900. Texas Agricultural Experiment Station Bulletin No. 56. (Austin, Texas: Von Boeckmann, Schutze & Co., November 1899.)

Territorial Horticultural Exhibition, Oklahoma City. (American Wine Press, 1899.) (It's unclear what Munson contributed to this, as no copy could be located.)

1900   "Munson on Grapes." *The National Nurseryman*, Volume 8, no. 9, October 1900.

1902   "Advantages of Conjoint Selection and Hybridization and Limits of Usefulness in Hybridization among Grapes." *Memoirs of the New York Horticultural Society: Volume I: Proceedings: International Conference on Plant Breeding and Hybridization, 1902.*

"Les Variations Asexuelles et le Deperissement de Certains Hybrides Greffes." *Revue de Viticulture*, Volume 18, page 460, October 25, 1902.

"Mission's [sic] Grape." Entry in *Traite General de Viticulture: Ampelo graphie: Tome III.* P. Viala and V. Vermorel, editors. (Paris.)

1903   "Les hybrides Munson." *Revue de Viticulture*, Volume 20, pages 411–415, 441–445, 473–476, and 558–562.

1904   "Grape Culture in Texas." *Texas Almanac and State Industrial Guide for 1904*, pages 105–106. (Galveston, Texas.)

1905   "Breeding Grapes to Produce the Highest Types." *Nebraska Farmer*, March 23, 1905.

"Breeding Grapes." *Proceedings of the American Breeders' Association*, Volume 1, pages 144–147. (Washington, D.C.: American Breeders' Association, 1905.)

"History of the Nursery Business in Texas." *The National Nurseryman:*
- Part 1: (missing)
- Part 2: Volume 13, no. 9, September 1905.

"Concord." "Ives." "Delaware." "Catawba." Entries in *Traite General de Viticulture: Ampelographie: Tome VI.* P. Viala and V. Vermorel, editors. (Paris.)

1906   "The Grape, the Commercially Neglected Fruit: Cause and Remedy." *Proceedings of the American Pomological Society, 1905.* (Lansing, Michigan, 1906.)

Length of life of vines of various species and varieties of grapes; profitableness; and by what diseases seriously affected. Texas Agricultural Experiment Station Bulletin No. 88. (1906)

The New Revelation. Written under pseudonym of Theophilus Philosophius. Privately printed. (Denison, Texas: Press of B. C. Murray, 1906.)

1907   "Can a Race of Commercial Seedling Pecans Be Evolved?" *Texas Fruits, Nuts, Berries and Flowers*, Volume 1, no. 5, December 1907.

1908   "Improvement of quality in grapes." *Proceedings of the Society of Horticultural Science, 1905*, Volume 3, pages 19–24. (College Park, MD: C. P. Close, 1908.)

"T. V. Munson's Theory: Cause of the Glacial and Warm Periods on the Earth." *Waco (Texas) Searchlight*, July 12, 1908.

1909   "Laying the Foundations of American Grape Culture." *The National Nurseryman,* Volume 17, no. 6, June 1909.

        Foundations of American Grape Culture. (New York: Orange Judd Company, 1909.)

1910   "Resistance to cold, heat, wet, drought, soils, insects, fungi in grapes." *Memoirs of the Horticultural Society of New York: Volume II: Proceedings of the International Conference on Plant Hardiness and Acclimatization, 1907.* (Horticultural Society of New York, 1910.)

1911   "Single-character vs. Tout-Ensemble Breeding in Grapes." *American Breeders' Magazine,* Volume 6. (Washington, D.C.: American Breeders' Association, 1911.)

1912   "Problems in Breeding Tree and Vine Fruits." *Annual Report: American Breeders' Association,* Volume 7. (Washington, D.C.: American Breeders' Association, 1912.)

        "Longavinbo and the mutation theory." *Annual Report: American Breeders' Association,* Volume 8, pages 444–448. (Washington, D.C.: American Breeders' Association, 1912.)

1925   "Grapes in the South." In L. H. Bailey's *The Standard Cyclopedia of Horticulture.* (New York: Macmillan Company, 1925.) (Published posthumously, probably written 1911.)

1929   "Horticulture in Texas." In L. H. Bailey's *The Standard Cyclopedia of Horticulture,* 1929. (Written circa 1911 and updated by Bailey for the 1929 edition.)

Beginning in 1886, Munson wrote dozens of articles for *Texas Farm and Ranch,* an agricultural newspaper published in Dallas, and its successor, *Farm and Ranch.* Below is a partial listing of those articles:

1886   "The McKee Grape." June 1.

        "The Lutie Grape." September 15.

1887   "Stamen characteristics of grapes: Extract from Mr. T. V. Munson's address before the American Pomological Society." January 15.

        "Horticultural surgery and propagation." April 1, April 15, May 15, August 1, September 15, October 1, and October 15.

        "The LeConte (pear) again." June 15.

        "Pedigreed LeContes." August 15.

1888   "Grafting peach and plum." December 15.

1890   Account of American Horticultural Society meeting in Austin, Texas. October 1.

        "The 'Whole Root Fraud.'" October 1.

1891   "The 'Whole Root Fraud.'" January 1.

        "The whole-piece root fraud dies hard." March 15.

        "Dr. Stell's insinuations." July 1.

1892   "Grapes and small fruits for the farm." January 1.

        "An ornamental evergreen hedge and windbreak." January 1.

        "The apple tree borer and its prevention." January 1.

"History of the Parker Earle strawberry." Reprinted from *Indiana Farmer*, March 12.

"What T. V. Munson is in Favor Of." March 26.

"How to make and apply the Bordeaux mixture." April 16.

"A sick orchard." May 7.

"Growing pears." May 14.

"Structural Botany." May 14.

"About grapes from cuttings." July 23.

"Prof. Munson (on horticulture) at Aransas Pass (Texas)." August 6.

"Experiments in root-grafting the apple." December 10.

1893   "Grapes and grape-growing." February 4, February 18.

"Horticulture at the World's Fair." July 8.

"Profitable fruit growing." August 26.

"Planting, pruning and training." October 14.

1894   "Why do trees die? The preventive." July 28.

"Berries in Texas." Reprinted from *Prairie Farmer*, September 15.

"Among the grape-growers some new varieties." Reprinted from *Prairie Farmer*, September 22.

1895   "Root tumor." April 20.

"Grapes in Texas." September 7.

"Stringfellow's criticism of Munson's paper." October 5.

"Different degrees of resistance to black rot in different varieties of grapes." October 19.

"Some late grapes." October 26.

"The 'science' of ignorance in horticulture." November 23.

1896   "Characteristics of southern grapes." February 8.

1906   "Longevity of grapes of various varieties and species in North Texas." February 17.

Appendix II.

PAPERS AND ADDRESSES GIVEN BY T. V. MUNSON
(Listed in chronological order)
(* indicates that paper was published)

1870   "Inductive Reasoning the Only Road to Advancement." Graduation oration. A&M College of Kentucky. Lexington, Kentucky.

1883   "Systematic Horticultural Progress." Mississippi Valley Horticultural Society. New Orleans, Louisiana.*

1884   "Trees Peculiar to Texas." Mississippi Valley Horticultural Society.*

1885   "Address on Native Grapes of the United States." American Horticultural Society. New Orleans, Louisiana.*

"American Grapes." American Pomological Society. Grand Rapids, Michigan.*

1888	"What Shall My Profession Be?" Graduation oration. Texas A&M College. College Station, Texas.*
1891	"Grapes in North Texas in 1891." Texas State Horticultural Society. Lampasas, Texas.
1892	(Relation of botany to scientific horticulture.) Texas State Horticultural Society. Dallas, Texas.
1893	"Progressive Grape Culture for the North." Michigan Horticultural Society.*
	"The Nurseryman's Position towards National, State and Local Horticultural Associations." American Association of Nurserymen. Chicago, Illinois.
	"Texas at the World's Fair." Texas State Horticultural Society.
	"Forecast of Better Thing Amongst Grapes." Congress on Horticulture. Chicago, Illinois.
1894	"Why Fruit Trees are Short Lived: Remedy." Texas State Horticultural Society. Houston, Texas.*
1895	"Grapes in Texas." Texas State Horticultural Society. Bowie, Texas.
1902	"Advantages of Conjoint Selection and Hybridization, and Limits of Usefulness in Hybridization among Grapes." International Conference on Plant Breeding and Hybridization. New York, New York.*
1905	"Breeding Grapes to Produce the Highest Types." American Breeders' Association. Champaign, Illinois.*
	"Improvement of Quality in Grapes." Society for Horticultural Science. New Orleans, Louisiana.
	"The Grape, The Commercially Neglected Fruit." American Association of Nurserymen. West Baden Springs, Indiana.
1907	"Resistance to Cold, Heat, Wet, Drought, Soils, Insects, Fungi in Grapes." International Conference on Plant Hardiness and Acclimatization. New York, New York.*
1909	"Single-character vs. Tout-ensemble Breeding in Grapes." American Breeders' Association. Omaha, Nebraska.*

Appendix III.

Grape Varieties Created by T. V. Munson
(Listed in alphabetical order)

The types of names Munson bestowed on his hybrids fall into six general categories: (1) family members and relationships, (2) friends and colleagues, (3) geographic locations, (4) parentage of the hybrids, (5) descriptors of the grape, and (6) whimsical to who knows? At one point, for example, he showed a great fondness for Oklahoma towns with unusual Indian names, which also likely were areas he had explored looking for grapes. Others, such as Eleala, make little sense to anyone but Munson. As Thomas Pinney wrote in *A History of Wine in America,* they are not only "graceless" but suggest "exhaustion of the poetic power" after having to name so many. Munson also did not bother to name some experimental hybrids; these are simply listed here by their crosses.

Munson's field notes have not survived and many of his grapes are now gone, so the task of assembling a list of his hybrids seemed at first daunting. Fortunately, just as this book was being completed, a letter emerged which proved to be, so to speak, "straight from the horse's mouth." On January 9, 1909, Volney wrote N. M. McGinnis of the New York Agricultural Experiment Station at Ithaca, complying with his request of December 30 to "name the varieties of grapes which I originated and introduced." With his usual thoroughness, Volney showed the names of each, parentage, color, size, season, and use (wine, table, market, etc.). He even noted those which he personally considered the most valuable; they are indicated here with a star (*) following the name. The authors have included those varieties which he eventually discarded for the same reason Volney did: "to explain the parentage of a number introduced and to show the scope of my [Volney's] work on grapes."

Additional sources for this list included: *The Grapes of New York* (Volume II) by U. P. Hedrick (1908); *Foundations of American Grape Culture* by T. V. Munson (1909); *Ampelographie* by Pierre Viala (1909); *A Practical Ampelography: Grapevine Identification* by Pierre Galet (1979); *Cepages et Vignobles de France: Tome I: Les Vignes Americaines* by Pierre Galet (1988); several published articles by Munson; and files compiled by W. Dancy in the 1970s and now in the collection of the T. V. Munson Viticulture and Enology Center.

Following the main list is a second one of wild varieties of grapes Munson discovered and named; many of these he used as breeding stock.

ADMIRABLE. (Also known as Munson's No. 76.) A hybrid of *Lincecumii* (Jaeger's No. 43) and *Aestivalis* (Norton), this black grape was introduced in 1894. Volney listed it as a wine grape but noted that it was "lost."

ADOBE GIANT OR ADOBE NUMBERS 1, 2, 3, 4, AND 5. These seedlings were hybrids of wild vines found in Hutchinson and Motley counties in the Panhandle of Texas. Munson used these black grapes for breeding and graft stock.

AGOMA. No parentage listed. Discarded.

AGONLA. No parentage listed. Discarded.

ALBANIA.* A cross between Ten Dollar Prize (*Lincecumii*), Norton, and Herbemont, Albania was a white grape created in 1896. Anthony J. Hawkins describes it as having "spicy aromatic and taste characteristics similar to Gewurztraminer."

ALBERS. A white Armlong/Pense (Malaga) cross, it was later discarded.

ALFERJON. No information listed.

ALFREDA. No parentage listed. Discarded.

ALMA. No parentage listed. Discarded.

ALUWE. A seedling of Lucky (one of Munson's Denison-area wild grape finds) pollinated by the hybrid Carman, this black grape was introduced in 1899.

AMAGAN. No parentage listed. Discarded.

AMALA. No parentage listed. Discarded.

AMBECON. Munson introduced this black grape, a hybrid of **America** and **Beacon**, in 1897.

AMERBA. No parentage listed. Discarded.

AMERBONTE. A hybrid of **Amer**ica and Herbe**mont**, this dark red grape derives its name from its parents. Volney considered it a good wine grape.

AMERICA.* A seedling of Hermann Jaeger's No. 70, Munson created this black grape in 1885 and commercially introduced it around 1891. U. P. Hedrick considered it a good grape for breeding since it combined *Rupestris* and *Lincecumii* but lamented in 1945 that it was no longer "largely grown." One of Munson's best known varieties, America was reputed to make a good dark red wine as well as a port. Volney also recommended it for jelly making and for juice, on account of its "peculiar but fine flavor."

AMERISA. A black America/Badart cross.

AMERSION. A black grape introduced in 1899, Amersion derived its name from its parents, **Amer**ica and Profu**sion**.

AMETHYST. A hybrid of Delago and Brilliant, first fruited in 1896 and commercially introduced six years later, Amethyst took its name from its soft red color. Hedrick considered it an excellent table grape, and Munson declared that it made a good white wine.

AMONTA. A seedling of *Monticola* and America, Amonta was a black grape introduced in 1899 but later discarded.

AMSION. A cross between America, Norton, and Ten Dollar Prize, Amsion was a black grape never introduced commercially.

AMUSEMENT. No parentage listed. Discarded.

ANITOAH. No parentage listed. Discarded.

ANSONYNE. No parentage listed. Discarded.

ANUTA. A black grape introduced in 1899, Anuta was a hybrid of America and Beacon, two other Munson hybrids.

ARBEKA. A cross between America and Profusion, this black grape was introduced in 1899 but later discarded.

ARBELA. No parentage listed. Discarded.

ARMADO. A black grape created in 1902, Armado was a hybrid of Armlong and Griesa de Piemonte (*vinifera*). It had not yet been introduced by 1909 and may never have been released commercially.

ARMALAGA.* This yellowish green grape was created in 1902 and introduced about five years later; it was a hybrid of **Arm**long and **Malaga** (*vinifera*).

ARMBRILONG. A seedling of **Armlong** crossed with **Bril**liant, this red grape was introduced in 1899.

ARMESA. No parentage listed. Discarded.

ARMLOCAR. No parentage listed. Discarded.

ARMLONG. Munson used this black grape, a hybrid of Ten Dollar Prize with Black Eagle (*vinifera*) (1886), for breeding only since it had "imperfect flowers."

ARONGA. No parentage listed. Discarded.

ATAVITE. This black Concord seedling, introduced in 1885, was later discarded by Munson because of problems.

ATOKA. A hybrid of America and Delaware, this purplish red grape was named for a town in south central Oklahoma. It was created in 1893 and introduced in 1899.

AUGUSTA. See Augustina.

AUGUSTINA. A cross between Delago and Brilliant, Munson created this close sister to Amethyst in 1899 and introduced it soon after as Augusta. However, he changed the name when he discovered there were already several Augusta grapes.

AUSTIN. A bronze San Jacinto seedling, Austin was named either for Stephen F. Austin (the "Father of Texas") or his namesake capital city.

BADART. Created and introduced around 1899, Badart was probably named for fellow grape-grower George Badart of Belton, Texas. It was a black hybrid of Ten Dollar Prize with Triumph.

BAILEY.* A cross of *Lincecumii* (Big Berry) and Triumph (a Concord), Bailey was planted in 1886 and first fruited in the 1889–1890 season. Munson thought it a promising wine grape and named the black-skinned creation for friend and leading American horticulturist Liberty Hyde Bailey of Cornell University.

BEACH. Munson eventually discarded this black grape, a hybrid of Post Oak Number 3 with Triumph and introduced in 1889.

BEACON.* An important Munson hybrid, Beacon, a black grape with a white bloom, was a cross of *Lincecumii* (Big Berry) and a *Labrusca* (Concord). Hedrick described it as "very vigorous" while the Kansas Agricultural Experiment Station declared it similar to Concord but with three times the yield. ("Grape Growing in Kansas," 1928.) (Planted 1886, first fruited 1889.)

BEAGLE. A black grape created in 1882, this was probably the result of an Elvira naturally pollinated by a nearby Ives and was likely named for Charles Darwin's famous research ship.

BEANUM. A black Beacon hybrid, Munson listed it as a wine grape.

BELL. (Also known as Munson's Number 21.) Munson honored his wife's Kentucky family with this greenish-yellow grape, a hybrid of Elvira and Delaware created in 1883. In *Foundations,* he wrote that he had "received very flattering testimonials of this grape." Among them was the Kansas State Agricultural College Experiment Station, which noted in 1902 that Bell was a "very good bearer," with a "peculiar sweet, agreeable flavor." (*Bulletin No. 110.*)

BELTON. A cross between De Grasset and Brilliant, Belton was a black grape named for the town in Bell County, Texas, near which Munson discovered his important wild rootstock, Dog Ridge.

BELVIN. A black grape with a bloom, this 1889 creation (Post Oak X Elvira) was later discarded.

BEN. Munson also eventually discarded this hybrid of Ten Dollar Prize and Norton's Virginia, a black grape introduced in 1889 and probably named for his brother Ben Munson.

BEN-HUR.* Munson created this black grape in 1899 by crossing a *Lincecumii* with Norton and Herbemont. He exhibited it at the American Pomological Society's 1903 meeting and, encouraged by its success, introduced it the following year. It was probably not named for Lew Wallace's best-selling 1880 novel, *Ben Hur: A Tale of the Christ*—since Munson didn't read novels—but for the Limestone County town. He considered it a good wine grape.

BERLAFERA. This black grape was probably a hybrid of **Berlandieri** and *Vitis vinifera.* It was later discarded.

BERLANDIERI BOUISSET. (?) A hybrid of *Berlandieri* and *Candicans*.

BERLAUSSEL. A black cross between **Berlandieri** and **Laussel** (another Munson hybrid).

BERQUANO.* A black *Berlandieri/Vinifera* cross, described in the 1909 list as a wine grape.

BICRABEL (?). A black cross between *V. bicolor, V. vulpina,* and Bell.

BIG BLACK. The vine died in 1898.

BIG EXTRA. Munson crossed *Lincecumii* and Triumph.

BIG HOPE. A dark red grape introduced about 1889 but later discarded. Big Hope was a hybrid of Big Berry and Triumph.

BLACK BEAR.

BLACK EAGLE X *LINCECUMII*.

BLACK HERBEMONT. This Herbemont seedling crossed with Norton (1893) was eventually discarded.

BLACK TAYLOR.

BLACKWOOD. Created in 1897, it was a Delago/Governor Ireland cross.

BLANCO. An Elvira/Triumph hybrid, this purple grape with a blue bloom was later discarded. It was named for the Texas county, town, and/or river of Blanco. Perplexingly, the Kansas State Agricultural College Experiment Station described it as "a pale, waxy yellow color, with just a shade of amber in the sun." (*Bulletin No. 73,* page 182)

BLONDELLA. A white grape, it was a hybrid of Ben with Green Mountain.

BLONDIN.* Pierre Viala described this translucent white as a complex hybrid of *Lincecumii, Aestivalis, Cinerea,* and *Vinifera.* Created in 1896 and introduced three years later, it was specifically a cross between Ten Dollar Prize, Norton, and Herbemont.

BLOOD. (Also known as Munson's No. 98.) This black grape was the result of a *Lincecumii* seedling fertilized with Herbemont. The Kansas State Agricultural College Experiment Station reported that Blood, Sweetey, and Letoney "give us a delicious juice for canning, and doubtless would make a fine wine." (*Bulletin No. 73,* page 185)

BLUE FAVORITE. A hybrid of *Aestivalis, Cinerea,* and *Vinifera.*

BOKCHITO. Introduced in 1899, this black grape was a cross between Early Purple and Brilliant. It was probably named for the town in Oklahoma, not far from the Red River, and was later discarded.

BOVERY. A *Candicans/Berlandieri* hybrid.

BRILLIANT.* One of Munson's favorite breeding stocks, this translucent dark red grape was a Lindley/Delaware hybrid planted in 1883 and commercially released in 1887. Even so severe a critic of Volney's Elvira/Delaware hybrids as the Kansas State Agricultural College Experiment Station declared Brilliant "one of the finest red grapes yet offered to the public." (*Bulletin No. 73,* page 183)

BUMPER. A black grape created in 1889 and later discarded, Bumper was a hybrid of Ten Dollar Prize with Norton.

BUSH (Isidore Bush/Munson's No. 104.) Named for Isidor Bush of Missouri's famous Bushberg Nursery, this black Herbemont/Post Oak cross was later discarded.

CAMPBELL. Originally named for colleague and grape originator George W. Campbell

(1883), this deep green to yellow grape was a Triumph seedling. Munson later (1894) re-named it Early Golden.

CAPTAIN. A black grape with a white bloom created in 1896, Captain was a hybrid of America and R. W. Munson.

CAPTIVATOR.* Munson modestly described this translucent red as the most attractive grape ever produced in the United States. Created in 1902, it was a Herbert/Meladel cross. Anthony Hawkins notes on his web site that it is "regarded by many as the most delicious of the sweet tablegrapes grown in humid summer regions of the southern U.S.A."

CARAMEL. A Delago/Brilliant hybrid, it was described in 1903 but later discarded.

CARMAN. One of Munson's most famous hybrids, this black grape, created in 1883 and commercially released in 1891, was a hybrid of Premier (*Lincecumii*) and Triumph. The name honored *Rural New Yorker* editor E. S. Carman. Gilbert Onderdonk wrote that it was a "good table grape, does well here [Texas], and we think it will prove a good shipper." It has, he continued, "a longevity beyond any pure *labrusca*." Carman was one of Munson Nursery's "most profitable varieties." (Also known as Munson's No. 129.)

CARNATION. A Delago/Brilliant hybrid, later discarded.

CATOOSA. A Lucky/Carman cross introduced in 1899, this black grape is probably named for a town outside Tulsa, Oklahoma. It was discarded.

CHAMBRIL. This purplish-black grape was a cross between **Champini** and **Brilliant**.

CHAMPANEL. Another Champini cross, this one with Worden, it was a black grape with a white bloom (1893). It has a high tolerance to heat and drought, is resistant to Pierce's Disease and cotton root rot, and can be used to make red wine.

CHAMPOVO. A black grape later discarded, Champovo was a hybrid of De Grasset and Worden.

CHOTEAU. A black Lucky/Carman hybrid introduced in 1899, Choteau was named for another Oklahoma town on the Missouri-Kansas-Texas railroad route which Munson used.

CLOETA.* This black America/R. W. Munson cross was created around 1899 and in-troduced in 1902.

COMPACTA. A white grape later discarded, it was a Herbemont/Triumph hybrid.

CONCORD X CYNTHIANA. A hybrid in which Cynthiana characteristics were dominant.

CONELVA. A black grape created in 1885 and later discarded, it was a cross between **Con**cord and **Elvira**. The Kansas State Agricultural College Experiment Station declared that it had "neither the tough pulp and [nor] the foxiness of the Concord." (*Bulletin No. 73*, page 183)

CONNELL. No parentage listed. Discarded.

CREAM. A white grape which was an Armlong/Malaga mix.

CROWN. This black grape was a cross between *Lincecumii* and Triumph and was not of-fered for sale in the Munson Nursery catalog after 1893.

CURTIS. A black grape introduced in 1889 and later discarded, Curtis was a hybrid of Post Oak Number 3 and Triumph.

CYNCON. A **Cyn**thiana seedling crossed with **Con**cord, this black grape (1885) was later discarded.

DAVKINA. A black America/Beacon cross introduced in 1899.

DELACON. No parentage listed. Discarded.

DELAGO. A hybrid of **Dela**ware and **Goe**the, this bronzy red grape was created in 1883 but not commercially released until 1896.

DELAKINS.* A red **Dela**go/Per**kins** hybrid.

DELBRILO. No parentage listed. Discarded.

DELGOETHE. This Delaware/Goethe hybrid may be the same as Delago.

DELICATESSEN.* A black, disease-resistant grape created in 1902, it was a hybrid of R. W. Munson and Delicious.

DELICIOUS. A black grape with a blue bloom (Big Berry crossed with Herbemont), it was created in 1887 and commercially released in 1894.

DELMAR. A hybrid of **Dela**ware and **Mar**tha.

DELMERLIE. Delago crossed with Governor Ireland (1898) produced this black grape, which was later discarded.

DELPRESY. No parentage listed. Discarded.

DENISON. A seedling of Moore Early and later discarded, this black 1885 creation was named for Munson's Texas hometown.

DE SOTO. A Scuppernong pollinated with *Vitis munsoniana*, this black grape was introduced in 1896 and later discarded because it was not freeze hardy in North Texas. It was likely named for the town south of Dallas, Texas.

DINKEL. A deep coppery red grape which resulted from a Catawba seedling. There is a Dinkel River in Holland, but it's hard to know if that's the source of the name.

DIXIE. Pierre Viala called this amber-colored hybrid of San Jacinto and Brilliant (introduced 1899) "curious." Munson listed it as a wine grape.

DR. COLLIER.* (Also called Collier or Big Red.) A dark red hybrid of Ten Dollar Prize and Concord (1885.) It was named for the director of the New York Experiment Station in Geneva.

DR. HEXAMER. A black grape introduced in 1893, it was a Post Oak/Triumph cross. It was named for Frederick M. Hexamer (1833–1909), a German-born doctor who left the medical field for horticulture. Hexamer served as editor of *American Garden* and later *American Agriculturist*. He also edited books for the Orange Judd Publishing Company in New York.

DR. KEMP. A dark red to purple grape, this *Lincecumii* and Herbemont hybrid was created in 1885 and commercially released in 1896.

EARLY BIRD. A black cross between Texas Highland and Brilliant, it was discarded.

EARLY GOLDEN. See Campbell.

EARLY MARKET. A black grape introduced in 1885 and later discarded, this was a cross between Elvira and Bacchus. (The Kansas State Agricultural College Experiment Station reported in 1891 that it was Elvira X Triumph and showed "a strong reversion to the *Riparia* ancestry of Elvira." (*Bulletin No. 28*, page 162)

EARLY WINE. Introduced in 1894 and later discarded, Early Wine was a black hybrid of Jaeger's No. 70 with a *Rupestris*.

EASY ROOTER. A Herbemont/Elvira cross, it was damaged by a blizzard in 1885.

EBONY. A black grape never commercially released by Munson.

EDNA.* A white Armlong/Malaga cross, it may be named for the Lavaca River town on Texas' Gulf Coast.

ELEALA. A Delago/Brilliant hybrid first described in 1903 but later discarded, it may be derived from the Spanish "el ala," or "wing."

ELEANOR. A cross between Rothenol and an unnamed Munson creation, Eleanor first fruited in the 1916–17 season, after Munson's death, and was named for granddaughter Eleanor Calvert, Olita's daughter.

ELLEN SCOTT.* Munson's beloved wife Nellie lent her name to this translucent violet grape, an Armlong/Malaga mix created in 1902. Munson believed that it had a promising commercial future and was "unexcelled as a dessert variety, even among foreign kinds." It has proved to be one of his most enduring varieties.

ELPO. An Elvira/*Lincecumii* hybrid, this pale green grape was similar to Elvira.

ELVIBACH. A black **Elvi**ra and **Bacch**us cross.

ELVICAND.* An accidental hybrid in 1885 of Elvira with a wild Mustang grape growing half a mile from the Munson vineyard, this translucent dark red combined three species—*Labrusca, Vulpina,* and *Candicans*—and was named for its two parents (**Elvi**ra and **Cand**icans). In *Foundations,* Volney noted that Elvicand was "the best of three accidental hybrids of Elvira with the native Mustang grape growing wild near my vineyard—about half a mile distant—and illustrates how readily hybrids between cultivated and wild grapes occur. . . . [Elvicand] is a base on which to build a distinct very successful class of high colored grapes for all the country south of Mason and Dixon's line."

ELVIN. A cross between **Elvi**ra and **Irvin**g, this white grape was one of the many 1885 introductions but was later discarded.

ENOLIAN (Oenolian).* A black grape (Amsion crossed with Malaga) created in 1902 but not yet introduced by 1909. The name may derive from "enology," or the science of winemaking.

EPURILL. A seedling of **E**arly **Pur**ple crossed with Br**ill**iant, this red grape was introduced in 1897.

EREBUS. No parentage listed.

ERICS(S)ON. A black hybrid of America and R. W. Munson (created 1897 and later discarded), it is likely named for early Norse explorer Leif Ericson, whose third landing in the Americas was at Vin(e)land, so called for its profusion of wild grapes.

ESTELLA. A white hybrid of Jaeger's No. 72 and Rommel, it was introduced in 1899.

EUFAULA. A red grape introduced in 1895 (America X Laura) but later discarded, it was named for a town in eastern Oklahoma.

EUMEDEL. A seedling of **Eume**lan crossed with **Del**aware, this black grape (1887) was later discarded.

EUMORELY. Another black **Eume**lan cross, this time with **More** Early, it was introduced in 1887, later discarded, and derived its name from its parents.

EXQUISITE. A Herbemont seedling, later discarded.

EXTRA. A dark purple/black hybrid of Big Berry and Triumph created in 1886.

FERN MUNSON.* A *Lincecumii* (Premier) pollinated with Catawba, this dark purplish red grape was planted in 1883 and released commercially a decade later. It was named for Munson's daughter Fern. Hedrick wrote that it was "vigorous, productive, and stand[s] drouth very well."

FERNOKE. No parentage listed. Discarded.

FOLROUGE. No parentage listed. Discarded.

GABRIEL. A complex red hybrid of *Rotundifolia, Lincecumii,* and *Aestivalis.*

GARNET.

GASPER (?). A white *Aestivalis/Lincecumii*/Herbemont cross.

GIANGO. No parentage listed. Discarded.

GOLD COIN. Named for its yellow color, this Norton and Martha mix was planted in 1885 and introduced in 1894. U. P. Hedrick called it "a handsome market grape for the South." The Kansas State Agricultural College Experiment Station declared in 1893 that it was "apparently equal in hardiness to anything [else] on our list." The *Bulletin* added that *Aestivalis/Labrusca* crosses such as Gold Coin were uncommon but felt, nonetheless, that it lacked the qualities to make it suitable for general cultivation. (*Bulletin No. 44,* pages 119 and 121.)

GOLD DUST. (Also known as Munson's Number 22.) A seedling of Lindley crossed with Delaware, this grape was introduced about 1880 but later discarded.

GOLDEN COIN. A hybrid of Norton's Early and Martha. Probably the same as Gold Coin.

GOLDEN GRAIN. A light green grape, this seedling of Linley crossed with Delaware (1883) was later discarded.

GOLDEN SEEDLESS. A hybrid of R. W. Munson and Rommel, this was listed as a table grape.

GOLNER. No parentage listed. Discarded.

GOODENUF. No parentage listed. Discarded.

GOUCH. (Gooch?) No parentage listed. Discarded.

GOVERNOR IRELAND. This black grape with a white bloom was a pure seedling of Moore's Early (1885). Munson named it for John Ireland, a Texas legislator and judge who served as governor from 1883 to 1887. He later discarded it.

GOVERNOR ROSS. Another pure seedling, this one of Triumph, this yellow green grape was introduced in 1894 and named for Lawrence Sullivan "Sul" Ross, who served as governor of Texas from 1887 until 1891. He later became president of the Agricultural and Mechanical College of Texas (now Texas A&M University) with which Munson worked extensively.

GRAYSON. Yet another seedling of Moore's Early, this one black, Grayson was introduced in 1885 but later discarded. It was named for the Texas county in which Munson lived.

GREAT CLUSTER. A black *Lincecumii*/Norton cross.

GROYNN. A variety of *Champini,* this grape may be named for the Groynn Mountains in Coryell County, Texas.

GULA. A black grape introduced in 1899, this is a cross between America and Beacon and was discarded. "*Gula*" is the Spanish word for gluttony.

GULCH. Munson introduced this black grape (Jaeger's No. 70 X a seedling of *Rupestris*) in 1888 but later discarded it.

HARTFORD SUR JACQUEZ.

HEADLIGHT.* One of Munson's most successful creations, Headlight derived from Moyer fertilized by Brilliant. Planted about 1895 or 1896, it was commercially released in 1901. Hedrick believed the dark, clear red grape to be "worthy of trial" in the northern United States. Headlight is one of the parents of the modern hybrid Beaumont.

HERBEMONT X MARTHA.

HERBEMONT X *LINCECUMII*.

HERBEMONT X NORTON'S EARLY VIRGINIA.

HERBEMONT X TRIUMPH.

HERMANN JAEGER.* (Also known as Munson's Number 81.) A hybrid of Herbemont and Post Oak No. 1, Munson planted this dark purple grape in 1883 and originally named it (1889) simply "Jaeger" for his colleague and friend in Missouri. A year later, he introduced it as Hermann Jaeger. In 1902 the Kansas State Agricultural College Experiment Station described the flavor as "spicy."

HERNITO. A pure seedling of Herbert, this black grape was created in 1900.

HEXAMER. (Also known as Dr. Hexamer. See entry.)

HIDALGO.* Probably named for the Texas town on the Rio Grande, this white Delago/Brilliant hybrid was created in 1889 and commercially released in 1902.

HIGBORN. No parentage listed. Discarded.

HIGHLAND. A seedling of Triumph crossed with Jura Muscat (described 1891), it may be named for a town west of Waco, Texas.

HILGARD. (Also known as Munson's Number 130.) This grape is known only from a reference in an 1891 report from the Experiment Station of Kansas State Agricultural College.

HOCK. A Herbemont/Norton hybrid introduced about 1890, it may be named for hock wine.

HOPEON. Introduced in 1899 and later discarded, this white grape was a cross between Big Hope and Carman.

HOPHERBE. A dark coppery red grape later discarded, this was a Post Oak (Ten Dollar Prize) and Herbemont cross.

HOPKINS.* Munson named this black grape (created 1888) in honor of John Hopkins, on whose property was found the original vine of Ten Dollar Prize. Munson crossed it with Norton.

HOPMONT. A cross between Ten Dollar Prize and Herbemont (1889), it was later discarded.

HOTPORUP. Munson used Porup to fertilize a seedling found in Hutchinson County, Texas, to produce this black grape.

HOWARD. This black grape was a seedling of San Jacinto.

HUSMANN. Munson crossed a seedling of Armlong with Perry (introduced 1892) and named the black grape for friend and colleague George Husmann of Missouri and later California. It was, Volney wrote in *Foundations,* "giving fine satisfaction in California, where tried, as a red wine grape."

ICTERIDA. A Gold Coin/Bell hybrid introduced about 1899, this grape's name may refer to its yellow color, for "icterus" is another word for jaundice.

JACINTOPAL. A black hybrid of *Rotundifolia* with *Lincecumii*.

JAEGER. (See Hermann Jaeger.)

JUDGE MILLER. (Also known as Judge Samuel Miller, Munson's friend and colleague in Missouri.) A greenish yellow grape, it is a Herbemont/Martha hybrid.

KEMP. A Post Oak and Herbemont cross introduced in 1885, it may be named for one of several towns in Texas or Oklahoma.

KENENA. (Kenema.) A purple America/Post Oak hybrid created circa 1898 and later discarded.

KIAMICHI. This purple Delago and Brilliant cross, introduced in 1899 and later discarded, is named for the Kiamichi Mountains in southeastern Oklahoma.

KIOWA.* A hybrid of *Lincecumii* (specifically Jaeger's No. 43) and Herbemont, this black grape was introduced in 1898 and likely named for Kiowa, Oklahoma, not far from Munson's Denison home.

KOSOMO. A purple Delago/Beacon hybrid introduced in 1899 and later discarded.

KRAUSE.* This pearly white grape was the result of a seedling Munson grew from a Herbemont/Niagara cross produced by Prof. E. W. Krause of Waco, Texas. It was created in 1893 and commercially released in 1908.

KRUGER. This black grape was a hybrid of America and R. W. Munson, introduced in 1899. The name may possibly refer to the pioneer ranching family (Krueger) in Blanco County, Texas.

KYLE. No parentage listed. Discarded. Likely named for horticulturist E. J. Kyle of Texas A&M University.

LABAMA.* A black seedling of San Jacinto, it may be named for Alabama.

LADANO.* A purplish red grape which first fruited in 1902, Ladano was a cross between Salado and Headlight.

LAGAMO. (Lagomo.) A red hybrid of Delago and Brilliant. Discarded.

LALOMA. A white hybrid of Armlong and Malaga.

LA REINE. Re Pierre Viala, this was a hybrid of America and R. W. Munson. U. P. Hedrick and, later, W. Dancy, however, call it an America/Beacon cross. The black grape was introduced in 1899 but later discarded.

LA SALLE.* Even Volney wasn't quite sure of the parentage here. In 1891, he wrote in *Foundations,* he saved seeds from a single Scuppernong vine which he found blooming near a group of Post Oak/Herbemont hybrids, the source, he believed, of the pollen. From these, he developed both La Salle and San Jacinto. There are several towns and a county in Texas named for the seventeenth-century French explorer.

LAST ROSE.* A rosy red grape created in 1902, it was a hybrid of Armlong and Jefferson.

LATANIA. A cross between Xlnta and Muench, this black grape may be named for a small ornamental palm. It was not offered for sale by Munson Nursery after the 1910–11 catalog.

LAUSSEL. Created in 1886, this was a Secundo/Gold Coin cross. The dark purple grape was named for Munson's daughter Viala Laussel, who was in turn named in honor of Pierre

Viala and his wife Laussel. Munson noted that his Laussel, Marguerite, and Neva hybrids were "commemorative of the three pretty[,] good girls for whom they are named."

LETOVEY (Letorey, Letoney, also known as Munson's No. 122.) A deep purple hybrid of *Lincecumii* and *Bourquiniana*. (See Blood.)

LINCEOLA. This *Lincecumii*/Elvira cross, according to U. P. Hedrick, was attributed to Munson but not catalogued by him.

LINDELL (Lindel). A **Lin**ley/**Del**aware hybrid of 1883, later discarded.

LINDHERBE (Linherbe). Another **Lin**ley hybrid of 1883, this time with **Herbe**mont, and also discarded.

LINELVI (Linilva, Linelva). (Also known as Munson's Number 45.) A **Lin**ley cross with either Humboldt or **Elvi**ra.

LINLEY. Munson named this yellow grape introduced in 1897 for his mother, Maria Linley Munson. It was a Rommel/Delaware cross.

LINMAR. A pale green hybrid of **Lin**ley and **Mar**tha from 1883 and later discarded.

LINRUPMENT. No parentage listed. Discarded.

LOBATA. A black Munson/Profusion hybrid introduced in 1897.

LOMALA. No parentage listed. Discarded.

LOMANTO.* This dark purple to black grape first fruited in 1902 and was a Salado cross with Pense.

LONGALA. No parentage listed. Discarded.

LONGESA (?). No parentage listed. Discarded.

LONGEFON. A black grape.

LONGFELLOW. A black hybrid of Armlong and Jefferson, it may be named for the poet.

LONG JOHN. A black Big Berry and Triumph mix, possibly named for the character in *Treasure Island.*

LOTNA. No parentage listed. Discarded.

LUKFATA. Munson probably named this black grape (1893) for the Oklahoma town. It was a hybrid of De Grasset (*Champini*) with Moore's Early.

LUCKYNE. A black cross between Lucky and Sweety introduced in 1897 but later discarded.

LYON. (See Presly.)

MADRE. A red hybrid of Delago and Brilliant and later discarded. "*Madre*" is Spanish for mother.

MAMLEAF. A black Lucky/Worden cross.

MANITO.* Munson planted this black America/Brilliant cross in 1895 and released it commercially four years later. The name probably derives from the town of Manitou in western Oklahoma near the Wichita Mountains.

MANNATEE. A form of *Vitis Simpsonii* isolated by Munson, the name may honor the Florida hometown of discoverer Simpson.

MANSON. This yellow grape, created in 1897 and introduced in 1906, uses the older, possibly Scottish form of "Munson." It was a hybrid of R. W. Munson with Gold Coin.

MARATHON. No parentage listed. Discarded.

MARCUS. One of Munson's posthumous introductions (1917–18) which were named for his grandchildren, this Ellen Scott hybrid honors Fern's son, Marcus Alexander Acheson.

MARFER. No parentage listed. Discarded.

MARGOLA (Margol.) A black grape, later discarded.

MARGUERITE. In 1886, Munson crossed a *Lincecumii* (Secundo) with Herbemont to create this dark purple grape named for his daughter. (See Laussel re name.) He called it "a splendid arbor grape" which made "a fine white wine."

MARVEL. A black hybrid of *Lincecumii* and *Rotundifolia,* later discarded.

MARVINA. A yellowish white hybrid of **Marvin**'s Laura and Brilliant (1897), later discarded.

MATHILDA. Another "grandchild" hybrid from the 1917–18 season (see Marcus, for example), this red grape was created by crossing a seedling of Violet Chasselas with Brilliant. The name honors Marguerite's daughter, Mathilda Raybourne Thompson.

MEANKO. A hybrid of Delaware and Brilliant (Dancy says Delago and Brilliant), this red grape introduced in 1899 resembled Amethyst. It was later discarded.

MEDALBA. No parentage listed. Discarded.

MEISSNER. Named for friend and colleague Gustave Meissner of the Bushberg Nursery in Missouri.

MELADEL. A red hybrid of a Delago seedling with Brilliant.

MELASKO. A black hybrid of Delago and Governor Ireland (1899).

MELASSUM.* A black hybrid of Delago and Worden.

MERICADEL.* A translucent purple grape created in 1893 and introduced five years later, Mericadel takes its name from its parents, **America** and **Del**aware.

MERMONTA. No parentage listed. Discarded.

MILLER. Probably the same as Judge Miller.

MINEKHA. A Delago/Brilliant hybrid of 1899.

MINNIE. Will Munson introduced this hybrid of his father's in the 1916–17 season and named it for his second wife, Minnie Secoy Munson. A white grape, it was a seedling of Edna crossed with Longfellow (?).

MODENA. A black hybrid of Delago and Governor Ireland (1899).

MONADA. No parentage listed. Discarded.

MONAGA. No parentage listed. Discarded.

MONLINTAWBA. Munson crossed a *Monticola* with Fern Munson to achieve this purple grape.

MONRIG. No parentage listed. Discarded.

MONTISELLA. A purple grape which took its name from its parents, **Monticola** and Laussel.

MONTUMUNOH. A purple *Monticola*/Muench grape, used as a breeder.

MORLAN. No parentage listed. Discarded.

MRS. MUNSON. A dark red Neosho/Herbemont cross (1889).

MUENCH.* Created in 1886 and released the following year, this purplish black grape was a hybrid of *Lincecumii* (specifically Neosho) and Herbemont. (Neosho was a native *Lincecumii* found in Missouri by Jaeger.) The name honored botanist Frederick Muench of Missouri. Gilbert Onderdonk called Muench a good late-ripening grape for the south of Texas.

MULONG. No parentage listed. Discarded.

MULTIPLE. (Also known as Munson's Number 107.) Eventually discarded, this purple grape was a cross between a Herbemont seeding and Triumph.

MYLITTA. A black America/Beacon hybrid (1889), later discarded.

NAMODEL. No parentage listed. Discarded.

NELL. This white Herbemont/Norton cross was later discarded, possibly because of damage from an 1886 blizzard.

NEVA MUNSON.* Munson crossed Neosho and Herbemont (1886) to achieve this dark purple grape named for his daughter. (See Laussel.)

NEWMANN (Newman). Introduced in 1894, this black grape was a hybrid of a *Lincecumii* (Big Berry, according to U. P. Hedrick) and Triumph.

NEWTONIA. A black America/R. W. Munson cross (1897).

NINEKAH. Munson probably named this red grape, introduced in 1899, for Ninnekah, Oklahoma. It was a hybrid of Delago and Brilliant.

NITODAL (Nitodel).* A cross between Salado and Pense (Malaga), a *vinifera*, this dark red grape first fruited in 1902.

NOGGER. No parentage listed. Discarded.

NOKNER. No parentage listed. Discarded.

NONPAREIL. A red Early Purple/Brilliant cross (introduced 1896).

NOSNOE (?). No parentage listed. Discarded.

NOSNUM. No parentage listed. Discarded.

NUMBER 6A. An embryo variety of late fall grapes created by crossing a *Berlandieri* with Fern Munson.

NUMBER 88. Known only from the Kansas State Agricultural College Experiment Station's *Bulletin No. 73,* this black Post Oak X *Rupestris* variety was dismissed as having "the least value" of Volney's five *Lincecumii*/Herbemont crosses tested there. It was, reported the *Bulletin* (pages 184–185), "too harsh and wild to be regarded as valuable."

OCTAVIA. Another red Early Purple/Brilliant mix (1896).

OENOFERA. No parentage listed. Discarded.

OENOLIAN. See Enolian.

OKTAHA. This black grape, introduced in 1898, was created by pollinating *Champini* with Delaware. It was probably named for the town near Muskogee, Oklahoma. Munson listed it as a wine grape.

OLD GOLD. (Also known as Munson's Number 29.) Munson crossed Elvira with Brighton in 1885 to achieve this yellow grape but never released it commercially and eventually discarded it.

OLITA (Oleta). Named for yet another Munson daughter, this yellowish white Delaware/Irving hybrid was released commercially in 1898.

OLITATOO. A white grape introduced in 1896, it was named for Munson's daughter, too. It was a cross between Armlong and Excelsior but was later discarded.

OLIVONTA. Munson crossed a *vinifera,* in this case **Oliv**ett de Cadinet, with Herbe**mont** to create this black grape (1899). The name derives from the parents.

ONALAGA. No parentage listed. Discarded.

ONDERDONK. Named for Gilbert Onderdonk, a pioneer South Texas horticulturist, this white makes an excellent wine, according to U. P. Hedrick. It was a Herbemont/Irving cross created in 1885 and introduced about five years later.

ONEONTA. No parentage listed. Discarded.

ONEOVEM. A hybrid of One Seed and Rommel (1897), this white grape may take its name from its parents or be a joke (One of 'em).

ONE SEED. Used as a parent stock but eventually lost, One Seed was a white grape created in 1886 by crossing Rommel and Humboldt.

ONYX. A dark red Delago/Golden Gem hybrid (1899), later discarded.

OPAL. Introduced about 1892 and later discarded, this yellowish white grape resulted from crossing a seedling of Linley with Martha.

ORIOLE. A black grape achieved by crossing a *Lincecumii* hybrid with an *Aestivalis.* (Hedrick noted the parents were Post Oak and Devereux.)

PALERMO. This yellowish green grape first fruited in 1899 after crossing a Delago hybrid with Brilliant. It was probably named for the town in California.

PAWHUSKA. No parentage listed. Discarded.

PERRY. Munson fertilized a *Lincecumii* (Post Oak No. 2) with Herbemont (1889) to achieve this black grape. Perry, the town for which it was named, was on the "Katy" rail line through Oklahoma. Fellow horticulturist Gilbert Onderdonk praised Perry for use in South Texas and as "an improvement on the Herbemont," a wine grape which was also good for the table.

PLANCHETTE. A white Herbemont/Triumph cross, later discarded.

PLANCHON. Munson crossed a wild hybrid of *Berlandieri* with *Candicans* and named the resulting grape for French ampelographer and entomologist Jules-Emile Planchon.

PONTOTOC. This red Delago/Brilliant hybrid (1899) was named for the county in southern Oklahoma.

PORUP. A seedling of **Post** Oak crossed with a ***Rupestris*** produced this oddly named grape.

PRESIDENT.* Munson called this black grape a sister of Hernito (see entry). It was introduced in 1900 and grew from a seedling of Herbert. It was long used by the now defunct Bright's Winery of Canada to make port.

PRESLY. (Originally named President Lyon, or Lyon.) This red Elvira/Champion hybrid (1886) honored T(heodatus) T(imothy) Lyon of Michigan, a friend and colleague from the American Horticultural Society (**President Ly**on). Lyon (1813–1900) worked with the Michigan Experimental Stations. The Kansas State Agricultural College Experiment Station called it a good quality table grape but recommended it for garden cultivation only because of its small size. (*Bulletin No. 44,* page 121)

PROFESSOR HILGARD (Professor Hilyard). A purple Post Oak/Herbemont cross.

PROFESSOR LYON. See Presly.

PROFITABLE. A greenish-red hybrid of Elvira and Perkins.

PROFUSION. Munson introduced this black grape in 1889. It was a hybrid of Ten Dollar Prize and Norton.

PUKWANA.* A black hybrid of *Rupestris* and *Monticola,* used for grafting.

PULPLESS. See Van Deman.

PURPUREA. A hybrid of Delago and Brilliant, later discarded.

QUINTINA. Munson crossed Early Purple with Jaeger to achieve this black grape, introduced in 1897.

RAGAN (Reagan). Another black grape, this one was the result of crossing an unknown *Lincecumii* hybrid with Triumph; it was introduced circa 1892.

RAGLAND. Friend and colleague Dr. A. M. Ragland of Pilot Point, Texas, was the namesake of this black grape, created by crossing an unknown *Lincecumii* hybrid with a *Rotundifolia.* It was later discarded.

RED BIRD. (Also known as Munson's Number 33.) This dark red grape was a hybrid of Linley and Champion, introduced about 1888. Munson felt it promised to be very valuable for the early market.

RED EAGLE. (Also known as Munson's Number 47.) A red seedling of Black Eagle, created circa 1887. The Kansas State Agricultural College Experiment Station reported in 1891 that it was "brisk and pleasant" in flavor. (*Bulletin No. 28,* page 162)

RED JACKET. No parentage listed. Discarded.

REDINA. No parentage listed. Discarded.

RENEAGUE. Another Black Eagle seedling.

REPHAIN. No parentage listed. Discarded.

RHAMY. Munson named this black *Rotundifolia/Lincecumii* hybrid for A. N. Rhamy, who had a nursery east of Denison.

RICASEL (?). No parentage listed. Discarded.

ROANOKE. A red *Cordifolia/Labrusca/Vinifera* cross used for breeding.

ROMBRILL. A yellow **Rom**mel/**Brill**iant hybrid (1897).

ROMMEL.* Munson crossed Elvira with Triumph to create this yellowish green grape in 1883; he released it commercially in 1889, naming it for viticulturist Jacob Rommel of Missouri. Munson declared it made a "sprightly" jelly but discontinued it in the 1905–06 catalog and replaced it with Wapanuka. U. P. Hedrick was more supportive; as late as 1945, he still recommended it as "worth growing in the South for a table grape and makes a very good white wine." (*Grapes and Wines from Home Vineyards,* page 170) However, the Kansas State Agricultural College Experiment Station dismissed Rommel: "While not with us a sort worthy of general introduction, it is very interesting in the study of hybrid effects." (*Bulletin No. 73,* page 182)

RONALDO (Ronalda.) A white Armlong/Malaga cross, later discarded.

ROSCOE. A pale green hybrid of Delaware and Martha (introduced circa 1888), Munson named this variety for his son.

ROTHENOL. (Also known as Eleanor Rothenol?) An Armlong/Jefferson hybrid, it was never introduced commercially.

ROTHVIN. No parentage listed. Discarded.

ROTUNDIFOLIA/LINCECUMII. Various hybrids.

RUBY. A red Elvira/Brighton cross introduced circa 1890.

RUPEL. This black grape, later discarded, was the result of pollinating a *Rupestris* with July Twenty-Fifth.

RUPERT.* A reddish black grape named for a Munson son who died as a baby, this was a hybrid of America and Brilliant (1894).

RUSTLER. Munson crossed a Linley seedling with Martha to produce this light green grape (introduced circa 1888).

R. W. MUNSON.* Munson pollinated Big Berry with Triumph in 1886 and introduced the black grape in 1894. It was named for his son.

SABINAL. Probably named for the town or river of the same name in the Texas Hill Country, this red grape was a Salado/Brilliant cross which first fruited in 1902.

SALADO. This black grape was a hybrid (1893) of De Grasset and Brilliant, named for the picturesque town north of Austin, Texas. Munson used it only as a mother stock and never introduced it commercially.

SALAMANDER. A translucent red hybrid of Salado, Delaware, and Linley.

SALMAL. No parentage listed. Discarded.

SALMANO. No parentage listed. Discarded.

SANALBA (San Alba).* Munson created this yellowish white grape in 1898 by crossing San Jacinto with Brilliant; he released it in 1906.

SAN JACINTO.* A hybrid of *Rotundifolia* (Scuppernong) and *Lincecumii X Herbemont*, this black grape was introduced circa 1908. (See La Salle.) It was named for the battlefield north of Houston where Texians gained their freedom from Mexico on April 12, 1836.

SANMELASKA (San Melaska).* A black seedling of San Jacinto, it was created in 1898 and commercially released in 1906.

SANMONTA (San Monta). A black grape created in 1898 and released in 1906, this is a San Jacinto/Herbemont cross.

SANRUBRA (San Rubra). A red San Jacinto/Brilliant hybrid created in 1898 and introduced in 1906.

SECUNDA (Secundo). Munson crossed Early Purple and Brilliant to create this red grape. He used it as breeding stock, never commercially introduced it, and eventually discarded it.

SEPTIMIA. A black Early Purple/Carman cross (1897).

SHALA. Pierre Viala wrote that this was a hybrid of America and R. W. Munson; Hedrick called it America and Beacon. Introduced in 1899, this black grape may have been named for the Assyrian goddess. It was later discarded; Munson did not name the parents on his 1909 list.

SHARP BEAK. This was a hybrid of *Rupestris* and Elvira; the black grape was later discarded.

SHERNAH (Sheruah). A black America/R. W. Munson (1899), later discarded.

SIBMAH. No parentage listed. Discarded.

SILKYFINE. A white hybrid of One Seed and Rommel introduced in 1898 but later discarded.

SOLINBRILA. (There's some question if this was a Munson hybrid.)

SOLINCRUP. A *Solonis/Lincecumii* hybrid.

SOLONIS MICROSPERMA. A variety of *Solonis*.

SOLRUPO. Munson crossed *Vitis Longii* with a *Lincecumii* seedling and *Rupestris* to create this black grape.

SPARKLER. A seedling of Delaware.

STRIPED RUBY. (Also known as Munson's Number 13.) It was on trial at the Virginia Experiment Station in 1893.

SUCCESS. A black Post Oak No. 3/Triumph cross.

SWEETEY. (Also known as Munson's Number 111.) A bronze *Lincecumii* (Post Oak No. 1)/ Herbemont cross (1889), it was later lost. Volney listed it as a wine grape, and Onderdonk recommended it as a table grape for the coastal South. (See Blood.)

TALALA. A red Elvicand/Brilliant hybrid, it may be named for a town near Vinita, Oklahoma.

TALEQUAH. An America/Herbemont cross introduced circa 1895 but later discarded, this dark red grape was probably named for the town east of Tulsa, Oklahoma.

TAMALA. A yellowish white grape which first fruited in 1899, this was a hybrid of Delago and Governor Ross. It was later discarded.

TECUMSEH. A yellow muscadine hybrid, this grape may be named for sister Louisa's hometown in Nebraska. It was later discarded.

TEMA. No parentage listed. Discarded.

TEXAS. (Also known as Munson's Number 181). A red Herbemont seedling.

TEXAS HIGHLAND. (Also known as Munson's Number 130.) Munson crossed Post Oak No. 1 (a *Lincecumii*) with Agawam and introduced the resulting black grape circa 1885. The Kansas State Agricultural College Experiment Station described the fruit as being "of fine size, black, juicy, rich, and pleasant." (*Bulletin No. 73,* page 184)

TEXAS PURE. A seedling of Herbemont, later discarded.

TINTERET. No parentage listed. Discarded.

TISHOMINGO. This was a black hybrid of Delago and Governor Ireland introduced about 1899 and probably named for the town in southern Oklahoma, not far from Denison. It was later discarded.

TONKAWA. A hybrid of Delago and Brilliant, this dark red grape was commercially released circa 1899. It was likely named either for the Tonkawa Indians of Texas or for the town of that name in north central Oklahoma.

TRYONE. A white hybrid of One Seed and Rommel (circa 1897).

TUSKAHOMA. Munson described this translucent red, a Brilliant/Delago cross, as a sister to his Amethyst. It first fruited in 1896 and was named for a town in Oklahoma's Kiamichi Mountains.

UFAULA. (See Eufaula.)

UNIVERSAL. A black America/Profusion cross (circa 1897), later discarded.

URIEL. This yellowish white grape was a cross between Dr. Collier and Marvina. Uriel was one of the four archangels of Hebrew tradition.

VALENCIA. Munson described this brownish red grape as a *Bourquiniana*. U. P. Hedrick wrote that it was grown from seed obtained from Valencia, Spain, hence the name.

VALHALLAH.* A translucent dark red, this grape, an Elvicand/Brilliant cross, was created in 1893 and commercially introduced in 1902. Valhalla was the mythological hall of heroes from the Munsons' Scandinavian heritage.

VALVERDE. Munson fertilized De Grasset with Brilliant but later discarded this hybrid. It was probably named for the Hill Country (Texas) town.

VAN DEMAN. (Also known as Munson's Number 92.) According to the Kansas State Agricultural College Experiment Station, this black *Lincecumii*/Triumph cross was originally named Pulpless but was later changed to honor USDA Division of Pomology chief Henry Van Deman.

VINITA.* Munson crossed Post Oak No. 2 with Herbemont to achieve this bronze grape, introduced circa 1885 and shown on his 1909 list as a wine grape. Since his Vinita Home had not yet been built, he probably named the grape for an Oklahoma town near Chetopa, Kansas, where he once owned property.

VINROUGE.* This black hybrid of America and Norton was introduced circa 1894. "*Vin rouge*" is French for red wine, and Munson listed it as a wine grape.

VOLNEY. Another of the "grandchildren" hybrids introduced in 1916–17 after Munson's death, this lavender grape was a hybrid of an Ellen Scott seedling and Muscat Rose. It was named for Fern's son, Volney Archibald Acheson.

WANETA. A red Delago/Brilliant cross (circa 1896).

WAPANUKA.* Munson called this yellowish white grape one of the best, if not *the* best, table and eating grapes produced in the United States. It was a hybrid of Rommel and Brilliant, created in 1885 and introduced in 1898, and likely takes its name from the town of Wapanucka near Atoka, Oklahoma. Volney considered that it was superior to and took the place of Rommel.

WARD. It may be a *Rupestris* seedling.

WARMITA. A black America/Beacon hybrid (circa 1896), later discarded.

WASHITA. This black grape, introduced circa 1896, was a cross between Delago and Governor Ireland. It was likely named for one of several landmarks in the Denison area: the Oklahoma fort, the Oklahoma river, or the bend in the Red River just west of Denison.

WATOVA. The town north of Tulsa, Oklahoma, likely gave its name to this yellow grape, a Gold Coin/Rommel cross, which was introduced in 1899 but later discarded.

WAUBECK. Munson fertilized Jaeger No. 43 with Laussel to create this black grape, introduced circa 1893 but later discarded.

W. B. MUNSON. Munson honored his son Will with this black grape (circa 1887), which was a hybrid of Post Oak No. 3 and Triumph.

WETUMKA.* This yellowish green grape was a hybrid of One Seed and Gold Coin. It was commercially released circa 1893 and likely was named for a town east of Oklahoma City.

WEWOKA. Yet another town east of OKC is the likely namesake of this black grape (circa 1893), a hybrid of America and Beacon.

WHATCOM. No parentage listed. Discarded.

WILLIE BELL. A pale green grape achieved by crossing Elvira and Delaware, it was named for one of Ellen Scott Bell Munson's brothers.

WINEDROP. A dark red grape introduced in 1884 and described as "lost" in 1909, this was a hybrid of Post Oak No. 1 and Herbemont.

WINE KING. A black Winona/America hybrid (1898).

WINNER.* There is some question whether Munson ever commercially released this black grape, created in 1902. (He had not done so by 1909.) It was an America /Badart cross.

WINONA.* A pure seedling of Norton, Munson created this black grape in 1889 and introduced it circa 1895. He noted that it was later accidentally destroyed. Wynona, Oklahoma, is near Tulsa.

WINTER WINE. A black *Simpsonii*/Marguerite cross (1898).

XENIA.* This white grape was a hybrid of Delago and Triumph (1896). "Xenia" is a word used to describe the effect of genes introduced by pollen.

XLNTA (Xylanta.)* A black hybrid of America and R. W. Munson (1893), Volney may have named it for a scientific term coined about the time he developed this grape. "Xylan" referred to a substance found in plant cell walls and woody tissue.

XYLOPHONE. No parentage listed. Discarded.

YOMAGO. A red Delago/Delaware hybrid introduced circa 1894.

ZENOBIA. No parentage listed. Discarded.

GRAPE VARIETIES FOUND AND/OR IDENTIFIED BY T. V. MUNSON AND USED BY HIM FOR BREEDING

ADOBE. This was a variety of *Vitis longii* which was found in Hutchinson County, Texas. It is found in the northwestern half of the state and extending into New Mexico, Oklahoma, and Kansas. Munson may have named it for the Adobe Walls, the site of several famous Indian battles in Hutchinson County.

ARIZONICA BESTFORM MUNSON. This was a variety of *Vitis arizonica* which is found in New Mexico, Arizona, and the southern regions of Nevada and California.

ARIZONICA GLABRA MUNSON. Munson named this variety of *Vitis arizonica* in 1890.

ARIZONICA HARDIEST MUNSON. Another variety of *arizonica*.

ARIZONICA WETMOORE MUNSON. Another variety of *arizonica*.

AUSTRALIS. This black grape, a variety of *Vitis longii*, was found on the Red River in Texas.

BARNES. This variety of *Vitis champini*—which is found only in Central Texas—came from Bell County; it was a black grape.

BERLANDIERI MUNSON. A variety of *berlandieri*.

BIG BERRY. Munson found this variety of *Vitis Lincecumii* four miles west of Denison on the Alkire Sand Hill about 1880. It was one of his favorite "mother" stocks but, unfortunately, was accidentally destroyed by the renters at Scarlet Oaks (the first Munson home in Denison) in 1889.

BIG BUNCH. A variety of *Lincecumii*.

BIG CLUSTER. This purple grape, a variety of *Vitis monticola*, came from Bell County, Texas.

BUNCOMBE. This black variety of *Vitis labrusca* came from North Carolina. Munson used it for breeding but later discarded it.

COLP. Another eventual discard, this white grape, a *labrusca* also, was found in Maryland.

CORDIFOLIA SEMPERVIRENS. This pure form of *cordifolia* grows in the chalky terrains of Texas.

DAVIS. Another Hutchinson County find, this black grape was a variety of *Vitis longii*.

DE GRASSET. Munson used this black *Champini* variety (found in Llano County, Texas) as the mother stock in several of his hybrids.

DOANIANA. Munson described it in 1890 and named it for Judge Doan of Wilbarger County, Texas, who raised grapes for many years and made wine from this variety. It is found primarily in the Texas Panhandle and in Greer County, Oklahoma. Today, Pierre Galet classifies it as *candicans*.

DOG RIDGE. Munson described this as a wild *Champini* hybrid and named the black grape for the area in Bell County, Texas, where he found it. Galet classifies it now as *Vitis candicans, Engelmann*. Jeff Cox (*From Vines to Wines*, 1999) notes that it is particularly useful in "the highly alkaline soils of the Southwest," where it supports "good grafts of vinifera and hybrids." (page 37) It remains popular in the Lower South because of its high resistance to Pierce's Disease, cotton root rot, and nematodes.

EARLY PURPLE. Another of Munson's preferred mother stocks, he found this black/purple variety of *Lincecumii* on a sand hill three miles west of Denison in 1884. On the 1909 list, it is shown as a wine grape.

GIANT. Munson eventually discarded this wild male *rupestris*, found in Missouri.

GIRDIANA. Formerly considered a variety of *Vitis californica*, Munson separated it out as a distinct species in his 1887 classification. (Today Galet classifies it as *Vitis girdiana* in his Arizona series.) It is found in southern Arizona and California, extending into the Baja.

GREER. A wild male *Vitis Doaniana* found in Greer County, Oklahoma.

GROTE. Munson secured this black variety of *Vitis vulpina* from Mauston, Wisconsin.

GROYNN MOUNTAIN. A variety of *Vitis champini* found in Coryell County, Texas.

HUTCHINSON. Probably a variety of *candicans* (Dancy believed it was *longii*), this black grape was found in Hutchinson County, Texas.

JOLY. This form of *Vitis Champinii, Planchon* (classified today by Galet as *V. Candicans, Engelmann*) was found in Lampasas County in the Texas Hill Country. It was a black grape.

JUDGE. A variety of *Vitis Doaniana Munson* (now *V. Candicans, Engelmann*), it was found in Greer County, Oklahoma, and may also have been named for Judge Doan.

LAMPASAS. It was found in the woods of Lampasas County, Texas, by nurseryman F. R. Ramsey and commercially released by Munson in 1889.

LARGE BERRY. This black variety of *Vitis longii* was found in Motley County, Texas.

LARGE LEAF. Another black *longii* found in the same area as Large Berry.

LIVE OAK. A wild hybrid of *rupestris* and *candicans,* it was found in San Saba County, Texas.

LUCKY. Munson found this black *Lincecumii* variety one of his favorite breeders, climbing a forty-foot tall blackjack oak three miles southwest of Denison around 1884.

LUDERS. A variety of male *Vitis riparia* found near Madison, Wisconsin.

MAUSTON. Another find from that town in Wisconsin, this black grape was a variety of *Vitis riparia.*

MILLARDET. Munson named this variety of *berlandieri* for French ampelographer A. Millardet. The black grape was found in Llano County in the Texas Hill Country.

MINNESOTA. Munson eventually discarded this white grape, a variety of *Vitis riparia* found near Carver, Minnesota.

MOTLEY. A variety of *Vitis Doaniana* (possibly *longii*) used for grafting, it was probably named for Motley County, Texas, northeast of Lubbock and the home of the famed Matador Ranch.

MUNSON'S BUCKSHOT. A variety of *rupestris* found in Hays County, Texas.

MUNSON'S PROLIFIC. A variety of *monticola.*

NOVO MEXICANA NUMBER 43. Galet now classifies this as a *candicans* hybrid. Munson discovered it in 1882 in Arkansas and Oklahoma.

NOVO MEXICANA D.

NOVO MEXICANA NUMBER 56.

NOVO MEXICANA MICROSPERMA.

POST OAK NUMBER 1. Munson found this black *Lincecumii* variety in 1881 or 1882 three miles west of Denison. He considered it the best of this species, but it was later lost.

POST OAK NUMBER 2. Another *Lincecumii* found in the same area as Number 1 in 1883 and also later lost.

POST OAK NUMBER 3. The third black *Lincecumii* found wild, west of Denison, in the early-to-mid-1880s and later lost.

POUROY (Ponroy, re U. P. Hedrick). A black variety of *Doaniana* found in Wilbarger County, Texas, along the Red River.

PREMIER. This was the first wild vine Munson transplanted into his vineyard—hence the name—and was one of his most heavily hybridized. He found it three miles west of Denison in 1880. Like Big Berry, this *Lincecumii* variety was destroyed by the renters at Scarlet Oaks.

RAMSEY. This black variety of *Vitis champini* was found in San Saba County, Texas. It was likely named for another well-known Texas horticulturist and plant originator, Frank Ramsey of Austin. Munson described it in 1909 as one of his most valuable, one he used for grafting stock.

RED LEAF. This black *rupestris* variety was found in Missouri; Munson eventually discarded it.

ROBUSTA. A variety of *longii* found in Motley County, Texas.

SALT CREEK. Though it was found in Greer County, Oklahoma, Munson named this black grape for the stream that flows through Lincoln, Nebraska, his home before moving to Texas. It is a variety of *Vitis candicans* and still widely used around the world as rootstock because of its resistance to cotton root rot, nematodes, and phylloxera.

SECUNDO. Grown from seed of the same vine that produced Big Berry.

SILVAIN. Used for grafting, this was another variety of *Doaniana* found in Greer County, Oklahoma.

SMALL LEAF. A black variety of *rupestris* found in Texas, it was later discarded.

SOLINIS. Variety of *Vitis longii.*

SPINOSA. This *labrusca* variety came from North Carolina but was eventually discarded.

TEN DOLLAR PRIZE. Munson used this black *Lincecumii* variety extensively in breeding. Found by John Hopkins east of Denison, Texas, it took its name from the award it won—given by T. V. Munson—at the 1882 Denison horticultural show for the best ripe, wild Post Oak grape. Unfortunately, the original vine was later lost.

TRELEASEI. Munson first observed it in 1887 in the Bradshaw Mountains of Arizona. He classified it as a wild hybrid of *Californica/Arizonica* but Pierre Galet now lists it as *Vitis Treleasei* in his Arizona series. It was named for and dedicated to Volney's friend, the director of the Missouri Botanical Gardens (St. Louis), William Trelease.

VERMOREL. A black variety of *champini* found in Williamson County, Texas.

VIALA. Some sources describe this a variety of *berlandieri,* others as a *champini.* It was a black grape, found in Coryell County, Texas, and named for Munson's friend and colleague Pierre Viala of Montpellier, France.

WILLIAMSON. A variety of *candicans* found in the Texas county of the same name, this black grape was later discarded.

GRAPE VARIETIES INTRODUCED BY T. V. MUNSON

RED GIANT. Originated in Pennsylvania by unnamed breeder and introduced commercially by T. V. Munson & Son in 1898.

Chapter Notes

Chapter 1: Ancestors and Childhood

1. Volney's notes are owned and have been expanded by his great-granddaughter, Joyce Acheson Maki and her husband John Allen Maki of Houston, Texas, who were kind enough to share the information with the authors. Her manuscript is entitled "The Genealogy and History of the Ancestors and Descendants of Thomas Volney Munson, B.S., M.S., D.Sc."

Volney contributed to two family histories: *The Munson Record* about the New Haven portion of the family and *Monsons-Mansons-Munsons.* The early portion of the genealogy chart which appears in Ben Munson's biography, *Ten Million Acres,* was also based on Volney's research.

2. Richard's children by his two wives, Mary and later Susannah, were Samuel (circa 1762–1765), William (1766–1830), Mary (circa 1768–1770), Sarah (circa 1769–1770), a son (Oliver?) (circa 1771–1790), Theodore (1775–1839), and Daniel (1778–1841).

3. Theodore and first wife Anna had no children. With Lydia, he fathered Sarah (1805–1865), William (1808–1890), Stephen (1811–1813), Daniel (1815–1893), Eliphalet (1817–1907), and Susanna (1819–1884).

4. Obituary of Maria Linley Munson. *Denison Sunday Gazetteer,* November 30, 1890, page 1. Maria's siblings were Joseph, Jr., Isaac, Thomas, Benjamin, Triphena, and Sibilla Linley. Isaac eventually followed his sister and brother-in-law to Grayson County, Texas.

5. T. V. Munson to his daughter Viala, 1906. Photocopy of inscription provided the authors by Joyce A. Maki.

6. This site was later known as the old John Fitz place. Fitz, the first German Baptist minister in the area, bought the property after the Munsons moved.

Astoria, on the Peoria-to-Quincy stage line, was first settled in 1828; the town was organized a decade later and named for John Jacob Astor. Fulton County is the home of the well-known archaeological site Dickson Mounds and of *Spoon River Anthology.*

7. *Volney's Ruins,* as it was commonly called, was an essay concerning the fall of so many ancient empires and was published in 1791 after the count's visit to the United States. In it, Volney predicted that, one day, the world's disparate religions would unite after finally recognizing the common truth(s) that lay at the basis of all of them. It concludes with these words, which Volney Munson appears to have taken to heart: "Preserve thyself; Instruct thyself; Moderate thyself; Live for thy fellow worthy citizens, that they may live for thee." Volney later gave a copy to his daughter Viala, with a moving but somewhat inaccurate description of his parents' early married life inscribed within it. (See footnote 5.) He also stated that *Volney's Ruins* "gave bent to his [Volney's] mind thru those of his parents. It was a blessed and elevating influence." (Photocopy of letter provided the authors by Joyce A. Maki.)

8. Trite was named after her mother's sister, Triphena Linley.

9. Woodland Township adjoined Astoria on the east and contained many native nut and fruit trees. Lewistown is the county seat and was settled in 1821; Edgar Lee Masters, author of *Spoon River Anthology*, grew up here.

10. Horticulture is a sub-category of agriculture and refers to the cultivation of plants, especially garden and orchard plants; i.e., fruits and vegetables.

11. Letter from T. V. Munson, quoted in Maki, page 25.

12. There is some evidence that he also studied medicine briefly with his namesake uncle in Livingston County, Kentucky, Dr. Thomas Linley, apparently testing the waters of various possible careers; this was probably just before he went to Kentucky. Linley was an 1833 graduate of Transylvania University, an institution crucial in Volney Munson's life as well. (See Chapter 2.)

Chapter 2: The Kentucky Years

1. Kentucky A&M separated from Kentucky University in 1878. For several confusing years, the two schools shared a similar name, which was remedied in 1908 when KU resumed its old name of Transylvania. In 1916 A&M became the University of Kentucky, the name it still retains. Ashland survives today as a house museum, but Woodlands was demolished around 1900 after most of its acreage was divided for residential lots.

2. *Ten Million Acres: The Life of William Benjamin Munson* by Donald Joseph, page 23.

3. The Ashland Institute, named for A&M's main building, was apparently short-lived, and the only evidence that it even existed may be T. V. Munson's surviving manuscripts. College archivists were unaware of the organization when questioned. Volney wrote in 1868 that the Institute "is one of the most important appendages to the College. ... Though young, she has gained an enviable reputation among older societies."

4. That Bible is now in the collection of the T. V. Munson Viticulture and Enology Center.

5. "Origin of Language" (May 23, 1867), "The Love of Glory (November 5, 1867), and "Patience" (December 15, 1867) are in the collection of the T. V. Munson Viticulture and Enology Center.

6. Article for the *Brass Button* written by T. V. Munson, June 8, 1868, and in the collection of the T. V. Munson Viticulture and Enology Center.

7. As with the Ashland Institute, archivists at the University of Kentucky were unfamiliar with the *Brass Button*.

8. *American Botany, 1873–1892: Decades of Transition* by Andrew Denny Rodgers, III.

9. One descendant says that Volney even considered entering the ministry as a young man, before he went to A&M.

The Campbellites evolved into Disciples of Christ and, eventually, the Christian Church.

For four generations, the Munsons had been Congregationalists, until Theodore left and took his family with him.

10. Munson may have been unfamiliar with the term "dinosaur." Coined in England in 1842, the word ("terrible lizard") did not gain common usage until discoveries in the western United States in the late 1870s precipitated a rush of paleontological research and popularity among the masses.

11. From *A History of Texas and Texans*. The author added that Ben's faith "is based on what may, perhaps, be called 'Natural Philosophy.'"

12. Ben was offered a faculty position in mathematics after graduation but declined it.

13. All trip quotes are from Leet's diary as published in *Ten Million Acres*, pages 43–50.

14. A velociped was an early three-wheeled bicycle.

15. Her siblings were George K. Bell (1846–1906), John Moffatt Bell (1848–1923), William Smith Bell (1852–1927), and Charles Stewart Bell, Jr. (1855–1917). Their mother, Margaret Bunyan Smith Bell, was a direct descendant of John Bunyan, author of *Pilgrim's Progress*; she was also distantly related to romantic novelist Sir Walter Scott.

16. Ironically, in light of his future son-in-law's career, grapes brought Bell to Lexington. He was hired by Henry Duncan, Sr. to supervise his new greenhouses of foreign grapes and to landscape the grounds at The Pines.

17. *A Daughter of Gardens,* a biography of Nellie by niece Linley M. Tonkin, page 11.

18. The school still preserves the original 1797 mansion where Nellie Bell attended classes. Sayre was one of the first American schools to offer its students a full college curriculum.

19. Biography of Dr. Robert Peter, Evans Collection, University of Kentucky Special Collections: Box 1, Folder 1.

20. During this time, Volney also sold land in Illinois given him by his father.

21. "What Shall My Profession Be?" Graduation address given by Munson at the Agricultural & Mechanical School of Texas, June 5, 1888, page 4.

22. According to Munson family genealogist Joyce A. Maki, Huxley was named for scientist Thomas Henry Huxley, whose books on evolution were among Volney's favorites.

23. Maria Linley Munson's brother Isaac and his second wife were among those who made the move to Denison.

24. According to Maki, his intentions to relocate to Texas were strong enough that he helped form the Grayson County Fair on one visit. Her assertion is based on material in *A History of Grayson County, Texas* by Mattie Davis Lucas and Mita Holsapple Hall, which states that Munson was one of the "moving spirits" in the organization of the second county fair in 1872. (page 203) It's entirely possible that this is true, his support voiced on one of his exploratory trips to visit Ben and check out the area for himself. However, in another chapter the authors state that Volney also promoted the first fair in 1867—while both he and Ben were still students at Kentucky A&M. (page 162)

25. Letter written by Munson to the *Kentucky Gazette,* December 20, 1873.

26. *Texas Agricultural Experiment Station Bulletin No. 56,* page 218.

27. Indeed, the Nebraska Horticultural Society had already received an award from the American Pomological Society for the quality of its fruit entries (1871) and would win two more during Volney's tenure there, in 1873 and 1875.

Chapter 3: On to Nebraska

1. Letter written by Munson to the *Kentucky Gazette,* December 20, 1873.

2. The Munson property was described legally as the south end of the east half of the northwest quarter of section 20 in township 10 (Lancaster County, Nebraska Deed Book M, pages 566–567 and Book N, page 433). The area is now largely residential and includes part of the Mount Forest Addition; it is bordered on the east and south by strip shopping centers.

3. "History of the Agricultural Farm" by Mrs. George Reeder in the University Archives of the University of Nebraska, page 1.

4. Manuscript map in the collection of the T. V. Munson Viticulture and Enology Center.

5. A similar lightning rod which Volney installed on Vinita Home in Denison was still in place during grandson Volney Acheson's time.

6. *Kentucky Gazette,* December 20, 1873.

7. This journal is in the collection of the T. V. Munson Viticulture and Enology Center.

8. Thompson (1833–1896) held the first Chair of Agriculture at the University of Nebraska.

9. Volney named one of his later wild grape discoveries "Salt Creek" in memory of their many excursions.

10. *Texas Agriculture Experiment Station Bulletin No. 56,* page 218.

11. His journal mentions another cultivator he invented "two years ago."

12. Manuscript copy of article in Nebraska journal in the collection of the T. V. Munson Viticulture and Enology Center.

13. Nellie had her own work, keeping house and raising the poultry Volney sold in town.

14. For some reason, the deed was not filed until July 24, 1878. Grimes was an attorney and justice of the peace.

15. A family of four could live on $30 to $40 per month, but "people are here in such a fix that they are glad to sell any thing for money." From "Nebraska in the Seventies" by K. Clive Jacob.

16. Railroad maps from this period are very sketchy. Available ones indicate that, had the Munsons moved by train, they would have taken a secondary line from Lincoln to Atchison,

Kansas, then used several more small lines to trace their way through Kansas and the northern Indian Territory (Oklahoma) to link up with the Missouri Kansas & Texas into Denison.

17. William C. Welch and Greg Grant, *The Southern Heirloom Garden*, page 173.

18. Gunter (1845–1907) came to Texas just before the Civil War and served in the Confederate Army. He afterward read law with Oran M. Roberts and practiced until 1878, when he and Ben Munson became busy buying land certificates and locating claims, especially for the railroads. With a third partner, they also ran the T-Anchor Ranch. The town of Gunter in Grayson County is on the site of Jot's 25,000-acre ranch there. He later moved to San Antonio and built the Gunter Hotel.

Chapter 4: Gone to Texas

1. The grape family is known scientifically as *Viticeae*. Of its fourteen genera, only one, *vitis*, contains food plants, and more than sixty species of it are recognized today. One of those, *Vitis vinifera*, the Old World grape which yields the classic wines, is the vine said to have been planted by Noah after the Great Flood and may have been a variety of *V. silvestris*, the common bunch grape. *Vinifera* probably originated in or near the Caucasus Mountains—in southern Russia near the modern borders with Turkey and Iran—and spread through the Mediterranean, then inland across Europe. Today there are some 14,000 cultivars (that is, a plant originated and grown under cultivation) and 5,000 named varieties of *V. vinifera* grown around the world. Twenty-five million acres of grapes are under cultivation, three-quarters of those in Europe. Twenty-six species of wild grapes are native to the United States, the most prolific center of wild grapes in the world, and half of those are indigenous to Texas.

Grapes were one of the first plants to be domesticated. Viticulture, or the cultivation of grapes, and winemaking (enology) are at least as old as ancient Egypt, Babylon, and China, where records from more than four thousand years ago have been found.

Seventeenth-century colonists—English, French, and Dutch—on the East Coast were the first in this country to attempt to cultivate *vinifera* for winemaking purposes. They surveyed the profusion of wild grapes they found growing here and believed that meant they could also easily grow the European varieties they were familiar with. In Virginia, for example, Lord Delaware, the royal governor of the colony, began urging the establishment of vineyards as early as 1616 to produce much-needed revenue. Three years later, both cuttings and experienced vignerons (grape growers) were shipped to Jamestown from France to further promote the industry; settlers were required to plant at least ten cuttings of grapes.

All these early colonists established hundreds of vineyards along the Eastern Seaboard; the Virginia effort alone resulted in 10,000 vines being planted. But they were mostly unsuccessful since the Old World grape is notoriously susceptible to diseases to which native American species east of the Rocky Mountains are immune.

Not until the late eighteenth and early nineteenth centuries did native American grapes begin to come into their own, largely by the impetus of Virginia statesman, inventor, and horticulturist Thomas Jefferson who traveled through Europe while ambassador to France, studying viticulture. He soon realized, however, that it would "take centuries to adapt [foreign vines] to our soil and climate." (Letter of 1809 quoted in USDA Circular No. 437, *American Grape Varieties*, page 2.) Still, even he persisted—unsuccessfully—in trying *vinifera* vines alongside American grapes in his Monticello vineyard.

Jefferson was a plant originator himself who once declared, "The greatest service which can be rendered any country is to add a useful plant to its culture." A sentiment which Volney Munson no doubt heartily endorsed.

2. Bishop Marin de Porras, writing in 1805, and quoted in "A Draught of Vintage: A History of the Texas Wine Industry" by Eva Crane in *Neil Sperry's Gardens* (January 1992). De Porras probably referred to the mustang grape which contains an extremely pungent, even acrid, juice just beneath the skin.

3. El Paso's grape industry was an essential in the city's economy. More than a quarter of a million vines had already been planted by 1755, producing hundreds of barrels of wine, brandy, and

vinegar. Less than a century later, the El Paso valley, watered by the then-mighty Rio Grande, was described as "one continuous orchard and vineyard." By 1858 local vintners were selling 200,000 gallons of wine each year at roughly $2 per gallon—a substantial revenue. Unfortunately, new irrigation projects on the Rio Grande in New Mexico and Colorado during the late 1880s and 1890s diverted much water from the river and caused the ruin of most of El Paso's vineyards.

4. *The Wines of Texas: A Guide and A History* by Sarah Jane English, page 7.

5. Quoted in Samuel Wood Geiser, *Horticulture and Horticulturists in Early Texas*, page 18.

6. *DeBow's Review: A Monthly Journal of Commerce, Agriculture, Manufactures, Internal Improvements, Statistics, etc. etc.* Volume XV, No. II, August 1853, page 194.

7. Nor did he ever lose his faith or vision regarding the state. "Texas," he declared, "a territory rather larger than France, with far greater areas of tillable surface, and soils, sites and climate equally favorable, should make grape growing one of its leading industries." He expounded on this theme in *Foundations of American Grape Culture* but concluded with a bluntly characteristic caveat: "America is the native home of many times more good species of grapes than all the world besides, and by natural rights ought to be the greatest grape country in the world, and will be, *when the fruit-growers wake up to the possibilities surrounding them.*" (Authors' italics) (page 236)

8. *Vitis aestivalis,* var. *lincecumii* was named for pioneer Texas naturalist and physician Gideon Lincecum (1793–1874). Born in Georgia and largely self-educated, Lincecum spent thirty years in Mississippi and made his first exploring trip to Texas in 1835. He moved there, to Washington County, in 1848 and subsequently became an authority on Texas grasses and ants. Among his correspondents were Charles Darwin (re ants) and George Engelmann (re Texas grapes). Like many other naturalists of the period, including Volney Munson, Lincecum held truth and logic in more esteem than religion and believed they would perfect the human race.

Vitis Lincecumii is commonly known as the pinewoods grape or the Post Oak grape (Munson preferred the latter designation) and became the basis of many of Volney's hybrids.

9. *Texas Agricultural Experiment Station Bulletin No. 56,* page 218. Much of this introduction was repeated verbatim in Volney's 1909 book, *Foundations of American Grape Culture,* pages 5 and 6.

10. Most of the following text is drawn from *Shinners & Mahler's Illustrated Flora of North Central Texas* (1999).

The fifty counties comprising the geographic region known as North Central Texas form a rough paisley shape extending west along the Red River from the county of that name to Montague, then cutting a ragged line southwest as far as Callahan County near Abilene. At its southernmost point, the region just touches Travis County, then curves back up to its starting point near the Red River. Its 40,000 square miles of area represent roughly one-sixth of the state's total land mass (making it about the size of Kentucky), within which can be found nearly half of *all* the plant species occurring in Texas—more than 2,200 in all.

The "modern" North Central region was formed in the Pennsylvanian period (320–286 million years ago) when North America collided with and joined the supercontinent of Pangaea; that uplifted the Ouachita (pronounced WASH-i-taw) Mountains, a range from southwestern Arkansas, across Oklahoma, and south to Austin, Texas. (Most of the Texas portion has long since eroded.) Over subsequent millennia, Pangaea coalesced, then began to break apart into the continental system we know today. Huge shallow seas covered much of Texas during the Cretaceous period and frequently connected the Gulf of Mexico with the Arctic Ocean by means of a trough just east of the Rocky Mountains. The numerous fluctuations of the water depth in this seaway account for the great varieties of soil now found in Texas. Moreover, the fossil-bearing sediments laid down during this period became the limestone and the chalky soils that would later prove critical in Volney Munson's work with the French (see Chapters 6 and 7).

The Hill Country counties where Volney found some of his most important native grape varieties lie within the Lampasas Cut Plain vegetational zone of North Central Texas. Here the hard white limestone has been cut into deep divides punctuated by buttes and mesas. This has exposed a tremendous variety of soils which, in turn, has resulted in a diversity of plant life, including grapes.

11. A hybrid, according to the dictionary, is the "offspring of two animals or plants of different races, breeds, varieties, species, or genera." In modern viticulture, a hybrid is considered specifically

a variety achieved by breeding *vinifera* with an American grape—i.e., two different species—and known as a French-American hybrid. A cross, on the other hand, results from using two varieties of the *same* species. According to Karen MacNeil in *The Wine Bible*, "most crosses occur spontaneously in nature" rather than through breeding. "Most modern grape varieties," she points out, "are natural crosses." (page 866)

In his own book, *Foundations of American Grape Culture*, Volney wrote poetically of these natural hybrids: ". . . we fall in love with nature for teaching us what wonderful development, the blind, haphazard selection by natural circumstances, aided doubtlessly by birds and other animals eating the best and carrying the seeds into new regions to start new and better families, have produced during several million years. . . ." (page 140) Hybrids continue to be important today as rootstock.

12. *Texas Agricultural Experiment Station Bulletin No. 56,* page 219.

Grayson County was already a large fruit-producing area when Volney arrived. On the 1870 Census, it ranked second in Texas in value of fruit sold or consumed.

13. Louis and Henri Bouschet devoted much time to experiments in grape breeding in France beginning in 1829, but most experimenters in Europe and the United States were simply curious amateurs. Many of the Americans used Catawba, the first native grape widely cultivated in this country, as parent stock.

14. I. W. Dix and J. R. Magness, *American Grape Varieties,* page 3.

15. That confusion continued for decades. U. P. Hedrick wrote thirty years later (1908), that "our knowledge of the genus is yet too meager to set those limits with authority." (*The Grapes of New York,* page 106) At the time Hedrick wrote, Munson was listing twenty-five American species and L. H. Bailey twenty-three while Planchon of France asserted there were only twenty-eight species in the entire world.

16. This land was part of a tract that Volney, Ben, and Jot Gunter had jointly purchased several years earlier. In 1878 they divided it and separated out Volney's farm, deeding the forty-three acres to him in return for his interest in another section. (Grayson County Deed Book 39, pages 486–487 and Deed Book 40, page 230)

The road on which Volney built his house led to Durant, across the Red River in the Indian Territory (now Oklahoma); it later became North Houston Avenue and then U.S. Highway 69. When the Munsons moved there, the house was close enough to the road that the children could watch the constant parade of Indians traveling to and from the Territory.

17. Theo lived with his parents until they died in 1890, when he moved into Denison. Trite and Jennie stayed on in the house for a time after William and Maria died, then moved to Point Loma, California (now part of San Diego), where Trite died; Jennie died in Pasadena, California. In 1918 Theo gave the property adjoining the old homeplace to the city of Denison for a park, which is still in use. Ben donated additional acreage for a lake and $10,000 for improvements.

18. According to Samuel Wood Geiser, author of *Horticulture and Horticulturists in Early Texas,* Willard Robison (1839–1917) was also associated with Munson in the nursery business during the period he (Robison) was in Denison, 1873–1883. He originated the blackberry that bears his name. Robison himself, in a condolence letter written to the family after Volney's death, remarked on "the little group of four of us pioneers of North Texas horticulture, Munson, Robison, [C. E.] Stephens, and [James] Nimon in order as we located around Denison in the dark ages, as it were."

19. Horticulturist Samuel Miller (1820–1901) developed the Martha grape, a variety Volney raised in his own vineyard. Born in Pennsylvania, Miller moved to Missouri in 1867 to enter business with another well-known viticulturist, George Husmann; they established Bluffton Winery, one of the first in the United States. Miller also wrote for *Colman's Rural World* and originated at least six varieties of grapes.

20. "Fifty Years of American Grapes: Part III: Brief Summary of T. V. Munson's Work with American Grapes," *American Gardening,* November 4, 1899, page 750.

He also brought home geological specimens from these trips, an indication of his wide-ranging interest in all the sciences. After his death, his collection was given to Texas A&M College. Unfortunately, the college's museum closed in the 1960s and the collections were dispersed. The whereabouts of Munson's specimens are currently unknown.

21. "Laying the Foundations of American Grape Culture," *The National Nurseryman*, Volume XVII, No. 6, June 1909, page 173.

22. A self-taught man, Husmann was later professor of pomology and forestry at the University of Missouri at Columbia. He moved to California's Napa Valley in 1882 and was well known for his many grape publications, especially *American Grape Growing and Wine Making*.

23. The extent of the Planchon/Riley collaboration is open to debate, and the answer largely depends on whether one is from France or Missouri. While on vacation in England, Riley, who was already famous in entomological circles, was asked to go to France, examine the devastation, and give his expert opinion. He immediately recognized the louse on the grapevine as similar to one he'd seen on oak trees in the United States and pronounced it to be a form of phylloxera. But it was through Planchon's writings that this information was disseminated, so he often receives the credit—at least in his country. Missourians firmly stand by their man as the discoverer.

Planchon (1823–1888) was editor of *La Vigne Americaine* and professor of pharmacy and of botany at the Universite de Montpellier but was more interested in his hobby of entomology. He was also director of Jardin des Plantes in Montpellier, established in 1593 as France's first botanical garden. Originally a study garden for the university's botany students, the Jardin today includes plants from around the world, including American grapevines, and is the oldest in France.

Pierre Viala was born in the Herault department of France in 1859 and graduated from l'Institut National Agronomique at Montpellier in 1881, where he concentrated on diseases of grapevines. He later taught viticulture at the school and founded *Revue de Viticulture* in 1894. He was named Inspector General of Viticulture in 1897 and elected to the French Academy of Sciences in 1919. Viala died in Paris in 1936.

Unfortunately, the Munson/Planchon/Viala correspondence has been lost. Viala's two remaining descendants in France have only a small collection of original material, which has never been published and only rarely exhibited.

24. George Engelmann (born Germany, 1809) immigrated to the United States in 1832 and settled in St. Louis as a physician. But his real love was botany, and he became an authority on several species of American trees. His garden of western plants at his St. Louis home and his herbarium were recognized as among the best in the country. Indeed, Engelmann's herbarium served as the basis of the collection when the Missouri Botanical Garden was established in 1889. His classification of American *vitis* was standard until it was replaced by Volney Munson's, and he wrote several monographs.

Engelmann was also known for his detailed weather observations, and he served as scientific adviser to many western expeditions. He was particularly interested in Texas and its role in the many new discoveries being made as the result of the expanded exploration of the American West after the Civil War. Among his contributions to Texas geological studies was his work on the 1858 Boundary Survey between the United States and Mexico.

25. *Texas Agricultural Experiment Station Bulletin No. 56*, page 220.

26. "Laying the Foundations of American Grape Culture," *The National Nurseryman*, June 1909, page 173.

27. The hoe allowed the user to stand erect while loosening soil and weeding and was particularly recommended for ladies. In a flyer Volney released in mid-1882 (printed on paper of a shocking pink hue), he enlisted testimonials from some of his better known horticultural friends:

• Samuel Miller (see footnote 19); Parker Earle, president of the American Horticultural Society for years; horticulturist George W. Campbell; John A. Warder (1812–1883), founder of the *Western Horticultural Review* (1850), United States commissioner to the Vienna World's Fair (1873), founder and president of the American Forestry Association, and one of the first to advocate landscaping of public grounds such as cemeteries; former governor of Nebraska Robert Furnas; and H. M. Stringfellow (1839–1912), a Virginia-born lawyer who became "one of the leading market gardeners" of Texas. As late as 1922, the *National Cyclopaedia of American Biography* attested to the continuing popularity of the diamond scuffle hoe, noting that it was "now widely known to horticulturists." (Volume XVIII)

28. Neva was probably named for her Aunt Ginevra. The origin of "Olita" is less clear. The

baby was born with a dark complexion—her siblings joked she'd been left on their doorstep by an Indian family—so Volney may have selected a derivation of a Spanish name, Lolita, which also translates as Louise or Louisa, the name of his sister in Nebraska. Or, family genealogist Joyce Maki speculates, it may be from the scientific term "Olitoria," that is, pertaining to a vegetable garden.

29. The group soon changed its name to the North Texas Pomological Society. Within a year of organizing, it had a membership of "about forty of the most active fruit men in this part of the State." As an officer, Volney wrote an article on "Fruit Culture in Northern Texas" for *Burke's Texas Almanac* of 1880.

30. Trelease had studied under Asa Gray and was considered one of the leaders of the "new botany" movement. While at the University of Wisconsin, he was selected to direct the new Washington University (St. Louis) school of botany founded by Henry Shaw in 1885. "Shaw's Garden," established in 1859 and later under the direction of William Trelease, is now the Missouri Botanical Garden, one of the oldest in the nation.

31. Thus, Texas had three separate Academies of Science, the first in existence from 1880 to 1887, the second from 1892 to 1912, and the third organized in 1929. Amazingly, the second incarnation—where almost half the initial members were immediately voted Fellows—never accorded Volney Munson that privilege.

32. Roughly two dozen letters were found in an old tin picnic basket in the summer of 1998 by the current owners of one of the Jaeger homes. (Hermann's house was razed some years ago.) With their kind permission, Kay Hively, a Neosho writer working on a biography of Jaeger, studied and transcribed the letters and shared them with the authors.

33. Volney elaborated on his feelings about amateur growers in an 1887 letter to Department of Agriculture botanist, A. (?) Crozier. The latter had written to ask Volney his opinion about "the effects of cross-fertilization on fruit." The Texan, in reply, acknowledged his lack of experimentation in this area and discussed who should take up such work. "It's certainly a very enticing field for investigation, but for certain results must be in the hands of the coolest, more critical experimenter, for other influences on the fruits … such as climate, weather &c. so readily affect them that it is a nice matter to sift out all influences save that of pollenization, and hence, I cannot even trust my own opinion in the matter with the lack, as at present, of the special course of careful scientific investigation. This indicates the value I place upon the opinion of the average fruit grower in the matter, who scarcely knows what part of the plant he is eating in an apple." (Letter from T. V. Munson to A. (?) Crozier dated October 19, 1887. Smithsonian Institution Archives, Record Unit 220, United States National Museum: Division of Plants, 1870–1893: Box 10, Folder 30.)

34. The *Botanical Gazette* had this terse comment on Hooker's theory: "Hence to state it all in one sentence, our Eastern flora has come from the North and our Western flora from the South." (Quoted in *American Botany, 1873–1892: Decades of Transition,* page 108.)

35. These papers represented the radical new approach which Munson and other "young lions" of the scientific movement advocated. American horticulturists were not then accustomed to gathering together for educational purposes since fruits and vegetables had long been regarded as less important aspects of farm production. The American Pomological Society had been organized in 1848, but not until the 1880s did growers begin to appreciate the need for more knowledge and to attend meetings in any numbers.

For this was the beginning of a new era, that of *commercial* agriculture, which lasted from roughly 1870 until the outbreak of World War I. During the seventeenth and eighteenth centuries, even up to the Civil War, land in America was so plentiful that farmers had little motivation to improve yields and efficiency. But the same scientific movement that had swept up Volney Munson in his college days was also sweeping away tradition. Farms were being mechanized and millions of acres of new lands brought to agriculture for the first time. From growing multiple crops, more farmers began to specialize.

So quick and so widespread was this revolution that one observer wrote as early as 1881, "the changes in farming in these latter times have been so marvelously great that it is probable that there are men now living who have seen in their lifetime greater changes in the methods and appliances of farming than took place in all the centuries before." (Quoted in *Agriculture in the United States: A Documentary History,* Volume 2, page 1227.)

36. In 1882 Volney wrote his former mentor, Dr. Robert Peter in Kentucky, and asked if he would be attending the MVHS meeting in New Orleans in February 1883. "It would be a supreme pleasure to meet you. ... What a treat it would be to listen to one of your characteristic lectures on the Chemical Constituents of Fruit Soils; besides this new Society needs some solid encouragement from our acknowledged authorities in scientific horticulture by their presence and papers." Since he wrote on MVHS letterhead listing him as vice-president for Texas, and he was presenting a paper himself, Volney understandably wanted to impress his old professor with how well he'd done. (Original letter in collection of T. V. Munson Viticulture and Enology Center.)

37. *Transactions of the Mississippi Valley Horticultural Society for the Year 1883,* Volume I, page 136.

38. Again, this was radical thinking for the time. Most naturalists and horticulturists from this period were more interested in variety than in specialization.

39. Volney's interest in and involvement with the American Forestry Association is yet another demonstration of his efforts to stay current with all that was new. Not until the late 1870s or early 1880s did forestry also begin to take the scientific approach; the first American Forestry Congresses were held in 1882.

40. James Nimon (1849–1905) was born in Ireland and, in 1879, moved to Denison, Texas, where he ran a nursery. He and Volney worked together on several hybridization projects such as the Parker Earle strawberry (see Chapter 8).

41. Few of these early catalogs have survived; one for fall 1885-spring 1886 is in the Leyendecker Family Papers at the Center for American History in Austin, Texas (Box 3R56, Folder 5). The headline declares that all transactions are strictly cash and all stock is from a home-grown supply. (Volney deplored traveling tree salesmen who took money for goods they knew they couldn't deliver.) Also in it, he listed a variety of fruit trees including Lone Star plums (15¢ each), Eureka apricots (15¢), and Early Richmond cherries (20¢). There were berries, figs, shade trees, and flowering shrubs such as althea, honeysuckle, and rose (12¢ each). Of course, he had a long list of grape stock—none of them his own creation yet—which ranged in price from 2¢ for a Concord vine to 20¢ for a mature Zinfandel.

Volney sometimes used his later catalogs as a platform for his personal views. In the 1908–09 edition he lashed out once more on the subject of traveling salesmen: "We employ no traveling agents. Those claiming to be our traveling agents are frauds and should be shunned. ... We never could see any benefit to the planter to be importuned by oily-tongued 'agents' ... with highly exaggerated pictures, samples, and deceptive speech ... we prefer DIRECT dealing." (From original in collection of T. V. Munson Viticulture and Enology Center.)

42. Living in a railroad boomtown such as Denison, which was on several main lines, made it easy for Volney to ship his nursery stock anywhere in the world and considerably eased the burden of his many travels. Generally speaking, the expansion of the American rail system in the 1870s and 1880s greatly increased the pace of botanical exploration in this country and in Mexico.

According to grandson Volney Acheson, Munson prepared his nursery stock for transport by dipping the roots into a red clay sludge to keep them from drying out; the clay was kept in a pit near the packing house. The plants were then wrapped in burlap, tagged, placed on a large mule-drawn flatbed wagon, and carried to the depot.

Volney's packing house was at the northwest corner of modern Hanna and Mirick streets, with the nursery's business office nearby. After his father's death, Will built a larger packing house.

43. *Foundations of American Grape Culture,* page 179.

44. We are accustomed today to thinking of the extinction of plant and animal species as strictly a modern problem. But in the late nineteenth century, many naturalists and horticulturists such as Volney Munson and George Engelmann bemoaned the number of species that were already gone or endangered.

Chapter 5: "No Little Work"

1. The fabulous Horticultural Hall was destroyed in a 1915 hurricane, and no remnants survive of any of the Expo's major buildings.

2. "Native Grapes of the United States." Published in *Transactions of the American Horticultural Society for 1885*, Volume 3, pages 128–140.

3. Volney wrote that science, the natural distribution of plants and animals, "and the results of recent deep sea dredgings," all made it clear that there had been such a continent in the past, possibly the legendary Atlantis for which the Atlantic Ocean was named. Many botanists of the period felt that the one-time existence of Atlantis explained the similarities between American and European plants.

4. That medal and the accompanying diploma are in the collection of the T. V. Munson Viticulture and Enology Center.

5. This is not the same organization as the modern American Horticultural Society.

6. Letter of March 29, 1885. Transcript in author collection.

7. Pomology is a sub-category of horticulture (garden plants) and refers to the cultivation of fruits specifically.

Volney was vice-president of the organization and took an active role in it.

Volney and the family visited Lexington, Kentucky, before traveling on to the meeting as Nellie's mother had died on July 24th.

8. Letter to Hermann Jaeger, September 24, 1885. Transcript in author collection.

9. The award, given for special merit, was named for Marshall Pinckney Wilder (1798–1886), a founder of the American Pomological Society and president for most of its first quarter century.

10. Letter to Jaeger, September 24, 1885. Transcript in author collection.

11. Born in Michigan in 1858 and raised in a wilderness area, Bailey graduated from Michigan Agricultural College and spent several years after school as a newspaper reporter. He studied taxonomy at Cambridge, then was selected by Asa Gray to arrange and name the materials in the university's gardens. He returned to his alma mater to teach and envisioned a new, more scientific horticulture; at Michigan Ag, he built one of, if not the first, distinctly horticultural labs in the country. In 1888 Bailey was chosen Cornell University's professor of horticulture and became dean of the College of Agriculture early in this century. (Cornell was the first American university to establish a separate horticultural department.) He eventually wrote some fifty books on American horticulture, several of which included biographical sketches of Munson.

12. None of these varieties was originated by Munson himself, though there were several by Jaeger. It was a complicated table, and Volney believed he was the first to attempt such a compilation. Listed were the common and botanical names of each variety, place of origin, rate of growth, hardiness in both North and South, ease of rooting the cuttings, time of year it bloomed at Denison, type of stamen, diseases to which it was prone, sizes of clusters and berries, color and date of ripening, skin and pulp type, number and size of seeds, and fitness for table use or in winemaking.

13. "Address on American Grapes." Published in the *Transactions* of the American Pomological Society, 1886, page 5.

14. Letter of Munson to Jaeger, April 4, 1886. Transcript in author collection.

15. Denison *Sunday Gazetteer*, March 28, 1886, page 3.

16. Henry Elias Van Deman (1845–1915) was born in Ohio and served in the Union Army during the Civil War. From 1878 to 1879, he was the first professor of botany and practical horticulture at Kansas State Agricultural College (now Kansas State University). From 1886 until 1893, Van Deman was founding director of the USDA's Division of Pomology. He later was associate editor of *Greens Fruit Grower* and *Southern Fruit Grower* and served as president of the American Fruit and Nut Company. He died at his home in Washington, D.C.

He first met Volney Munson in St. Louis in 1882 at a meeting of the Mississippi Valley (later American) Horticultural Society. They remained friends until Volney's death; in a condolence letter written to the family at that time, Van Deman remarked that he had last seen Volney when they roomed together at the American Pomological Society meeting the previous year.

17. *Report of the Commissioner of Agriculture, 1886*, pages 261–262.

The Denison vineyards eventually included about sixty *vinifera* varieties obtained from "Japan, Persia, Greece, Italy, Spain, Hungary and France." Volney noted in *Foundations of American Grape Culture* (1909) that eighteen of them, grafted onto American rootstock and treated with Bordeaux mixture, were already being profitably grown in West Texas, thus foreshadowing the successful wine industry there which began in the 1970s. (See Chapter 15.)

18. Letter from Munson to Van Deman, dated February 6, 1887. (National Archives II, Record Group 54, General Correspondence/Pomology, Box 1, Folder 3 of 4, January 1885-June 1887.)

19. The Central Texas Horticultural Society was the first to call for reorganizing the state society, which had been established in Houston in 1875 but subsequently lapsed. The Texas State Horticultural Society eventually merged with the Texas Nut Growers Association (organized circa 1906).

20. Letter from Munson to Jaeger dated April 11, 1887. Transcript in author collection.

This is likely the same collection Volney referred to in an 1888 letter he wrote to George Vasey of what is now the Smithsonian Institution (see Chapter 8). The collection at the modern school in Montpellier still contains specimens of Munson's.

21. Additional acreage was planted in vines in 1887 (total of 6 acres), 1894 (total of 1.12 acres), 1904 (total of 5 acres). As a whole, the south nursery totaled 109 acres of sandy loam with yellow and red clay subsoil. This clay is likely the same used as wet sludge to coat the roots of nursery plants before shipping. (See Chapter 4, footnote 42.)

Volney continued to use the old location north of town for several years, too, but eventually moved all the operations to Vinita.

22. Among the designs Volney played with in the pages of his journal was a circular house: "The most Economical, Comfortable, Durable, Beautiful & Convenient," he wrote. The center of the house was a vestibule containing the staircase; all the rooms were of equal size and radiated from the vestibule.

23. Videotaped interview with Marcus Acheson, conducted by Dr. Roy Renfro at the Acheson home in Palo Alto, California, and in the collection of the T. V. Munson Viticulture and Enology Center.

24. Volney later built a separate office building near the other nursery structures. The original room in the house was then converted to storage and an indoor bathroom. By the time the new office was built, probably in the 1890s, Will was working for his father and moved his residence from the main house to two rooms on the second floor of the office.

25. After moving to Vinita Home, Volney rented out Scarlet Oaks, or The Old Place, as he sometimes called it. Unfortunately, the house and property did not fare well under the "tender mercies" of the renters, who destroyed several of his prized grapevines.

26. Frank Lamson-Scribner was born in Massachusetts in 1851. Orphaned as a child, he was adopted and took his new parents' name in hyphenated form. He became interested in botany at the age of fifteen and published his first paper at eighteen. He received his B.S. in 1873 and was an officer of Philadelphia's Girard College from 1877 to 1884. In 1885 he was appointed an assistant botanist with the USDA, Special Agent in charge of the Mycological Section the next year, and chief of vegetable pathology in 1887. In 1888 he moved to Tennessee to teach botany and horticulture and ran the university's agricultural experiment station until 1894, when he returned to the USDA. He served three years in the Philippines for the Department and was later given charge of all government exhibits at expositions around the world. He published extensively and was known for his work on American grasses and on fungal diseases.

27. Volney wrote H. E. Van Deman on March 19, 1887, and thanked him for recommending that he work with Lamson-Scribner on the Bordeaux mixture project. "I beg of you," he continued, "to have Herman [sic] Jaeger of Neosho, Mo, appointed by Prof. Scribner for a similar work. He is thoroughly capable, has the vineyards, and is read up on the subject as treated by the best French authorities. He is a very close observer." (N. A. II, RG 54, Gen. Corr./Pom., Box 1, Folder 3 of 4, January 1885-June 1887.)

28. The recipe for Bordeaux mixture was 6.5 pounds of copper sulphate, 3.25 pounds of lime, and 22 gallons of water. Some modern writers suggest also adding table salt.

Bordeaux mixture is said to have originated with a French vineyardist who used the blue color to make his grapes look sickly. It seems that some of his grapes grew alongside the road, tempting passersby to grab a handful.

Lamson-Scribner spent some time trying to find spray pumps made in the U.S. that were similar to those used in France but was unsuccessful and concluded by importing them. The USDA furnished all four experimenters with chemicals and apparatus since, as Van Deman wrote Volney, "uniformity of experiment is quite as necessary as to make the experiment at all." (N. A. II, RG 54, Gen. Corr./Pom., Box 1, Folder 2 of 4, January 1885-June 1887.)

29. There are two pieces of evidence that document a stipend was paid. One is Lamson-Scribner's letter of April 9, 1887, to Volney in which he states, "Your terms I consider reasonable for the work to be done. Satisfactory arrangements can be made after the first of July—have already told you that we could not enter into any contract before that date." (Probably a reference to the government's fiscal year.) (N. A. II, RG 54, Box 1, Book 1, page 409.)

The second is a letter written January 6, 1888, in which Acting Commissioner L. C. Nesbitt spent some time describing the final printed form of the report and told Volney that his section "will not exceed six pages of printed matter, and this is all that can be approved under the commission of July 5th. If you can conveniently wait until the report is actually in type before presenting your vouchers there will then be no doubt as to the number of pages covered." (Box 1, Book III, page 38.)

30. Feeling that the 1887 tests were encouraging but not conclusive, Lamson-Scribner repeated them the following year. But Volney declined to participate again, probably because he was extensively involved in other work and had no time.

31. Andrew Denny Rodgers, III, *American Botany, 1873–1892: Decades of Transition,* page 130.

32. Norman J. Coleman (1828–1896) was a lawyer before he moved to Missouri, where he started the newspaper *Coleman's Rural World.* He served as lieutenant governor of that state and helped organize the Missouri State Horticultural Society before moving to Washington to become secretary of agriculture.

33. Letter from Munson to Van Deman dated August 24, 1887. (N. A. II, RG 54, Gen. Corr./Pom., Box 1, Folder 1 of 2, July-December 1887.)

34. *Ibid.*

35. Letter from Munson to Van Deman dated July 4, 1887. (N. A. II, RG 54, Gen. Corr./Pom., Box 1, Folder 2 of 2, July-December 1887.)

36. This inaugural meeting of the new Texas Horticultural Society attracted visitors from California and Louisiana and was hailed as "the first glad cry of the fairest industrial infant born to Texas in many years." (*Texas Farm and Ranch,* August 15, 1887.) Volney won accolades for his hybridized grape display, which viewers felt "entitles that learned gentleman to the front place in advanced horticultural experimentation." He also exhibited the Bordeaux mixture sprayer he'd used in Lamson-Scribner's black rot experiments earlier that year. In fact, Scribner—doubtlessly as a result of his work with Volney—attended and discussed "Fungous diseases of the grape and other fruits, and their treatment." At the end of the conference, Volney was reelected to a second term as president of the Society.

Primary material from the TSHS is very rare. The E. W. Kirkpatrick Collection at the Collin County Historical Society (McKinney, Texas) contains a copy of the original "Charter and Organization" of the Society from its Houston meeting in September of 1875. Established to promote "the advancement of the science of Horticulture and Pomology, rural adornment, and landscape gardening" among other lofty goals, the group pledged to meet annually in Houston and to sponsor an exhibition of members' products.

Also, a preliminary program for the 1887 conference referenced in the text can be found in the Leyendecker Family Papers at the Center for American History in Austin, Texas (Box 3R56, Folder 3), and a more complete program is in the National Agricultural Library in Beltsville, Maryland.

37. Letter from Munson to Van Deman dated August 24, 1887.

The Red River actually arises in New Mexico and has four main channels. Munson was probably referring to the Prairie Dog Town Fork which flows, through several streams, from the source.

38. Modern ampelographer Pierre Galet classifies this as *Vitis Candicans Engelmann,* or Mustang Grape, and believes it to be a natural hybrid. Munson felt it was distinct enough to be a separate species.

39. Letter from Van Deman to Munson dated August 19, 1887. (N. A. II, RG 54, Gen. Corr./Pom., Box 1, Folder 1 of 2, July-December 1887.)

40. Prestele (1838–1895) was the son of another famous botanical illustrator, Joseph Prestele. His appointment as first artist of the Pomological Department came on August 1, 1887.

41. Letter from Munson to Van Deman dated August 24, 1887.

Chapter 6: "A Great Pest"

1. Station Viticole du BNIC, *Dossier: Il Y A 100 Ans . . . Le Phylloxera*.

The importation of American grapevines was particularly prevalent in the late 1850s. Some Missouri wine historians claim that was due to the increasing number of medals won by Missouri wines at international competitions; the French, say the historians, decided to see what they could do with these rootstocks and began importing them. (See "Missouri's Rhineland" by Archie Satterfield, *Chicago Tribune*, May 8, 2000.)

2. "The Story of a Great Pest," *London Society*, Volume 36, October 1879, page 328.

3. The Gard borders the Rhone River west of Marseille and just touches the Mediterranean; also known as the southern Rhone Valley, it includes the city of Nimes. One of the points of origin for the epidemic there was Chateau de Clary, which experimented with what proved to be phylloxera-laden vines from California (1863).

Some wine historians theorize that phylloxera had not reached Europe earlier because of the slowness of sailing ships crossing the Atlantic: the bug simply couldn't live long enough. New steam-powered ships and rapid inland rail transport, they argue, carried phylloxera to the vineyards fast enough to enable it to survive. Tim Unwin, author of *Wine and the Vine: An Historical Geography of Viticulture and the Wine Trade* (1991), disagrees. He believes that the sudden appearance of the pest, after years of importing American vines, had to do with a change in the source of the vines and a more virulent strain of phylloxera. The plague of oidium which preceded phylloxera may also have weakened the vines and made them more susceptible. Lastly, Unwin points out that Europe enjoyed generally warmer weather from 1857 to 1875, which may have helped the aphid survive and spread.

4. H. M. Jenkins, "Notes on Market-Gardening and Vine-Culture in the Northwest of France" (1880), page 103.

5. "The Story of a Great Pest," page 326.

6. "The Grape Phylloxera in California" in *The American Entomologist*, Volume III, page 3.

7. Julian Jeffs, *The Wines of Europe* (New York: Taplinger Publishing Company, 1971), page 74.

8. "Insects Injurious to the Grape-Vine" in *The American Entomologist and Botanist*, Volume 2, No. 12, December 1870, page 354.

9. According to H. M. Jenkins (*op. cit.*, page 98), "The best grapes are grown on hill-sides or elevated ground with a southern aspect. . . . The difference of a foot or two in the elevation of the land has a marked effect on the quality of the grape for wine-making purposes."

10. "The Story of a Great Pest," page 333.

11. The debate over using American vines raged for years. One French correspondent wrote *The American Entomologist* early in 1880 that "the great news of the day is the conversion of Prof. Dumas, at Paris, to the belief in American vines as the salvation of our Phylloxera-ridden vineyards." (Volume III, page 77.)

12. "The Story of a Great Pest," page 333.

The pre- versus post-phylloxera taste argument continues today. Some declare that modern wines made from grafted vines mature more quickly and have shorter lives, although this may be due also to twentieth-century changes in the fermentation process. The vines themselves become less vigorous at an earlier age and live about half as long.

13. "The Story of a Great Pest," page 334.

14. The cost of replanting in American vines was estimated at 5,000 francs per hectare.

15. Greece would pay the price of this in later years when crop failures forced many to emigrate. A surprising number of Greeks wound up in the coalfields of the western United States.

16. Jenkins, *op. cit.*, page 103.

17. Letter of April 12, 1898, from J. Leenhardt-Pomier of Montpellier, France, to *La Vigne Americaine et la Viticulture en Europe* (3rd Series, Tome II [Tome XXII of the entire work], 1898) re-

garding the recent death of Gustav Meissner. (Photocopy in author collection, courtesy of Mr. and Mrs. Joseph Gleason.)

By 1891, phylloxera had finally traveled far enough north to invade France's champagne district. But by that time, most areas—even those adamantly opposed to it—were using American rootstock. C. V. Riley, quoted in the *Kew Gardens Bulletin* of 1891, had this to say: "The work [eradicating phylloxera] is practically at an end in such Departments as Herault, Gard, and Gironde, where the American resistant vines have most effectually been used; while the wine-growers of Algeria, Spain, Italy, Portugal, Hungary, Austria, and Switzerland are all [still] battling against it. . . . [Even in Medoc and Sauterne, where . . .] opposition to the use of American resistant stocks has been greatest . . . [they] have finally vanquished prejudice." (pages 44–45)

18. Jenkins, *op. cit.*, page 103.

19. In his work, much of which he conducted at the Bushberg Nursery south of St. Louis (1871–1873), Riley had found that the American and European phylloxera were different forms of the same insect.

In the same letter referred to in footnote 17, Leenhardt-Pomier stated that it was he who asked C. V. Riley, in 1871, to request Gustav Meissner (see footnote 21) to send the first shipment of American vines to France. "It consisted of vines from direct producers, such as, Clinton, Taylor, Concord, Herbemont, Cunningham, Norton Virginia, Cynthiana and some other varieties discontinued since then." However, later in the letter, he wrote that Meissner had the "happy and lucky thought" of sending vines while on an 1876 trip to the Indian Territory (Oklahoma) searching for wild varieties he thought would help.

20. Guilbeau (1813–1879) was born in Brittany and immigrated to San Antonio in 1839; about a decade later, he opened a wine and import business on the Plaza. Among his city real estate holdings were fifty acres across from the modern south entrance of Brackenridge Park, which he rented to Knox about 1876 on the latter's moving his nursery from Brenham, Texas. Knox selected most of the vines from an area just southeast of San Antonio, where the mustang grapes were prolific. They were packed in Guilbeau's warehouse, carted to the coast by ox train, then shipped from the Texas coastal ports of Indianola or Galveston. (*San Antonio Express*, May 28, 1933)

21. Isidor Bush (1822–1898) was born to a wealthy Jewish family in Prague and privately educated. His father was an investor in the world's largest Hebrew publishing house, so young Isidor apprenticed as a printer. His liberal politics forced him to flee to the United States in 1848, and he soon made St. Louis his home. An ardent Unionist, he became friends with viticulturist and fellow German immigrant George Husmann when both served on the committee which drafted Missouri's emancipation laws. Bush invested in Husmann's new Bluffton Winery in Hermann and apprenticed his son, Raphael Bush, to Husmann to learn the nursery business. In 1865 he bought 241 acres along the Mississippi River in Jefferson County, south of St. Louis, and established the Bushberg Nursery, where he specialized in native American grapes. He also built a wholesale wine business in the city to distribute American wines.

Bush published the first of his four nursery catalogs in 1869. These became quite successful among viticulturists for they eventually included a complete listing of American grapes and a classification by George Engelmann. Volney Munson himself would write in a later edition that "your modesty in simply calling it a 'Catalogue' does the great work an injustice. It is a most complete and valuable treatise on American grapes." The catalog became required reading in agricultural schools and was translated into French and Italian.

In 1871 Bush customer and nurseryman Gustav E. Meissner (born in Germany in 1843, the same year as Munson) joined the firm. Meissner was an accomplished linguist whose family had moved to the United States in 1851. He worked several years for Isidor Bush as a foreman before becoming a partner in the business. Meissner made several trips to France studying the phylloxera problem and later established his own 600-acre vineyard across the Mississippi from Bush. He died in 1898, a few months before Isidor.

The homes of both Isidor Bush and Gustave Meissner still survive at the old site of Bushberg and have been restored, but only terraces and cellar ruins remain of the nursery.

22. Letter of Leenhardt-Pomier, *op. cit.*

23. Some vineyards refused to graft. The famous Domaine de la Romanee-Conti in Burgundy, for example, maintained original *vinifera* vines until World War II, when the shortage of labor made it impossible to keep phylloxera at bay any longer. Mouton-Rothschild and Haut-Bailly held out, too, until 1907. Clearly, those who resisted grafting were the vignerons known and respected for the quality of their wines, reputations which they were loathe to endanger.

German vineyards also used fewer graftings since the cold winters there considerably slowed the spread of the aphid and the Germans exercised meticulous care—at some expense—to keep the vines clean. As with the last holdouts in France, however, World War II disrupted the labor supply and caused a scarcity of necessary materials.

Around 1879, to combat phylloxera, the Turkish government banned the importation of plants of *any* type. The *Gardener's Chronicle* noted caustically that "this wholesale edict (supposing it were practicable to enforce it) has at least more logic and sense about it than the rules which permit a wagonload of hay to cross the frontier, but confiscate a rose in the buttonhole of a tourist." (Reported in *The American Entomologist*, Volume III, page 248.)

Chapter 7: Mr. Munson to the Rescue

1. The Ardeche and Gard in southern France are marked by rolling, tree-covered limestone hills and upthrust ridges with exposed sides nearly as white as snow. In some places, subsidiary streams flowing into the main rivers are actually cream colored from the quantity of chalky soil and limestone that has washed into them. (Geologically, chalk is simply a softer, crumbly limestone.)

Such soils were laid down by prehistoric oceans that covered much of the continent millions of years ago. They are more permeable than clay, percolating and draining water quickly and thus aiding root growth. Many of France's most famous vineyards are in either chalky or limestone soil; the Cote d'Or in Burgundy, for example, is on a fossil-laden limestone ridge and Champagne is in chalk.

2. Quoted in *The Winegrowers of France and the Government Since 1875* by Charles K. Warner, page 6.

3. The Montpellier-based journal, *Le Progres Agricole et Viticole* published by the agricultural school, also provided funding, and the French government paid the remaining costs.

4. Texas State Geological and Scientific Association, Bulletin No. 3, *The French Viticultural Mission to the United States* by M. Pierre Viala, page 31.

Marl is a crumbly mixture of clay and calcium carbonate; it's often used as a fertilizer in lime-deficient soils.

In the foreword to this English translation, Robert T. Hill of the U.S. Geological Survey lamented that it had taken foreigners to conduct a good geological survey of Texas. Viala's report, he declared, "presents for the first time an intelligible discussion of our wild vines, and recognizes the true geologic character of our soils, and their relation to plant growth." (page 30)

5. The map accompanying Viala's final report indicates that he primarily visited places along America's new and still expanding railroad routes. From Washington, he went up the East Coast to Vineland, New Jersey; Hammondsport, New York; and Boston, where he visited W. G. Farlow at Cambridge University. Traveling south, he stopped in Charlottesville, Virginia, and Fayetteville, North Carolina, where he likely visited the nurseries used in Lamson-Scribner's black rot/Bordeaux mixture experiments.

He then turned his attention west, making a small exploratory loop outside Nashville before continuing on to St. Louis and the Bushberg Nursery south of the city, conferring with William Trelease, Isidor Bush, and Gustav Meissner. Viala searched in the Neosho-northeastern Oklahoma area with Hermann Jaeger, then cut straight down into Texas (see text). From there, he traveled with few stops across the Southwest to Los Angeles, San Francisco, San Jose, and the Napa Valley. To return east required a torturous rail journey through Nevada to Salt Lake City, across the Rocky Mountains to Pueblo, then north to Denver and across Nebraska. From Chicago, Viala headed to Ohio where he visited several sites before concluding his trip back in Washington.

6. "This formation occurs in north Texas, from the Panhandle to the Rio Pecos, and from New Mexico eastward, where it is limited by a line drawn through Sherman, Dallas, Austin, San Antonio

and Eagle Pass; it rejoins the Rio Pecos in the latitude of San Antonio." (Viala, pages 34–35) Viala was describing the western half of Texas, including part of North Central Texas, the Edwards Plateau, the Panhandle, and West Texas.

7. Lamson-Scribner confirmed to Viala on April 25 that Volney was "one of the best informed grape growers in this country." (Letter quoted in Stephens, "Thomas Volney Munson: A Case Study in the Professionalization of American Horticulture, 1875–1913," page 10; original letter now missing from National Archives II USDA collection.)

8. Letter from Munson to Van Deman dated September 23, 1887. (National Archives II, Record Group 54, General Correspondence/Pomology, Box 1, Folder 2 of 2, July-December 1887.)

9. Johnson managed a nursery at McKinney Avenue and Pearl Street in Dallas; he was also vice-president for Texas of the Nurseryman's Association of America, an organization in which Volney, too, was active. Both Johnson and his wife were members of the Texas State Horticultural Society, of which she was secretary for many years.

About the time of the Viala mission, Volney and Mrs. Johnson co-authored a pamphlet for the TSHS on organizing a local horticultural group. In addition to a model constitution, it contained practical and hard-learned experience such as "Do not expect to have everyone belong, for they will not." (Center for American History, University of Texas-Austin, Leyendecker Family Papers, Box 3R57, Folder 6.)

10. In El Paso, Viala and Lamson-Scribner stopped to examine the Mission grapes and reported favorably on prospects for the industry in that area. In an interview with the local *Times*, the Frenchman reported that the lush valley made him homesick, for it was "the first locality in this country where I have seen the vines pruned as we see them in France." He also declared his belief that such fertile land, if located in France, "would be worth at least $2,000 per acre for grape growing."

After leaving El Paso, the two men headed for California, where the USDA wanted Lamson-Scribner to look at new problems in the vineyards there. According to both the *Dallas Morning News* and the *El Paso Times* (October 8, 1887), growers were "pretty badly scared" about a new disease "which is sweeping away whole vineyards" in the Los Angeles area, probably Pierce's Disease. Viala, meanwhile, occupied himself by conducting "various experiments" in Napa Valley on *Vitis californica*. They convinced him that that variety would "never possess great power of resistance to the Phylloxera" though he thought it would still "have a future in the vineyards of France."

11. "History of Bell County," no author. (Center for American History, University of Texas-Austin, WPA Historical Records Survey, Bell County, 4H193, Folder 4.)

Interestingly, the United States Agricultural Census of 1887–88 shows no grapes cultivated in Bell County.

12. The Dog Ridge formation lies between Belton and Killeen in Bell County and is said to have been named for the number of wild dogs living in the area when it was first settled. The rootstock continues to be commercially important today in both Texas and California vineyards, but the difficulty in rooting it eventually led to its being discarded in France. Galet classifies it in the *Candicans* family (the same as Salt Creek), a natural hybrid of *rupestris* and *candicans*.

13. *Berlandieri* had already been imported to France some years earlier but, until the release of Viala's report, it was little known except by a few researchers and hybridizers. The variety was named for French botanist Jean Louis Berlandier (1805–1851), who collected plant specimens in Texas from 1828 to 1834.

Volney had recommended, in addition to these three, *Vitis rupestris Scheele*.

14. Viala, *op. cit.,* page 39.

15. Hermann Jaeger had good luck selling *Rupestris/Lincecumii* hybrids to the French. Among the most successful was his No. 70, which was used to create a number of successful hybrids.

The French report concluded: "Everywhere, the Americans helped me in my studies with a devotion above all praise, and I'm sorry to not be able to tell you the names of all those who helped me. I cannot end this note without thanking Mr. T. V. Munson, Mr. Hermann Iaeger [*sic*], Mr. Bush, Mr. Meissner." He then repeated his comment about Munson and Jaeger quoted in the text here.

16. Letter from Munson to Van Deman dated September 23, 1887. (N.A. II, RG 54, Gen. Corr./Pom., Box 1, Folder 2 of 2, July-December 1887.)

17. And they were indeed desperate. By 1889, production of wine in France had dropped from an 1875 high of 1.8 billion gallons to little more than half a million.

18. Many of these early hybridizers bought their seeds from Jaeger or Munson, who advertised them in *Le Revue de Viticulture*. Descendants of some of those Munson imports are reported to still be in the viticultural collection of Ecole Nationale d'Agriculture in Montpellier.

19. 41B was replaced in the 1960s with Ruggeri 140 (itself a later *Berlandieri* hybrid) because it proved sensitive to drought and phylloxera when planted in shallow soil.

Modern names are composed of a number and the last name of the breeder.

20. C. V. Riley, for example, recommended as a rootstock Viala, a black grape Volney had found in Coryell County, Texas. Viala, noted Riley, "gives the greatest per-centage of successful grafts, and is admirably adapted for grafting on cuttings." (Paper presented by C. V. Riley to the Association of Economic Entomologists and reprinted in *Kew Gardens Bulletin* for 1891, pages 44–46.)

21. There have been two Legion of Honor recipients from the small railroad town of Denison, Texas. Dwight D. Eisenhower, who was born there in 1890, would also receive the award for his leadership of Allied forces in World War II.

22. In 1874, Riley was presented with "a grand gold medal" in appreciation of his research on phylloxera, but the authors have found no other identification.

Chapter 8: Back to Work

1. All the children had specific chores around the nursery and house. Neva, for example, helped bake the breads and cakes, pounding out the dough on a special section of the back porch set aside for that purpose. As she grew older, Viala helped her father with his correspondence and typed the manuscript of his 1909 book. Will, of course, helped in the nursery as did Fern after a brief college experience. All of them worked to one extent or another in their father's business, where every pair of hands was important at planting and harvest times.

The children often traveled with Volney on his shorter grape-hunting jaunts, carrying a picnic basket and collecting rocks and fossils. (The Denison-Sherman area in prehistoric times was on the edge of what would eventually be called the Gulf of Mexico and is replete with marine fossils.) Sometimes they attended special events or programs where Volney was speaking. Fern later recalled to her own children that the whole family was once packed off to Gainesville, Texas, thirty miles away, to see a public hanging. (Maki, *op. cit.*, page 54.)

Volney and Nellie raised their children to be creative, respectful, and philanthropic. Theirs was a financially comfortable life but, again following the example of William and Maria Munson, each child received only one year of college tuition. They were expected to work and earn the money for additional education, just as Volney and his siblings had done.

2. Letter from Munson to "My Dear Friend," probably H. E. Van Deman, dated October 2, 1887. (National Archives II, Record Group 54, General Correspondence/Pomology, Box 2, Folder 1 of 2, July-December 1887.)

3. Letter from T. V. Munson to George Vasey dated July 26, 1888. (Smithsonian Institution Archives, Record Unit 220, United States National Museum: Division of Plants, 1870–1893: Box 10, Folder 30.)

4. *Ibid.*

5. University Archives, University of Nebraska-Lincoln: Charles Bessey Papers: December 7, 1910.

6. Letter from George Vasey to T. V. Munson dated October 1, 1888. (S. I. A., R. U. 220: Div. of Plants, 1870–1893: Box 18.)

7. Letter from T. V. Munson to George Vasey dated October 5, 1888. (S. I. A., R. U. 220: Div. of Plants, 1870–1893: Box 10, Folder 30.)

The new grape referred to in the letter is listed as *Vitis blancoi Munson* in the New York Botanical Garden's list of herbarium specimens.

8. Letter from T. V. Munson to George Vasey dated October 16, 1888. (S. I. A., R. U. 220: Div. of Plants, 1870–1893: Box 10, Folder 30.)

9. Letter from T. V. Munson to George Vasey dated November 2, 1888. (S. I. A., R. U. 220: Div. of Plants, 1870–1893: Box 10, Folder 30.)

10. Munson's collegiate sets consisted of these grapes (taken from his handwritten list):

SECTION I: EUVITIS, PLANCHON

SERIES 1: RIPARIE, MUNSON	*Vitis rupestris, Scheele*
	V. riparia, Mich(au)x
	V. Solinis, Engelm(ann)
SERIES 2: LEUCOBRYAE, PLANCH(ON)	*V. Turnerii, Munson*
	V. Arizonica, Engel(mann)
	V. Cencobrya (?), Munson
SERIES 3: CORDIFOLIAE, MUNSON	*V. cordifolia, Mich(au)x*
	V. rubra, Michaux
	V. Monticola, Buckley
SERIES 4: LABRUSCAE	*V. Labrusca, L.*
	a V. Romaneth, Rom.
SERIES 5: LABRUSCOIDEAE, PLANCH(ON)	*V. Doaniana, Munson*
	V. candicans, Engelm(ann)
	V. coriacea, Shutt.
SERIES 6: VINIFERAE, MUNSON	*(b) V. vinifera, L.*
	(c) V. Bourquinana, Munson
SERIES 7: AESTIVALES, PLANCH(ON)	*V. Linsecomii, Buckley*
	V. aestivalis, Mich(au)x
	V. Floridana, Munson
	V. Simpsonii, Munson
SERIES 8: CINERACENTES, PLANCH(ON)	*V. cinerea, Eng(elmann)* (North Texas)
	V. cinerea, Engelmann (Southwest Florida)
	V. Berlandieri, Planch(on)

SECTION II: LENTICELIS, MUNSON

SERIES 1: TENDRONAE-SIMPLEXAE, MUNSON	*V. Munsoniana, J. H. Simpson*
	V. rotundifolia, Mich(au)x

(Letter from T. V. Munson to George Vasey dated January 5, 1889. S.I.A., R. U. 220: Div. of Plants, 1870–1893: Box 10, Folder 30.)

11. *Foundations of American Grape Culture*, page 226.

Other sets were sent to Columbia, Philadelphia Academy of Sciences, the National Herbarium, and the Botany Department of the USDA. Eleven additional sets went to Dr. Vasey at the latter to be distributed among agricultural colleges; according to Liberty Hyde Bailey, these included Mississippi Agricultural & Mechanical College and the University of Florida.

Unfortunately, few of these sets could be found intact or even extant by the authors. In many cases, they had been integrated into larger herbaria collections within the institutions or even lost or discarded over the years. A further complication is that there is no complete list surviving of all the schools and individuals who received them.

12. Liberty Hyde Bailey wrote in his *Gentes Herbarum* (Volume III) that he had recorded a list of those living plants in his *Annals of Horticulture* for 1889. According to that list, Munson sent the following plants:

Vitis Arizonica, Engelmann	*V. aestivalis, Michaux*
V. Berlandieri, Planchon	*V. bicolor, LeConte*
V. Bourquina, Munson	*V. Californica, Bentham*
V. candicans, Engelmann	*V. cinerea, Engelmann*
V. cordifolia, Michaux	*V. coriacea, Shuttleworth*
V. Doaniana, Munson	*V. Labrusca, Linnaeus*
V. Linsecumii	*V. monticola, Buckley*

V. Munsoniana, Simpson	*V. riparia, Michaux*
V. rubra, Michaux	*V. rupestris, Sheele*
V. Simpsonii, Munson	*V. Solonis, Engelmann*
V. vulpina, Linnaeus	

In the same article, Bailey also listed another Munson creation introduced that year, the Riverside pecan.

13. *Foundations of American Grape Culture*, page 226.

14. Letter from Munson to Van Deman dated July 25, 1888, and quoted in Stephens, *op. cit.*, page 9. The author was unable to find the original of this letter in the USDA collection at National Archives II.

15. "What Shall My Profession Be?" Murray's Print (House), Denison, Texas, no date, pages 4–5.

16. *Ibid.*, page 6.

17. *Ibid.*, pages 12–13.

18. *Ibid*, page 16.

19. Theophilus Philosophius (a.k.a. T. V. Munson), *The New Revelation*. Press of B. C. Murray, Denison, Texas, no date, page 22.

20. Neva taught in the Denison school system from 1902 until 1920; she was also well known in the Texas art community and had several exhibits of her work. Viala taught music privately, was supervisor of music classes in Denison's public schools, and sang professionally. Fern worked for several decades at the Denison Post Office after her husband died at a young age, leaving her with two sons to support. Marguerite attended Kentucky State University for two years; Olita graduated from high school but did not continue her schooling.

21. T. V. Munson, "History of Parker Earle Strawberry," *Texas Farm and Ranch*, March 12, 1892, page 11. Reprinted from the *Indiana Farmer*.

22. For some reason, he withdrew Jaeger shortly after its introduction and issued a new grape the following year which he named Hermann Jaeger.

23. Will, Warder, and Fern also received acreage outside the addition around the turn of the century.

24. One descendant recalled that the only black people allowed on the grounds of Vinita Home were washerwomen.

25. Family genealogist Joyce Maki believes that Theo was in Company I of the 36th Virginia Infantry (the Second Kanawha); but there are so many Joseph and J. Munsons on the Confederate rolls that it's hard to be sure.

26. *Denison Sunday Gazetteer*, June 16, 1889, page 4.

Van Deman also planned to issue a special report on the results of the western trip, but it apparently was never published.

27. Before leaving the last week of June, Volney prudently made out his will. He never changed it, and this was the same document probated on his death in 1913.

The microscope was German-made. Housed in a mahogany box, most of the brass parts are still wrapped in the original paper and show little sign of wear.

Volney later described C. L. Hopkins as "spicy enough not to be monotonous and dull, and with notions of his own, without which I could not have long held him in cherished memory." Hopkins, originally from Florida, was forced to resign from the Department in 1892 after just five years of service because of ill health. (Letter from Munson to Van Deman dated November 11, 1889. N. A. II, RG 54, Gen. Corr./Pom., Box 5, Folder 4 of 4, October-December 1889.)

Letter of May 14, 1889, from B. Galloway to Munson re a microscope he's using that he thinks Volney would like, too, since he's interested: "If it were for Government use I could of course obtain it free of duty. . . . But as I have bought considerable from the firm I have no doubt that they will give me all the reduction possible."

28. Nellie allowed the letter to be reprinted in the *Denison Sunday Gazetteer*, September 22, 1889, page 2.

29. Letter from Munson to Van Deman dated October 12, 1889. (N. A. II, RG 54, Gen. Corr./Pom., Box 5, Folder 3 of 4, October-December 1889.)

Van Deman later had to write the hotel in St. Louis and ask for government property Volney had left there, "some pails and other camping outfit in an old broken trunk that he had abandoned on account of its poor condition. ... I am especially anxious," Van Deman continued, "to secure a copy of a French work on grapes [probably Viala's report on his 1887 mission] that it is possible that he left with these things. It is his own personal property, but he had to use it during his trip ... and I am especially anxious to secure and return it to him." (Letter from Van Deman to the proprietor of Hurst's Hotel dated November 11, 1889. N. A. II, RG 54, Gen. Corr./Pom., Box 5, Folder 4 of 4, October-December 1889.)

30. Letter from Munson to Van Deman dated July 24, 1888. Quoted in Stephens, *op.cit.* This letter is now missing from the USDA collection in National Archives II.

31. Letter from Munson to Van Deman dated October 28, 1889. (N. A. II, RG 54, Gen. Corr./Pom., Box 5, Folder 4 of 4, October-December 1889.)

32. Letter from Munson to Van Deman dated August 20, 1890. (N. A. II, RG 54, Gen. Corr./Pom., Box 7, Folder 2 of 4, August-December 1890.)

At least two of Volney's grandsons, Volney and Marcus Acheson (Fern's sons), suffered from macular degeneration. Among its causes is excessive exposure to sunlight.

Chapter 9: "A Valuable Treatise"

1. Letter from Munson to Van Deman dated October 2, 1887. (National Archives II, Record Group 54, General Correspondence/Pomology, Box 2, Folder 1 of 2, July-December 1887.)

2. Letter from Munson to Van Deman dated August 9, 1890. (N. A. II, RG 54, Gen. Corr./Pom., Box 7, Folder 2 of 4, August-December 1890.)

3. Not everyone agreed with Volney's conclusions on these questions. The Kansas State Agricultural College Experiment Station, for example, planted twenty of his "interesting seedlings" in its 1890 and 1891 test vineyards. In *Bulletin No. 44* (December 1893), the editors noted that "Mr. Munson, after extensive research, believes he has traced them all [cultivated varieties of *Vitis Aestivalis, Michaux* or the Fox Grape] to an early importation from Europe into the South, and erects [creates] the species Bourquiniana to accommodate them." However, the editors disagreed, remarking austerely that "as there is nothing known in Europe, either wild or cultivated, from which these could have been derived, they are retained in the oestivalis [sic] class in these records for the present." (page 117)

4. Letter from Van Deman to Munson dated December ?, 1889. (N. A. II, RG 54, Gen. Corr./Pom., Box 5, Folder 4 of 4, October-December 1889.)

5. Letter from Munson to Van Deman dated December 9, 1889. (N.A. II, RG 54, Gen. Corr./Pom., Box 5, Folder 4 of 4, October-December 1889.)

6. Letter from Munson to Van Deman dated October 2, 1887. (N.A. II, RG 54, Gen. Corr./Pom., Box 2, Folder 1 of 2, July-December 1887.)

Pierre Marie Alexis Millardet (1838–1902) was a doctor who turned to botany after serving as a surgeon in the Franco-Prussian War. He taught at Nancy several years before moving to the university at Bordeaux in 1876, where he remained until his 1899 retirement. He co-discovered downy mildew in 1878 along with Jules-Emile Planchon and studied both it and the phylloxera louse. Millardet was an advocate of using American grape stock and is credited with discovering "Bordeaux mixture."

7. George Washington Campbell (1817–1898) lived in Delaware, Ohio, and originated or disseminated several famous varieties of grapes including the Delaware and Campbell's Early.

Munson elaborated on this in an undated letter to Hermann Jaeger: "If I was unable to spend but a short time with you[,] yet I shall try to make up in writing at leisure moments, trying not to consume time uselessly. I do read your letters with great pleasure. There is one advantage in writing, we can take notes of little discoveries along [the way], and of thought, and queries, and embody their all in a letter. ... Do not be timid in asking any questions which maybe I can aid in answering. Thus we can throw together our knowledge and be mutually benefited [sic]." (Jaeger cache, transcript in authors' collection.)

8. Letter from Munson to Van Deman dated October 12, 1889. (N.A. II, RG 54, Gen. Corr./Pom., Box 5, Folder 3 of 4, October-December 1889.)

9. Letter from Munson to Van Deman dated September 23, 1887. (N.A. II, RG 54, Gen. Corr./Pom., Box 1, Folder 2 of 2, July-December 1887.)

10. Letter from Hopkins to Munson dated August 15, 1890. (N.A. II, RG 54, Gen. Corr./Pom., Box 7, Folder 2 of 4, August-December 1890.)

11. Letter from Munson to Van Deman dated August 20, 1890. (N.A. II, RG 54, Gen. Corr./Pom., Box 7, Folder 2 of 4, August-December 1890.)

The others he felt had "taken credit" for his discoveries were Pierre Viala and Sereno Watson; the latter was completing a major book on American flora begun by Asa Gray. Volney had shared his new classification with them, hoping for constructive criticism.

12. *Texas Agricultural Experiment Station Bulletin No. 56*, page 226.

13. Despite Van Deman's pessimism about the probable fate of the second publication, Volney continued to send him detailed instructions on its layout. He was especially concerned about the plates depicting the wood portions of the vines at various stages since that component was critical for identification of the species when leaves and fruits were not available.

14. Letter from Van Deman to Munson dated September 23, 1890. (N. A. II, RG 54, Gen. Corr./Pom., Box 7, Folder 1 of 4, August-December 1890.)

15. William Munson's will left only $500 to each of his three sons since he had earlier given them land in Illinois. All the personal property except money and stocks went to Trite and Jennie, and everything else, including real estate, was divided among the three daughters, including Louisa in Nebraska.

16. Letter from Munson to W. A. Taylor dated July 1, 1893. (N.A. II, RG 54, Gen. Corr./Pom., Box 15, Folder 2 of 4, July-October 1893.)

17. Van Deman had lost his position when the administration changed in Washington.

18. Letter from Taylor to Munson dated July 14, 1893. (N.A. II, RG 54, Gen. Corr./Pom., Box 15, Folder 2 of 4, July-October 1893.)

Secretary of Agriculture Sterling Morton (whom, coincidentally, Munson knew from his Nebraska days) asked Van Deman to resign in June 1893, ostensibly because he would not cut his divisional budget (see footnote 17). Van Deman retired to Parksley, Virginia, but continued to be active in horticultural circles.

19. Letter from Munson to Taylor dated September 13, 1893. (N.A. II, RG 54, Gen. Corr./Pom., Box 15, Folder 3 of 4, July-October 1893.)

20. Letter from T. V. Munson to J. W. Rose dated March 29, 1899. (Smithsonian Institution Archives, Record Unit 221: Division of Plants, 1886–1928: Box 16.)

21. "Laying the Foundations of American Grape Culture," *The National Nurseryman*, June 1909, page 174.

USDA records contend that the manuscript was returned to Volney to use for *Foundations of American Grape Culture.*

Chapter 10: The World Goes to Chicago

1. Liberty Hyde Bailey, *Annals of Horticulture in North America for the Year 1893*, page 33.

2. At the time, according to Bailey, there were ten major national horticultural societies, making it too expensive for anyone to belong to more than a few.

The *Southern Horticultural Journal* of December 15, 1889, reported on a development at the last meeting of the American Horticultural Society. Volney was chair of its committee on societies and organizations, and he "made an extended report giving the views of many prominent men of the country on the advisability of uniting the six or seven National Horticultural Societies into one society covering the ground." (page 4). (The *SHJ* had recently moved from Tyler, Texas to Denison and changed its name from the *Texas Journal of Horticulture*. Volney was listed as one of the paper's special contributors.)

3. Letter from Munson to Van Deman dated November 15 (?), 1890. (National Archives II, Record Group 54, General Correspondence/Pomology, Box 7, Folder 2 of 4, August-December 1890.)

4. *Report of the Texas State Horticultural Society, for the Year 1891,* page 85.

Volney also gave a paper at the meeting on "Grapes in North Texas in 1891" in which he reported on "my vineyards of some eight acres, containing altogether standard and newly introduced varieties, *with hundreds of my own production,* aggregating probably 1,000 bearing varieties, representing every species and mixture." (Authors' italics)

In addition, he was drawn once more into a controversy concerning whole root versus piece root stocks, about which he had written on several occasions. The discussion debated whether it was best to propagate apples and pears from stock four to eight inches long (whole root) or something shorter. According to Volney, fraudulent agents were traveling the country decrying nurseries that did not use the whole-root system, all in a swindle to sell unsuspecting customers less valuable trees. When pressed for his opinion at the Lampasas meeting, Volney stated that "the longer your cutting, the further you get from Nature ... use the smallest piece."

This seemingly simple question had subjected Volney to much criticism, especially from Dr. W. W. Stell of Paris, Texas, a longtime opponent of his. However, Volney could take comfort from Luther Burbank's support of his position. "I am greatly pleased," wrote the great California horticulturist, "to observe the stand you take on the whole root fraud. ... I do not care how small a piece of root is used, as the result depends wholly on the quality not the size of root." (Letter from Burbank to Munson dated March 30, 1891. Copy in collection of the T. V. Munson Viticulture and Enology Center.)

5. *Denison Sunday Gazetteer,* October 9, 1892, page 3.

6. *Texas Farm and Ranch,* December 3, 1892, page 10.

7. One such piece of publicity was entitled "A Day with Munson" and appeared in *Texas Farm and Ranch.* The article related a visit by the editor of *Orchard and Garden* to Munson's nursery to discuss with him pruning and trimming grapes. Volney, the author wrote, was "giving to southern horticulture a new race of grapes, which in time will supercede [*sic*] older varieties." Along with Onderdonk and Kerr, he continued, Volney was helping to provide "a cheerful view of the coming renaissance of Southern horticulture."

8. Letter from Blue Mound Nurseries in Kansas to Munson dated November 29, 1888, and inquiring about land in Milam City, Texas. (W. B. Munson Archives, Box 1, Letters 1886–1891.)

9. Munson Archives, Box 6, Letters, 1899, W.

10. Munson Archives, Box 10, Letters January-June 1904, K.

11. To be fair to Stell, one must note that the editor of *Garden and Forest* also thought Volney should "reduce the price on some of his best known varieties ... which would put them in reach of hundreds who do not feel they can afford to pay [the] present price. We believe thousands would be planted at a popular price." ("A Day with Munson," *Texas Farm and Ranch,* n.d.)

12. Samuel Wood Geiser, *Horticulture and Horticulturists in Early Texas,* page 81.

13. T. V. Munson, "History of the Nursery Business in Texas," *The National Nurseryman,* September 1905, page 157.

14. Volney could be very scornful of *labrusca,* which he referred to in print as "coarse" and a "miserable Skunk Grape." He thought it "a pity that so much effort has been wasted on it." (*Garden and Forest,* Volume III, page 475.)

15. The Agriculture Department's annual pomology report for 1891 mentioned yet another plant originated and released commercially by Volney that year, an apple he named Bradford's Best.

16. Letter from Munson to Van Deman dated November 15, 1890, *op. cit.*

17. Trite and Jennie were in Chicago at the time and visited with their brother. In an incomplete letter dated March 18, 1893, one of the sisters wrote back to Denison that "his is the first exhibit put in place so far and received commendatory press notices in the papers here." (Munson Archives, Box 4.) Fern, then sixteen, also traveled with her father to help set up the display.

Coincidentally, brother Ben Munson had been appointed one of seven commissioners from Grayson County to the fair.

18. The *Bulletin* of London's Kew Gardens had a much less complimentary description than appeared in fair literature. The Horticultural Building, declared the editors, was "altogether an unsuitable structure in which to grow plants, some of the latter showing unmistakeable signs of distress very soon after being placed in it." (*Bulletin for 1894,* page 55.)

19. *Texas Agricultural Experiment Station Bulletin No. 56,* page 226.

The Prestele drawings are not mentioned in any press coverage of the exhibit, so that idea apparently failed to materialize.

The photographs were made by J. Swartz of Denison. Mounted on cardboard, they remain in excellent condition today. Forty-two of the original fifty-five have survived; they depict:

Vitis aestivalis, Michaux	*V. arizonica, Engelmann*
V. Baileyana, Munson	*V. Berlandieri* (2 views)
V. bicolor, LeConte	*V. california, Bentham*
G. W. Campbell	*V. candicans, Engelmann*
V. champini, Planchon	*V. cinerea*
V. cordifolia, Lamarch, Michaux	*V. coriacea*
V. Doaniana, Munson	*V. Girdiana, Munson*

V. labrusca, Linnaeus (2 views)
V. Lincecumii: America, Bailey, Dr. Collier, Extra, Dr. Hexamer, Husmann,
   Hermann Jaeger, Lincy, Laussel, Mrs. Munson, and
   Ozark (There were separate plates of each.)
V. Lincecumii, Buckley
V. Lincecumii, var. Glauca, Munson (2 views)

V. monticola, Buckley	*V. Munsoniana, Simpson*
Onderdonk	*Opal*
Rommel	*V. rotundifolia, Michaux*
V. rubra, Michaux	*V. rupestris, Scheele*
V. simpsonii	*V. solonis, Souays*

V. vulpina, Linnaeus

20. The certificate's artwork featured an improbable mix of classical figures, "Red Indians," and American allegories. In one scenario, for example, an elegant Roman lady rested against a buffalo. And across the bottom of the document, Christopher Columbus appeared in a canoe of eastern American Indian design, its sides embellished with classical motifs.

21. Reprinted in the *Denison Sunday Gazetteer,* April 2, 1893, page 2, and headlined as "A Stinging Letter from our Fellow Citizen T. V. Munson."

22. In later nursery catalogs, Munson even sold the "juice of the vine ... fresh grape juice ... one of the most appetizing, wholesome foods and remedial preparations known." He prepared it "by a special process"; the juice was available in either pint or quart bottles which had been "hermetically sealed."

Chapter 11: The Waning Century

1. *Texas Agricultural Experiment Station Bulletin No. 56,* page 237.

2. Will was also active in military organizations. At the age of seventeen he enrolled in the Denison Zouave Volunteer Guards and was later captain of the Denison Rifles Volunteer Guard (Texas State Militia). He served with the Guards, a sort of late-nineteenth-century National Guard, until 1898. This state militia had been organized in 1879 and provided its 2,000 to 3,000 members with an extra, though small, income.

3. Obituary of Will Munson, *Sherman Daily Democrat,* September 14, 1931, page 1.

4. Letter from M. Ginevra Munson to J. T. Munson dated September 11, 1895 from Oakland, California. (W. B. Munson Archives: Box 4.)

5. According to *The New Handbook of Texas,* "in April 1898, Congress allowed soldiers in existing organized militia units to volunteer for federal service. Under this law, state troops formed the First Texas Volunteer Infantry Regiment, which sailed to Havana in late 1898." (Volume 4, page 729)

6. Some family members believe that Will, a quiet and reserved man, never really "had his heart" in the nursery, though his love for plants was obvious, and, shortly after the turn of the twentieth century, he started his own business as a florist. One cannot help thinking that, as matters worked themselves out, Will had the drudge job and Volney the more prestigious and exciting one.

The circumstance of the nursery's name change to Munson & Son in 1895—just as Will was forced to leave college—also has the air of a concession, though that may be pure coincidence.

Volney's passion was for the creation of these new varieties, but he felt that further development and refinement of them, as well as their broad dissemination, belonged to others. "I am breeding for stocks ... I am only an introducer," he wrote the editor of *Pacific Tree and Vine* in 1896. "[I] cannot go into extensive growing of stock vines. Your people [California vine growers] will soon get a start and then can grow their own supplies." Perhaps he felt there wasn't enough profit potential in it for the nursery, or perhaps the day-to-day production of his precious vines seemed mundane and mechanical when compared to the excitement of the hybridization "chase."

7. Photocopy of original speech in possession of Joyce Maki.

8. Munson referred to the monopolies he perceived as choking American economy, some of which Teddy Roosevelt would soon hunt down as president.

9. Letter from M. Ginevra Munson to J. T. Munson dated August 22, 1895, from Oakland, California. (Munson Archives: Box 4, Letters 1895)

10. Jennie usually handled the correspondence, at least that which has survived, and, consequently, the demands for money. In the summer of 1895, she felt the funds weren't coming fast enough and petulantly wrote Theo, "Can't you manage the estate business so that it will not be necessary to go to the expense of *you three boys* having to sign release papers every time a lot is sold?" (Authors' italics) (Letter from M. Ginevra Munson to J. T. Munson dated August 22, 1895, from Oakland, California. Munson Archives: Box 4, Letters 1895, M-Z)

11. Letter from Louisa Douglas to J. T. Munson dated November 2, 1895. (Munson Archives: Box 4, Letters 1895, A-L)

12. The authors were unable to locate coverage of the speech, which was entitled "The Superiority of Natural over Supernatural Moral Codes," in the *Dallas Morning News*. The *Denison Gazetteer*, however, reported that it was "well attended and received." (October 25, 1896)

13. Indeed, it could justifiably be said that Munson and Jaeger between them put *Lincecumii* on the viticultural map. U. P. Hedrick wrote that both "considered it one of the most, if not the most, promising form from which to secure cultivated varieties for the Southwest." (*The Grapes of New York*, page 142) And nurseryman J. W. Stubenranch of Mexia, Texas, wrote the *Rural New Yorker* his opinion: "It was a lucky hit on his [Volney's] part to strike upon the *Vitis Lincecumii* race of grapes for a basis of his new productions ... there can be no question as to their adaptability to our section of country and climate." (*Texas Farm and Ranch*, August 13, 1898.)

14. Volney's ego was well known, apparently, to his friends but tolerated. In 1898, John S. Kerr of Sherman Commercial Nurseries wrote E. W. Kirkpatrick, another nurseryman from McKinney, Texas, regarding his most recent "spicy article" in *Texas Farm and Ranch*. The issue containing Kirkpatrick's piece is missing but it appears that he wrote a tart letter concerning the frequency with which Munson was publicized and heard from in that newspaper. "I want to say you are mighty hard," wrote Kerr humorously, "on our weekly correspondent of O & Garden [the Orchard and Garden section] & the Editor thereof [Dr. A. M. Ragland of Pilot Point, a friend of Volney's] for giving us so much <u>interesting</u> Munsonian reading matter. Yes it has been 'rammed down' pretty regularly & we have become used to it, but had you thought what a shock it must produce upon the nerves of <u>such modest people</u>. Truly Consistency is a jewel." Both men would serve as pallbearers at Volney's funeral, and Kerr would write a moving eulogy on his longtime friend and colleague. (Letter from John S. Kerr to E. W. Kirkpatrick, January 11, 1898. Kirkpatrick Collection, Collin County (Texas) Historical Society.)

15. Bush & Son & Meissner, *Illustrated Descriptive Catalogue of American Grape Vines*, 1895, page 159.

16. *Ibid.*, page 160.

17. Letter from T. V. Munson to J. W. Rose dated March 29, 1899. (Smithsonian Institution Archives, Record Unit 221: Division of Plants, 1886–1928: Box 16.)

18. Parras is located in the state of Coahuila in northeastern Mexico and is the center of one of that country's major grape-producing areas. "*Parras*," in fact, is Spanish for vines, a tribute to the profusion of wild grapes found growing there. The first wine produced in the New World was made

in Parras from those vines sometime around 1560, and the first true "Mission" grapes were developed there in the seventeenth century when friars crossed Spanish grapes with the native vines. From thence, they were carried to Texas, California, and other parts of the Spanish Southwest. Volney Munson reported after his trip that he had also found native *Vitis cinerea* in abundance there.

Parras was the home of Mexico's first winery, La Hacienda de Santa Maria de las Parras, established in 1593. Four years later, Don Lorenzo Garcia received a land grant from King Philip II and founded Casa Madero Winery, which remained for centuries the oldest winery in the Americas in continuous operation; unfortunately, however, it closed in 1989.

19. "Inauguration du Monument Planchon a Montpellier," *Le Progres Agricole et Viticole*, Volume 22, 1894, page 621.

20. There is another statue on the grounds of the agricultural school which also commemorates American help in fighting phylloxera. It depicts a young woman—America—holding up an old, infirm one—France. Both, *naturellement*, are nude. The statue is dedicated to Gustave Foex, one of the early presidents of the school. However, one book, *Vineyards in Bloom*, which discusses the Missouri wine industry, declares that statue to have been erected in honor of Jaeger, Husmann, Munson, and others. There are also statues attributed to Munson in Bordeaux and Aix en Provence.

21. The statue is in Planchon Square, just across from the train station, in Montpellier.

22. *American Gardening*, November 4, 1899, page 750.

Munson referred to E. S. Rogers, creator of the Rogers hybrids such as Goethe; E. W. Bull, disseminator of the Concord grape; and his friend Hermann Jaeger.

23. The agricultural experiment stations in the U.S. were created in 1887 under the Hatch Act, which provided federal funds to land grant colleges such as Texas A&M.

Munson's *Bulletin No. 56* was not the first in the series to deal with grapes; two other writers provided discourses on black rot (No. 23, 1892) and the grape itself (No. 48, 1898). This series of bulletins began in 1888.

24. *Investigation and Improvement of American grapes at the Munson experimental grounds near Denison, Texas, from 1870 to 1900*, page 218.

25. *Ibid.*, page 224.

26. George W. Campbell. Reprint of his paper on "The Agency of Crossing and Hybridizing in the Improvement of the American Grape," published in *The National Nurseryman*, July 1896, page 77.

27. *The Grape Culturist*, page 4.

28. U. P. Hedrick, *Grapes and Wines from Home Vineyards* (1945), page 151.

29. Volney wrote: "Allow me to thank you most heartily for a copy of your unique presentation, *New Creations in Fruits and Flowers*. It is a rare, rich feast, and I congratulate you upon your marvelous accomplishments." (Quoted in *A Gardener Touched with Genius*, page 152.)

30. U. P. Hedrick noted that the original modification of the Kniffin plan was devised by a Long Island grower but was "afterwards improved and brought into prominence" by Munson. (*Manual of American Grape-Growing*, page 136.)

31. R. H. Price and H. Ness, *Texas Agricultural Experiment Station Bulletin No. 48: The Grape*, page 172.

32. Charles L. Sullivan, "Two Frenchmen in America," *The Wayward Tendrils Newsletter*, March 1992, page 2.

33. Vincent P. Carosso, *The California Wine Industry, 1830–1895: The Formative Years*, page 119.

34. Geiser, *op. cit.*, pages 65–66.

Peach varieties listed by Geiser were Alton, Bonanza, Crimson Beauty, Dulcey, Family Favorite, Munson's Free, Red River, Rupley, Shipler, Success, and Superb. Munson Cling is listed in a 1910 nursery price list. Volney also originated Minnie and introduced it in 1886; he had stopped selling it by 1912. At the time of his death, he was planning to introduce the Ben Munson (named for his nephew), which he found growing in his brother Ben's Denison yard.

Plums were Bestovall, Chicrigland, Clifford, Minco, Nimon, Topsy Turvy, and Ward's October Red. Drs. D. V. Fisher and W. H. Upshall, in their 1976 *History of Fruit Growing and Handling in*

United States of America and Canada, 1860–1972, wrote that Volney's "Munson variety" was "an excellent plum, but because of its yellow color, soft texture and susceptibility to brown rot, it never attained widespread importance." (page 130)

Persimmon varieties were American Honey and Kawakami. American Honey, Volney wrote the USDA in 1906, was descended from a flat-fruited persimmon tree sent him by Judge Sam Miller of Bluffton, Missouri, in 1883 or 1884. Volney grafted it on a native Texas persimmon and named the resulting tree American Honey because it originated in this country, rather than Japan. William A. Taylor, pomologist in charge of field investigations, declared it "the best persimmon in flavor that I have yet tasted." Since it eventually proved to be similar to, if not the same as, another persimmon, the Department suggested changing the name to Munson. But this Volney declined emphatically, saying it should be Miller if it was changed at all. In fact, the name Josephine was found to have precedence.

About 1893 Volney selected one from a group of Josephine seedlings and planted it in his persimmon orchard; he believed the other parent to be Among. He later propagated some 200 additional seedlings by grafting onto native persimmons. In 1902 he named it Kawakami in honor "of one of our highly esteemed customers in Japan," nurseryman Zembei Kawakami.

35. Volney had been doing this type of work for the USDA for several years. In 1888, for example, he was sent a Blackman plum for examination. His letter to H. E. Van Deman described its flowering habits and anatomy, as he had observed them, then offered this conclusion: "this Blackman bud seems a sort of compromise between the single peach bloom and half dozen plum blooms but <u>wanting the petals and stamens</u>. ... Nothing could show more unmistakably the hybridity of plum and peach than this mulish Blackman." (Letter from Munson to Van Deman dated June 6, 1888, National Archives II, Record Group 54, General Correspondence/Pomology, Box 2, Folder 2 of 5, January-August 1888.)

36. An 1893 report, for example, lists the following:

—from Gonzales County: Big Acorn, Big Nutmeg, Crown Prince, Madam, Maj. Harwood, Riverside, San Marcos, Smith, Wild

—from Grayson County: Grayson, Red River

—from San Saba: Competitor, Gem, Jumbo, Longfellow, Prize, Superb.

(List in NA II, RG 54, Gen. Corr./Pom., Box 14, Folder 3 of 4, October 1892-June 1893.)

37. Letter from Munson to Van Deman dated December 20, 1889. (NA II, RG 54, Gen. Corr./Pom., Box 5, Folder 4 of 4, October-December 1889.)

38. *Texas Fruits, Nuts, Berries and Flowers,* December 1907, page 13.

Chapter 12: Meet Me in St. Louis

1. Unfortunately, archivists at the New York Botanical Garden no longer have a record of the collection and are unaware of any such photographs still existing.

Nellie accompanied Volney on this trip, and they traveled to New York on the Mallory Steamship Line vessel *San Marcos.* Once there, the couple stayed at the Murray Hill Hotel at the corner of 41st Street and 4th Avenue. During the conference, they both joined in a pleasure excursion up the Hudson River which was arranged for the attendees; the party visited several houses, including Hyde Park, the home of F. W. Vanderbilt. The Munsons may also have crossed over into Canada so that Volney could research grapes in that area.

2. The paper was published by the Horticultural Society of New York in Volume I of its *Memoirs.*

3. *Proceedings of the Society for Horticultural Science,* 1905, page 13.

4. At the time, Volney apparently was not a full member. He joined in 1910 and served as vice-president in 1911.

5. *Proceedings of the Society for Horticultural Science,* 1905, 1908, pages 21 and 24.

6. Undated partial manuscripts in the Gilbert Onderdonk Collection at the University of North Texas, Box 2, Folder E.

7. Peter Dreyer, *A Gardener Touched with Genius: The Life of Luther Burbank,* page 164.

8. Archy died in Oklahoma at a young age, leaving Fern with two sons to raise alone. They returned to Vinita Home for two years, and she took a job with the Denison Post Office, retiring in 1941.

9. Calvert worked for the Houston & Texas Central Railroad. After the wedding, the couple moved to Sherman.

10. In the midst of this family crisis, in 1903, Theo had a serious accident when he was hit by a train while trying to cross the railroad tracks in his horse-drawn wagon. He was unconscious for many days.

11. Letter from Tryphena Munson to W. B. Munson, Sr. dated May 25, 1900. (Munson Archives: Box 7, Letters 1900: M.)

12. Letter from Tryphena and Ginevra Munson to Mr. Handy, dated April 3, 1902 (but apparently 1904). (Munson Archives: Box 11, Letters March-November 1904, L.)

13. Daughter Viala's initials, for example, appear on several typed letters in 1905, identifying her as the typist. By 1902, Will had also opened his floral shop, "Munson's," at 1315 S. Myrick, where one could purchase floral designs, cut flowers, and potted plants.

14. *A Twentieth Century History and Biographical Record of North and West Texas*, Volume I, page 606.

15. There is some evidence that Headlight received the Wilder Award from the American Pomological Society in 1903 as a champion grape. (Several letters from the authors to the current Society for additional information went unanswered.) The "Index to General Letters Received, 1891–1908" of the Records of the Bureau of Plant Industry, Soils, and Agricultural Engineering (in National Archives II) lists a letter from Will on July 13; the very brief description of the contents indicates that the award was the subject of the letter. However, the actual letter, like dozens of other Munson letters in this collection, is missing.

16. George C. Husmann wrote Volney early in 1907 of his problems in obtaining good stock for that station, noting that he was giving cuttings and plants "from inexperienced parties nursery treatment before attempting to plant them in the vineyard. This will, of course, not be necessary with vines received from you." (Letter from Husmann to Munson, dated February 12, 1907. Reproduced in Renfro's *The Correspondence of T. V. Munson with the United States Department of Agriculture*.)

17. *Denison Sunday Gazetteer*, June 26, 1904, page 1.

18. H. A. Ivy, *Rum on the Run in Texas: A Brief History of Prohibition in the Lone Star State*, page 57.

19. Nor did he change his mind on this subject. Five years later, in *Foundations of American Grape Culture*, Volney discussed his feelings on liquor laws. "It is a fact that the interdiction cuts off from the most of our country one great industry and source of wealth, that the country might otherwise enjoy to as great [an] extent as even France." (page 239)

Naturally, Volney was not the only viticulturist to extol the fruits of the grape over that of the grain. Isidor Bush of Missouri addressed "The Temperance Question" in his 1883 and 1895 catalogs. "Wine itself," he wrote, "is an apostle of temperance. ... Let wine and beer drinking be prohibited, and the use of opium, the secret tippling of strong drinks, the increase of vice and intemperance, would be the consequence." (Quoted in "Isidor Bush and the Bushberg Vineyards of Jefferson County," page 50.)

20. Letter from Don A. Bliss to J. T. Munson, dated July 7, 1904. (W. B. Munson Archives: Box 11, Letters March-November 1904, B.)

21. Don Bliss felt that Theo's absence in Minnesota on a fishing trip was a critical factor in the loss: "I was sorry you went away. With your assistance we could have elected him with ease." (Letter from Bliss to J. T. Munson, dated July 12, 1904. Munson Archives: Box 11, Letters March-November 1904, B.)

22. Louisiana Purchase Exposition Company Collection, Series III, Subseries VII, Folder 2: Report of the Departments of Agriculture and Horticulture, page 402. (Missouri Historical Society.)

23. Sam H. Dixon, *Texas Fruits at the World's Fair, 1904*, page 61.

Jurors were compensated for their travel costs and paid $7.00 per day.

24. The Texas exhibit in the Palace of Horticulture comprised 3,500 square feet of floor space between Virginia and Michigan and was touted as the only state exhibit where visitors were treated to a glass of cool water. Volney donated arbor vitaes to decorate the booth, and Will added a "fine collection of ferns."

As with the Columbian Exposition, the Texas Legislature once again refused to appropriate any funds for a state building. The Texas Pavilion, in the form of the five-pointed "Lone Star," was built with private monies: the only non-state funded building at the fair. There were also Texas exhibits in six "palaces," or sections, including Horticulture. Texas Week began September 12 and drew an estimated 50,000 Texans, many of whom took advantage of the special railroad excursions arranged by the commissioners.

25. There's some confusion as to whether there were six or seven Texas gold medal winners; the authors can find only six, though there are references to one more. In addition to Volney, they included two other entries from Grayson County: John S. Kerr of Sherman, an old friend of Volney's, and a countywide exhibit entered by the World's Fair Fruit Committee of Sherman. Also taking home gold for their fruits were S. D. (S. N.?) Thompson of Bowie and C. W. Wood of Swan in Smith County. Sam Dixon, secretary of the Texas State Horticultural Society, was awarded a gold for his service to horticulture.

Altogether, Texas fruits claimed 238 awards at the St. Louis Fair, a number exceeded only by California, New York, and Missouri.

26. Leyendecker Family Papers, Box 3R57, Folder 2.

27. *Denison Sunday Gazetteer* carried a reprint on July 12, 1908.

28. Mendel's work was not appreciated or understood at the time it was first published. Several decades passed before new developments in botany and biology led to the rediscovery of his revolutionary work.

And while Volney acquainted himself quickly with the "new" material, others did not. Luther Burbank, for example, when interviewed by scientists, not only knew nothing of the rediscovery but was equally ignorant of other recent scientific advances. (*Luther Burbank: Gardener to the World.*)

29. Texas Agricultural Experiment Station Historical Files, Folder 1–7, page 3. (Texas A&M University Archives.)

30. Quotations in the text below are from photocopies of these letters, courtesy of the Arnold Arboretum.

Charles Sprague Sargent (1841–1927) traveled through Europe for several years before returning to the United States to continue his botanical studies. He was professor of horticulture at Harvard from 1872 to 1873 and director of the university's botanical garden from 1872 to 1879. He served as director of the new Arnold Arboretum from 1873 until his death in 1927. In the late 1870s and early 1880s, he compiled a forestry census of the United States and later was one of the advocates for establishing Glacier National Park.

31. Letter from Sargent to Munson, dated August 17, 1909: "I am going to try and get down to see you myself this autumn as Texas seems to be a most interesting and still unexplored botanical field."

32. Sargent visited on at least these occasions: March 1909; October 1909; March 1910; and March 1911.

Letter from Sargent to Munson, dated May 6, 1910: "We have no fruit or fruiting specimens of any of the Texas Plums and I am very anxious to get as many of them as possible. Do not hesitate to make as many journeys to these as is necessary, or rather as you feel like making. We are interested in this investigation and I do not think there is any more useful piece of work that the Arboretum can undertake, so far as the money goes you need not give the matter a thought."

33. This would have been a trip of approximately 600 miles, most likely by rail.

Chapter 13: Munson's Magical Flying Car

1. Volney was not the only famous horticulturist to create a stir with his beliefs. In 1926, taking the controversy several steps further, Luther Burbank declared himself an infidel. (According to Webster, an infidel is one who is not a Christian or who opposes Christianity, an unbeliever with respect to a particular religion, or one who acknowledges no religious belief.) Burbank's "remarks were seized upon by the national press, and enraged an America already torn between Biblical fundamentalists and Darwinian scientists." (*Luther Burbank: Gardener to the World.*)

2. Letter from John Leet to W. B. Munson, dated February 1, 1913. (W. B. Munson Archives: Letters, L.)

3. Ahura-Mazda was the supreme god of the late Persian Empire. Mithraism was a more ancient Persian cult which became popular once again under the Romans; it shared many commonalities with early Christianity.

4. Born in Greece, Lafcadio Hearn (1850–1904) was educated in England. He immigrated to the United States in 1869 and settled for a time in New Orleans, where he began writing editorials, essays, and translations. His first original work was *Chita: a Memory of Last Island,* which told of a horrific tidal wave destroying an island town in the Mississippi River. Hearn spent several years in Martinique writing and eventually (1890) moved to Japan, where he taught and wrote twelve books about that country and its folklore.

5. Volney no doubt enjoyed a book written in 1911 by Elbert Hubbard, published by Handmade Books, and still in the family's possession. It was a new version of the Bible which included The American Bible and The Gospels According to Lincoln, Jefferson, Etc. There are ample notations by Volney scattered throughout, one of which declared that Hubbard's work was an "improvement over the Bible of 2,000 years ago." (Courtesy of Joyce A. Maki.)

6. That paper was entitled "Resistance to Cold, Heat, Wet, Drought, Soils, Insects, Fungi in Grapes." It was published in Volume II of the organization's *Memoirs* in 1910.

Volney had used the phrase "foundations of American grape culture" in letters and published articles for many years.

7. "Laying the Foundations of American Grape Culture," *The National Nurseryman,* June 1909, page 174.

8. Volney wrote G. W. Brackett, USDA pomologist, on September 17 that he had been unable to attend the American Pomological Society meeting that year because he was just getting his book out and couldn't get away from Denison throughout the summer or fall. (Quoted in Renfro, *The Correspondence of T. V. Munson with the United States Department of Agriculture.*)

9. *Science,* November 19, 1909, page 714.

10. Both *Foundations of American Grape Culture* and *Ten Million Acres,* a biography of Volney's brother Ben Munson, have been recently reprinted (2001) and are available through the T. V. Munson Viticulture and Enology Center at Grayson College.

11. The drawings bear the notation "Invented Oct. 1911."

12. Sikorsky would invent the first single rotor helicopter in 1939.

13. Isaac Newton's law, that every action has an equal and opposite reaction, means that the fuselage, or car, of the helicopter would want to whirl in the same direction as the rotor blades. The problem was to overcome that.

14. Quotations are from Munson's patent application; photocopy courtesy of Mr. and Mrs. John Maki.

Chapter 14: "A Great and Good Man"

1. Viala studied in New York City (1919) and, in the summer of 1923, traveled to Europe to pursue her musical studies in Germany, Switzerland, and Vienna.

2. Roy Miller, "The Texas Industrial Congress," *The Texas Magazine,* October 1910, page 11.

3. *Denison Sunday Gazetteer,* February 25, 1912, page 4.

4. Letter from Ginevra Munson to Ben or Theo Munson, dated January 20, 1913. (W. B. Munson Archives.)

5. Warder was then working at a Houston bank; he arrived at 11:00 that night. Fern made it home from Hugo, Oklahoma, just shortly before Volney died. Marguerite, Volney and Nellie's youngest child, was a sophomore that year at Kentucky State University (as it was then known); she arrived by train the following day. The rest of the family was at Volney's bedside.

6. There is a recurring family story that Volney died in St. Vincent's Hospital in Sherman, but all the official documentation notes his death at Vinita Home.

7. Photocopy of handwritten notes (probably those of Fern Munson Acheson) in the collection of the T. V. Munson Viticulture and Enology Center.

8. *Denison Herald,* January 23, 1913, page 1.

Ben and Theo paid for the funeral out of love and their "appreciation of his whole life," but Nellie insisted on reimbursing them for the cost of the family's expenses. (Letter from E. S. Munson to Munson & Bro., dated March 17, 1913, Munson Archives.)

9. The XXI Club was a ladies' study club which Theo had helped organize and of which he was a benefactor. In 1896 they built a handsome brick structure—the first women's clubhouse in the state of Texas—with funds he gave them. It was located at the corner of Scullin Avenue and Gandy Street, next door to Fern's house on Scullin.

10. Pallbearers were: Dr. Alexander Acheson, a founder of Denison and uncle of Fern's husband Archy; Dr. Elaunus R. Birch; Godwin L. Blackford, president of State National Bank of Denison; Samuel J. Boldrick, owner of Boldrick & Swan Shoes; James Boyd, owner of a men's clothing store; Patrick J. Brenham, cashier of National Bank of Denison; Nathaniel Decker, attorney; Walter S. Faires, businessman; A. M. Ferguson; John R. Handy, insurance agent and owner of fuel and feed store; Howard Hanna, druggist; Frank B. Hughes, superintendent of Denison public schools; A. D. Jackson; John S. Kerr, Sherman nurseryman; E. W. Kirkpatrick, Collin County nurseryman; John B. McDougall, president of Denison Bank & Trust Co.; Brudett M. Murray, owner of Murray Printing House; A. N. Rhamey, Denison nurseryman; Erwin J. Smith, attorney; Charles E. Stephens, nurseryman; and Isaac Yeidel, owner of a stationery shop.

11. Unless otherwise noted, all tributes are taken from the family scrapbooks in the collection of the Munson Viticulture Center.

12. *Proceedings of the Thirty-third Biennial Session of the American Pomological Society,* 1913, page 296.

13. All of Volney's original probate documents are missing from the Grayson County Clerk's Office. The condensed material presented here was found in Probate Minute Book 29, pages 481–484.

14. Many of these are now on long-term loan to the Munson Viticulture Center.

15. Despite Volney's insistence that he wanted "no cold stone" on his grave, the family later placed a granite column entwined with grapevines on it. A long slab of granite connects it to the stone rose urn under which Nellie was later buried. Once decorated with fresh-cut roses from her garden, the urn is now filled with stone roses.

16. Some family members disliked Minnie Secoy Munson. She was a Christian Scientist, and they believe she kept Will from taking the medication that would have prolonged his life; he was only fifty-seven at the time of his death.

17. Minnie added India to the list of places growing Munson hybrids after she shipped a large order to the Punjab Agricultural College at Lyallpur, for the government experiment station there.

18. In 1930, Texas A&M College had opened its Winter Garden Research Center in Crystal City, southwest of San Antonio. Station manager Ernest Mortenson, who had corresponded with Will and Nellie for years, secured a large collection of Volney's grapes for the facility, seeming to ensure that they would survive. Tragically, Winter Garden closed in 1961 and the vineyard was bulldozed.

19. The 1940–41 Wolfe Nursery catalog listed the following varieties as "Wolfe's Munson Grapes." However, those in italics were not Volney's creations: Albania, America, Armalaga, *Bachman's Early,* Bailey, Beacon, Bell, *Black Spanish, Brighton,* Brilliant, *Campbell's Early,* Carman, *Catawba,* Champanel, Cloeta, *Concord, Edna,* Extra, Fern Munson, *Fredonia,* Headlight, *Herbemont,* Last Rose, Lomanto, Longfellow, Manito, Mericadel, Minnie, *Mission, Moore's Early,* Muench, *Niagara, Portland,* President, and R. W. Munson.

Chapter 15: A New Beginning

1. Pierce's Disease is an insect-transmitted bacterium which prevents the plant from absorbing water and nutrients.

2. Dr. Roy E. Renfro, co-author of this book, is one of the founding members of the Texas Grape Growers Association. Now known as the Texas Wine and Grape Growers Association and

based in Grapevine, the group allied in 2002 with the Wine Society of Texas (founded 1996). The new organization is the only one in the United States which joins grape and wine producers with those who enjoy and appreciate the results. Increased promotion of Texas wines is the goal.

3. Texas had such public lands because it joined the United States in 1845 as a separate nation—the Republic of Texas—after belonging for centuries to either Spain or Mexico.

4. Eloise Munson (1889–1969) was president and general manager of Munson Realty (1937–1955), president of the Denison Cotton Mill (1936–1955), and a director of Citizens National Bank (1936–1969), all businesses founded by her father. Like him and like her Uncle Theo, she was an important benefactor to the Denison community, though she often insisted on anonymity. "Miss Eloise" was also an avid sportswoman and competed at the national level in golf.

Since its formation, the W. B. Munson Foundation has funded many Texas projects and is especially supportive of Grayson County College and the Munson Viticulture Center. Its directors instigated this biography and underwrote the research expenses for it as part of the Center's 25th anniversary celebration.

5. "The Wine Rack," *Denison Herald,* August 7, 1974.

6. Grayson County College opened its main facility, the East Campus, in 1965. When nearby Perrin Air Force Base closed in 1971, the college acquired 325 acres and more than twenty buildings there, too, and operates it as the West Campus.

7. Brochure on "The Thomas Volney Munson Memorial Vineyard and Viticulture-Enology Center."

8. Foster's father had also worked there, and he stayed on for a few years after Volney's death before starting his own grape business. Horace recalled his employer in a later interview: "I remember T. V. Munson as a tall, lanky man. He always had a sparkle in his eye. He was very active and would think nothing of walking miles. He had a wonderful personality, too. . . . I remember a motto plaque he had hanging in his office. It went 'Nothing great is ever achieved without enthusiasm.'" ("Horticulturist Recalls Working for Munson," an interview with John Clift for "The Wine Rack," *Denison Herald,* May 26, 1974.)

9. Galet's classification of *vitis* is the one currently in use today, and it is based in part on Volney Munson's research.

10. Volney considered grapevines as useful for landscaping purposes as for edible ones. In the final chapter of *Foundations of American Grape Culture,* he discussed this. "If there be some stiff-looking, ungraceful trees about the yard, they can, in a few years, be made graceful and charming by training up their bodies among the branches, grapevines planted near their roots . . . [these varieties] have such climbing powers that they can ascend to the tops of tall trees, and then throw out pendant arms that . . . give a charm of which the Babylonian willow may well be jealous. . . . Every farm having such objects about it will marry the boys and girls perpetually to farm life, resolved to have such charming features in their own homes. . . . The humblest cottager and the millionaire may engage with pleasure and success in producing handsome clusters, luscious berries and comforting shade." (pages 242–243)

11. Renfro received another award from the Texas Wine and Grape Growers in 2001 when he was given the John E. Crosby, Jr. Award for lifetime, meritorious service to the industry. He also served as vice-chair and, later, chairman (1995–1996) of the Wine Grape Growers of America.

Bibliography

I. Archival Collections

Collin County Historical Society: E. W. Kirkpatrick collection. (McKinney, Texas.)

Dallas Historical Society (on long-term loan to the T. V. Munson Viticulture and Enology Center): T. V. Munson collection. (Dallas, Texas.)

Ecole Nationale Superieure Agronomique: viticultural library and archives. (Montpellier, France.)

Grayson County College: T. V. Munson Viticulture & Enology Center: T. V. Munson collection, including Munson Memorial Vineyard reports. (Denison, Texas.)

Harvard University: Arnold Arboretum: Records of the Director, Charles Sprague Sargent, 1893–1927.

Kentucky Historical Society collections. (Frankfort, Kentucky.)

Lexington Cemetery archives. (Lexington, Kentucky.)

Lexington Public Library: Kentucky Room collection. (Lexington, Kentucky.)

Library of Congress: (1) Frank Lamson-Scribner collection (2) Luther Burbank collection. (Washington, D.C.)

Missouri Botanical Garden: (1) George Engelmann collection (2) William Trelease collection: Director's Office: Series 1: Administration. (St. Louis, Missouri.)

Missouri Historical Society: (1) Hermann Jaeger collection (2) Special Collections: Louisiana Purchase Exposition (3) Meissner Family Papers. (St. Louis, Missouri.)

National Agricultural Library: Prestele collection. (Beltsville, Maryland.)

National Archives II: United States Department of Agriculture collection (Record Group 54): (1) General Correspondence: Pomology (2) Correspondence: Bureau of Plant Industry, Soils, and Agricultural Engineering. (College Park, Maryland.)

Nebraska State Historical Society: photograph collection. (Lincoln, Nebraska.)

Neosho Public Library: Hermann Jaeger collection. (Neosho, Missouri.)

New Orleans Public Library: Louisiana Division. (New Orleans, Louisiana.)

Newton County Historical Society: Hermann Jaeger collection. (Neosho, Missouri.)

Panhandle-Plains Historical Museum: W. B. Munson Archives. (Canyon, Texas.)

Smithsonian Institution Archives. (1) Record Units 220 and 221: United States National Museum: Division of Plants, Records (2) Frederick Vernon Coville Papers. (Washington, D. C.)

Texas A&M University: Records of the Texas Agricultural Experiment Station: Historical files. (College Station, Texas.)

Texas A&M University at Commerce: University Archives: Skip Steely collection, Ed McCuistion Scrapbook. (Commerce, Texas.)

Texas State Library: Vertical files. (Austin, Texas.)

University of Kentucky: Special collections. (Lexington, Kentucky.)

University of Nebraska at Lincoln: University Archives.

University of North Texas: Special Collections: Gilbert Onderdonk Collection. (Denton, Texas.)

University of Texas at Austin: Center for American History: (1) Leyendecker Family Papers (2) T. V. Munson vertical file (3) Works Progress Administration: Historical Records Survey: Bell County, Texas (4) Texas newspaper collection. (Austin, Texas.)

Williams Research Center: Special collections. (New Orleans, Louisiana.)

II. Articles

Ackerman, Diane. "America's First Gardener." *Parade Magazine,* July 15, 2001.

The American Entomologist and Botanist. 1870 and 1880. Various articles.

"The American Horticultural Union." *Southern Horticultural Journal,* Volume II, No. 18, December 15, 1889.

Appleby, J. Gavin and Pierquet, Patrick. "T. V. Munson: American Grape Hybridizer." *American Wine Society Journal,* 1979.

Asher, Gerald. "The Return of the Native: Missouri's Vintage Grape." *Gourmet,* April 1993.

Atkins, Ken. "The Munson Grapes: The Grapes That Lived." *Texas Garden.* Volume 11, No. 1, November-December 1982.

Bailey, Liberty Hyde:
- "The Species of Grapes Peculiar to North America." In *Gentes Herbarum: Occasional Papers on the Kinds of Plants.* 1933–1935. Volume III, No. 4.
- "Hybridisation [*sic*] in the United States." *Journal of the Royal Horticultural Society.* Summer 1900, pages 209–213.

Banks, Suzy. "Vine Times in Texas." *Texas Journey,* September-October 2000.

Ciesla, Thomas and Regina. "The Texas 3 Step: dancing through a history of Texas viticulture and wine: Part II." Wineskinny.com: December 1999-January 2000.

Cirani, Richard. "Guide to the Selection of Phylloxera Resistant Rootstocks." Phylloxera and Grape Industry Board of South Australia, 1999.

Crane, Eva. " 'A Draught of Vintage:' A History of the Texas Wine Industry," *Neil Sperry's Gardens,* January 1992.

Dickson, Gordon. "Texans Turn Up Their Noses at State Wines, Study Finds." *Fort Worth Star-Telegram,* September 2, 2001.

"France Will Honor Early Texas Pioneer: Phylloxera Fighter Munson Finally Recognized." *Vinifera Wine Growers Journal,* Spring 1988.

Gaudette, Merton. "The Little World That Almost Was." *The Reader: San Diego's Weekly,* Volume 7, No. 14. April 20, 1978.

Geiser, Samuel Wood:
- "Thomas Volney Munson." *Dictionary of American Biography,* Volume 7.
- "The First Texas Academy of Science (1880–1887)." Reprinted from *Field and Laboratory,* Volume XIII, No. 1, January 1945.

Giordano, Frank, Jr. "Texans' Venture in the Vineyards." *Texas Highways,* January 1984.

Jacobs, Julius L. "The Saga of T. V. Munson, 19th Century Texas grape breeder of 300 crosses." *Wines and Vines,* September 1975.

Jenkins, H. M. "Notes on Market-Gardening and Vine-Culture in the Northwest of France." *The Journal of the Royal Agricultural Society of England,* 2nd Series, Volume 16. London: John Murray, 1880.

Kerr, John S. "The Life and Labors of the Late T. V. Munson." Texas Department of Agriculture *Bulletin No. 33,* September-October 1913.

Ladwig, Tom. "Herman Jaeger's Legacy." *Missouri Life,* December 1982.

"The Legion of Honor." *Temple Bar: A London Magazine for Town and Country Readers.* London: Richard Bentley & Son., Volume 89, May–August 1890.

Le Progres Agricole et Viticole. Montpellier: Imprimerie Grollier Fils:
- "Mission de M. Viala en Amerique." 5e. Annee, No. 3, 15 Janvier, 1888.

- "Sur les resultats de la mission de M. P. Viala en Amerique." 5e. Annee, No. 3, 15 Janvier 1888.
- "Decorations pour services rendus a l'Agriculture." Tome XI, 1889.
- "Trois Hybrides de Munson." Translated from an article in the *Rural New Yorker*. Tome XVI, 1891.
- "Inauguration du monument Planchon a Montpellier." 11e. Annee, No. 50, 16 Decembre 1894.

Lowry, Jack. "Roots in Texas." *Texas Highways,* July 1990.

MacEachern, Dr. George Ray:
- "About the Vine: A Texas Grape and Wine History." Part I. *Texas Horticulturist,* Volume 23, No. 6, November 1996.
- "About the Vine: Researching Grapes in Mexico." *Texas Horticulturist,* Volume 24, No. 4, September 1997.
- "About the Vine: Texas Grape Rootstock." *Texas Horticulturist,* Volume 24, No. 5, October 1997.

McClurg, Avery. "A Day in Munson County [*sic*]." *The Texas Horticulturist,* Volume 4, No. 2, Summer 1977.

Miller, Roy. "The Texas Industrial Congress." *The Texas Magazine,* Volume 2, No. 6, October 1910.

Muehl, Siegmar:
- "The Wild Missouri Grape and Nineteenth-Century Viticulture." *Missouri Historical Review,* July 1997.
- "Isidor Bush and the Bushberg Vineyards of Jefferson County." *Missouri Historical Society Review,* Volume XCIV, No. 1, October 1999.

"Munson a Hero, but Not the 'Savior.'" *Vinifera Wine Growers Journal: Correspondence,* Spring 1988.

"Note on a Centenary." *Baltimore Sun,* September 27, 1943.

Onderdonk, Gilbert. "Pomological Possibilities of Texas." Texas Department of Agriculture *Bulletin No. 9.* Austin, Texas: Von Boeckmann-Jones Company, September-October 1909.

Roddy, Ann. "T. V. Munson, Who Found 'Grape Paradise' in Denison, Recalled as World's Chief Vineyard Expert on Centennial of Birth." *Denison (Texas) Herald,* October 4, 1943.

Rodriguez, Rebecca. "Bearing Fruit." *Fort Worth (Texas) Star-Telegram,* September 16, 2001.

Schenck, Anita A. "San Diego's Debt to the Point Loma Community." *Listener of San Diego,* June 1969.

Stein, Dr. Larry A. "What's New With Grapes." *Texas Gardener,* Volume XX, No. 6, September-October 2001.

"The Story of a Great Pest." *London Society,* October 1879.

Sullivan, Charles L. "California's Early Response to the Dreaded Phylloxera Root Louse." *Vinifera Wine Growers Journal,* Spring 1988.

Summers, Floyd G. "Charles V. Riley, Benefactor of Agriculture." *Missouri Historical Review,* July 1925.

Tarara, Julie M., and Edward W. Hellman. "The Munson Grapes—A Rich Germplasm Legacy." *Fruit Varieties Journal,* Volume 44, No. 3, 1990.

"The Telegraph." *De Bow's Review: A Monthly Journal of Commerce, Agriculture, Manufactures, Internal Improvements, Statistics, etc. etc.* Volume. XV, No. 11, August 1853.

"Texas Mustang Grape Once Saved the Great Vineyards of Europe." *San Antonio (Texas) Express,* May 28, 1933.

Texas Wine Marketing Research Institute. "Texas Wine and Wine Grape Industry." Lubbock: Texas Tech University, n.d. (circa 1997.)

"Thomas Volney Munson." *The Texas Horticulturist,* Volume 1, No. 1, February 1913.

"Two Frenchmen in America." *The Wayward Tendrils Newsletter.* Volume 2, No. 1, March 1992.

Unzelman, Gail. "The Bushberg Catalogue." *The Wayward Tendrils Newsletter,* Volume 5, No. 2, April 1995.

"A Visit to the New Orleans Exposition." *Demorest's Monthly Magazine,* Volume XXI, No. 5, March 1885.

Walker, R. R. "Thomas Volney Munson—The Burbank of the Vineyards." *East Texas,* November 1928.

Ward-McLemore, Ethel. "The Academies of Science of Texas, 1880–1987." *The Texas Journal of Science,* Volume 41, No. 3, August 1989, supplement.

West, Richard. "Root de France." *American Way,* September 15, 1988.

III. Government Documents

Bowman, J. B. Legislative Document No. 11. *Report of the Agricultural and Mechanical College of Kentucky.* Frankfort: S.I.M. Major, Public Printer, 1869.

Fayette County, Kentucky: tax lists, deeds.

Grayson County, Texas: marriages, deeds, wills, probate records.

Lancaster County, Nebraska: deeds, assessment records.

Leslie, Gov. P. H. Legislative Document No. 17. *Report Concerning the Agricultural and Mechanical College of Kentucky.* Frankfort: S.I.M. Major, Public Printer, 1872.

Smithsonian Institution: American Museum of Natural History: Herbarium collections. (Washington, D.C.)

United States: Census Records:
- Illinois, 1820–1860.
- Kentucky, 1810–1860, including Slave Schedules.
- Nebraska, 1870.
- Texas, 1880.

United States Department of Agriculture:
- *Report upon Statistics of Grape Culture and Wine Production in the United States for 1880.* Special Report No. 36. 1881. Washington, D.C.: Government Printing Office.
- *Report of the Commissioner of Agriculture* for the years 1886–1888. Washington, D.C.: Government Printing Office.
- *Report on the Experiments Made in 1888 in the Treatment of the Downy Mildew and Black Rot of the Grape Vine.* Botanical Division, Bullein No. 10: Section of Vegetable Pathology. Washington, D. C.: Government Printing Office, 1889.
- *Report of the Secretary of Agriculture* for the years 1889–1893. Washington, D.C.: Government Printing Office.
- *Yearbook of the United States Department of Agriculture* for the years 1900, 1903, and 1937. Washington. D.C.: Government Printing Office.
- *American Grape Varieties.* (I. W. Dix and J. R. Magness, authors.) Circular No. 437. September 1937.
- *Mexico's Grape Industry: Table Grapes, Raisins, and Wine.* Foreign Agricultural Service, 1979.

United States: Patent Office. "Improvement in Scuffle-Hoes." Patent No. 213,584. 1879.

IV. Interviews/Correspondence

Acheson, Marcus. (Palo Alto, California.)

Bye, Beatrice. (Bibliotheque Agro, Ecole Nationale Superieure Agronomique. Montpellier, France.)

Coates, Claire. (Bureau Nationale Interprofessionel du Cognac. Cognac, France.)

Geraci, Dr. Thomas A. (Las Vegas, Nevada.)

Grunwald, Thomas A. (Sayre School. Lexington, Kentucky.)

Hively, Kay. (Neosho, Missouri.)

Maki, Joyce Acheson and John Allen. (Houston, Texas.)

McEachern, Dr. George Ray. (Texas A&M University.)

Shrader, Dorothy Heckmann. (Hermann, Missouri.)

Steely, Skipper. (Paris, Texas.)

Stevens, Linda Waler. (Hermann, Missouri.)

V. Manuscripts

Hardy, Donald Clive. "The World's Industrial and Cotton Centennial Exposition." M.A. thesis, Tulane University, 1964.

Jacob, K. Clive. "Nebraska in the Seventies." Unpublished manuscript, Nebraska State Historical Society.

Letters, T. V. Munson to Hermann Jaeger. Transcribed by Kay Hively of Neosho, Missouri. Courtesy of Mr. and Mrs. John Brock.

Maki, Joyce Acheson: T. V. Munson collection. (Houston, Texas.)

Maki, Joyce Acheson, and John Allen Maki. "The Genealogy and History of the Ancestors and Descendants of Thomas Volney Munson, B.S., M.S., D.Sc." Unpublished manuscript, 1988.

McGrew, J. R. "The W. H. Prestele paintings of the American species of grapes." Unpublished manuscript, National Agricultural Research Library.

Morgan, Melanie. "The History of Pilot Point, Texas." Unpublished manuscript, Pilot Point (Texas) Public Library, 1975.

Olmo, H. P. "Some Contributions of Thomas Volney Munson." Speech presented at the dedication of T. V. Munson Viticulture and Enology Center, September 10, 1988.

Roth, Bryan Gregory. "Rapid Graft Propagation of Dormant Native Texas Grapevines." M.S. thesis, Texas A&M University, 1992.

Stephens, Carlene E. "Thomas Volney Munson: A Case Study in the Professionalization of American Horticulture, 1875–1913." M.A. thesis, University of Delaware, 1976.

VI. Newspapers, Magazines, Etc.

City Directories:
• Denison, Texas.
• Lexington, Kentucky.
• Lincoln, Nebraska.
Daily Picayune. (New Orleans, Louisiana.)
Daily State Journal. (Lincoln, Nebraska.)
Dallas Morning News. (Dallas, Texas.)
Denison Herald. (Denison, Texas.)
Denison Sunday Gazetteer. (Denison, Texas.)
Kew Gardens: Bulletin. (London, England.)
Lexington Daily Press. (Lexington, Kentucky.)
Lexington Herald. (Lexington, Kentucky.)
Lexington Leader. (Lexington, Kentucky.)
The National Nurseryman magazine.
New York Times. (New York, New York.)
Palestine Daily Herald. (Palestine, Texas.)
The Rural New Yorker magazine.
Sherman Daily Register. (Sherman, Texas.)
Sherman Democrat. (Sherman, Texas.)
Texas Farm and Ranch. (Dallas, Texas.)

VII. Published Books, Reports, Etc.

Adams, Leon D. *The Wines of America.* 2nd edition. New York: McGraw-Hill Book Company, 1978.

Allen, H. Warner. *The Romance of Wine.* New York: Dover Publications, Inc., 1932, reprinted 1971.

Bailey, L(iberty) H(yde):
• *Annals of Horticulture in North America for the Year 1889.* New York: Rural Publishing Company, 1890.

- *Annals of Horticulture in North America for the Year 1893.* New York: Orange Judd Company, 1894.
- *Sketch of the Evolution of Our Native Fruits.* New York: The Macmillan Company, 1898.
- *The Survival of the Unlike: A Collection of Evolution Essays Suggested by the Study of Domestic Plants.* New York: The Macmillan Company, 1911.
- *The Standard Cyclopedia of Horticulture.* Volume II. New York: The Macmillan Company, 1925.

Barnhart, John Hendley, compiler. *Biographical Notes Upon Botanists.* Boston: G. K. Hall & Co., 1965.

Barns, Cass G. *The Sod House.* Lincoln: University of Nebraska Press, 1970.

Bennitt, Mark, editor in chief. *History of the Louisiana Purchase Exposition.* St. Louis: Universal Exposition Publishing Company, 1905.

Biographical Cyclopedia of the Commonwealth of Kentucky. Chicago: John M. Gresham Company, 1896.

Briggs, Asa. *Wine for Sale: Victoria Wine and the Liquor Trade, 1860–1984.* Chicago: University of Chicago Press, 1985.

Bush & Son & Meissner. *Illustrated Descriptive Catalogue of American Grape Vines: A Grape Growers' Manual.* 4th edition. St. Louis: R. P. Studley & Co., 1895.

Carosso, Vincent P. *The California Wine Industry, 1830–1895: A Study of the Formative Years.* Berkeley: University of California Press, 1951.

Catalog of H. B. Bryant's (Bryant & Stratton) Chicago Business College, 1873. Chicago Historical Society, Chicago, Illinois.

Catalogue of the Officers and Students of Kentucky University for the years 1865–1871. Transylvania University, Lexington, Kentucky.

Charter and Organization of the Texas State Horticultural and Pomological Association, Chartered September 17, 1875. Houston: W. M. Hamilton, 1876.

Clark, Helen Hollandsworth, editor. *A History of Fulton County, Illinois in Spoon River Country, 1818–1968.* Astoria: Stevens Publishing Company, 1969.

Coats, Alice M. *The Plant Hunters: Being a History of the Horticultural Pioneers, Their Quests, and their Discoveries.* New York: McGraw-Hill Book Company, 1969.

Coleman, J. Winston, Jr. *A Centennial History of Sayre School, 1854–1954.* Lexington, Kentucky: Winburn Press, 1954.

Copple, Neale. *Tower on the Plains: Lincoln's Centennial History, 1859–1959.* Lincoln Centennial Commission Publishers, n.d.

Cox, Jeff. *From Vines to Wines.* Storey Books, 1999.

Crawford, Robert Platt. *These Fifty Years: A History of the College of Agriculture of the University of Nebraska.* University of Nebraska College of Agriculture, 1925.

Creigh, Dorothy Weyer. *Nebraska: A Centennial History.* New York: W. W. Norton & Company, Inc., 1977.

Dick, Everett. *Conquering the Great American Desert.* Nebraska State Historical Society, 1975.

Diggs, George M., Jr., Barney L. Lipscomb, and Robert J. O'Kennon. *Shinners & Mahler's Illustrated Flora of North Central Texas.* Fort Worth, Texas: Botanical Research Institute of Texas, 1999, 2nd printing, 2000.

Dixon, Sam H. *Texas Fruits at the World's Fair, 1904.* Houston, Texas: Texas State Horticultural Society, 1905.

Dreyer, Peter. *A Gardener Touched with Genius: The Life of Luther Burbank.* Revised edition. Berkeley: University of California Press, 1985.

English, Sarah Jane. *The Wines of Texas: A Guide and a History.* Austin, Texas: Eakin Press, 1986.

Fairall, Herbert S. *The World's Industrial and Cotton Centennial Exposition, New Orleans, 1884–1885.* Iowa City, Iowa: 1885.

Federal Writers' Project of the Works Progress Administration for the State of Kentucky. *A Guide to the Blue Grass State.* New York: Hastings House, 1939, revised 1954.

Fisher, Dr. D. V., and Dr. W. H. Upshall. *History of Fruit Growing and Handling in United States of America and Canada, 1860–1972.* Kelowna, British Columbia: Regatta City Press, 1976.

Foex, M. Gustave and M. Pierre Viala. *Ampelographie Americaine.* Montpellier: E. Isard, 1885.

Francis, David R. *The Universal Exposition of 1904.* St. Louis: Louisiana Purchase Exposition Company, 1913.

Fraser, Samuel. *American Fruits: Their Propagation, Cultivation, Harvesting and Distribution.* New York: Orange Judd Publishing Company, 1931.

Frolik, Elvin F., and Ralston J. Graham. *The University of Nebraska-Lincoln College of Agriculture: The First Century.* Board of Regents of the University of Nebraska, 1987.

Fuller, Andrew S. *The Grape Culturist: A Treatise on the Cultivation of the Native Grape.* New York: Orange Judd Company, 1899.

Fuller, Robert. *Religion and Wine: A Cultural History of Wine Drinking in the United States.* Knoxville: University of Tennessee Press, 1996.

Galet, Pierre. *A Practical Ampelography: Grapevine Identification.* Translated and adapted by Lucie T. Morton. Ithaca: Comstock Publishing Associates, 1979.

Geiser, Samuel Wood. *Horticulture and Horticulturists in Early Texas.* University Press in Dallas: Southern Methodist University, 1945.

Geiser, S. W., and Claude C. Albritton, Jr. *Field and Laboratory.* Volume XXVII, No. 3, July 1959. Dallas: Southern Methodist University Press.

Gibson, Melissa R. *Index of Lexington Deaths, 1894–1907.* Lexington (Ky.) Public Library, 1997.

Giordano, Frank. *Texas Wines and Wineries.* Austin, Texas: Texas Monthly Press, 1984.

Hardy, D. Clive. *The World's Industrial and Cotton Centennial Exposition.* The Historic New Orleans Collection, 1978.

Hayes, A. B., and Sam D. Cox. *History of the City of Lincoln, Nebraska.* Lincoln: State Journal Company, 1889. Reprinted 1995.

Hedrick, U(lysses). P.:
- *The Grapes of New York.* Albany: J. B. Lyon Company, 1908.
- *Manual of American Grape-Growing.* New York: Macmillan Company, 1919, reprinted 1924.
- *Cyclopedia of Hardy Fruits.* 2nd edition. New York: Macmillan Company, 1938.
- *Grapes and Wines from Home Vineyards.* London: Oxford University Press, 1945.
- *A History of Horticulture in America to 1860.* New York: Oxford University Press, 1950.

Heintz, William F. *Wine Country: An [sic] History of Napa Valley: The Early Years, 1838–1920.* Santa Barbara: Capra Press, 1990.

Hyde, William, and Howard L. Conrad. *Encyclopedia of the History of St. Louis.* New York: Southern History Company, 1899.

Illinois Guide & Gazetteer. Chicago: Rand McNally & Company, 1969.

Ivy, H. A. *Rum on the Run in Texas: A Brief History of Prohibition in the Lone Star State.* Dallas, Texas: Temperance Publishing Co., 1910.

Janick, Jules, and James N. Moore. *Fruit Breeding: Volume II: Vine and Small Fruits.* New York: John Wiley & Sons, Inc., 1996.

Jeffs, Julian. *The Wines of Europe.* New York: Taplinger Publishing Company, 1971.

Jensen, Richard J. *Illinois: A Bicentennial History.* New York: W. W. Norton & Company, Inc., 1978.

Johnson, Frank W. *A History of Texas and Texans.* Volume V. Chicago: The American Historical Society, 1916.

Johnson, Hugh. *Vintage: The Story of Wine.* New York: Simon and Schuster, 1989.

Joseph, Donald. *Ten Million Acres: The Life of William Benjamin Munson.* Denison, Texas: Privately printed, 1946.

Kansas State Agricultural College Experiment Station: Manhattan, Kansas:
- *Bulletin No. 28*, December 1891.
- *Report for 1891.*
- *Bulletin No. 44*, December 1893.
- *Bulletin No. 73*, July 1897.

- *Bulletin No. 110*, May 1902.

Kansas State University. *Grape Growing in Kansas.* 1928.

Kaufman, William I. *Encyclopedia of American Wine.* Los Angeles: Jeremy P. Tarcher, Inc., 1984.

Kerr, Bettie L. *Lexington: A Century in Photographs.* Lexington, Kentucky: Lexington-Fayette County Historic Commission, n.d.

Kleber, John E., editor in chief. *The Kentucky Encyclopedia.* The University Press of Kentucky, 1992.

Lichine, Alexis, in collaboration with Samuel Perkins. *Alexis Lichine's Guide to the Wines and Vineyards of France.* New York: Alfred A. Knopf, 1986.

Lipe, William N., and David Davenport. *Grape Cultivar Performance on the Texas South Plains, 1974–1986.* Texas Agricultural Experiment Station, Texas A&M University.

Lipe, William N., and Robert Eddins. *Grape Cultivar and Rootstock Evaluations for the Texas South Plains, 1975–1976.* Texas Agricultural Experiment Station, Texas A&M University.

Loubere, Leo A.:
- *The Red and the White: A History of Wine in France and Italy in the Nineteenth Century.* Albany: State University of New York Press, 1978.
- *The Wine Revolution in France: The Twentieth Century.* Princeton, New Jersey: Princeton University Press, 1990.

"Luther Burbank: Gardener to the World." Pamphlet, Luther Burbank Museum. City of Santa Rosa (Cal.) Press, 1983.

Lukacs, Paul. *American Vintage: The Rise of American Wine.* Boston: Houghton Mifflin Company, 2000.

MacNeil, Karen. *The Wine Bible.* New York: Workman Publishing, 2001.

Maguire, Jack. *Katy's Baby: The Story of Denison, Texas.* Austin, Texas: Nortex Press, 1991.

Masson, G., editor. *Le Phylloxera: Comites d'Etudes et de Vigilance: Rapports et Documents.* Paris: Janvier 1879.

McKee, James L.:
- *Lincoln: A Photographic History.* Lincoln: Salt Valley Press, 1976.
- *Lincoln: The Prairie Capital.* Woodland Hills, California: Windsor Publications, Inc.

Milward, Burton. *A History of the Lexington Cemetery.* The Lexington (Ky.) Cemetery Company, 1989.

National Cyclopaedia of American Biography. Volume XVIII. (New York: James T. White & Company, 1922.)

Official Catalogue: Cotton States and International Exposition, Atlanta, Ga. 1895. (Atlanta Historical Society.)

Onderdonk, G., et al. *Facts and Figures for Farmers, Fruit-Growers and Florists.* Houston, Texas: (Southern Pacific) Sunset Route, circa 1905.

Paddock, Capt. B. B., editor. *A Twentieth Century History and Biographical Record of North and West Texas.* Volume I. Chicago: The Lewis Publishing Company, 1906.

Peter, Robert, M.D. *History of Fayette County, Kentucky.* Chicago: O. L. Baskin & Co., 1882.

Pinney, Thomas. *A History of Wine in America: From the Beginnings to Prohibition.* Berkeley: University of California Press, 1989.

Premium List of the Department of Horticulture of the World's Exposition, New Orleans, Louisiana, USA. Revised edition. (Tulane University.)

Premium List Offered by the Citizens of Houston, Texas with Programme for the Eighth Annual Meeting and Fair of the Texas State Horticultural Society to be Held at Houston, Texas. Austin, Texas: Eugener Von Boechmann, 1894. (Texas State Archives.)

Price, R. H. *Black Rot of the Grape: Life History, Treatment.* Texas Agricultural Experiment Station Bulletin No. 23. Bryan, Texas: Cox, 1892.

Price, R. H., and H. Ness. *The Grape.* Texas Agricultural Experiment Station Bulletin No. 48. Austin: Ben C. Jones & Co., 1898.

Ranck, George W. *History of Lexington, Kentucky.* Cincinnati: Robert Clarke & Co., 1872. Reprinted 1989.

Rasmussen, Wayne D., editor. *Agriculture in the United States: A Documentary History.* Volume II. New York: Random House, 1975.

Renfro, Dr. Roy E., Jr.:

• *The Thomas Volney Munson Memorial Vineyard and Viticulture-Enology Center.* Denison, Texas: Grayson County College, 1980.

• *The Correspondence of T. V. Munson with the United States Department of Agriculture.* Denison, Texas: Grayson County College, 1981.

• *Viticulture and Enology Degree Program.* Denison, Texas: Grayson County College, 1987.

• *Grayson County College T. V. Munson Viticulture and Enology Center.* Denison, Texas: Grayson County College, 1988.

Report of the Texas State Horticultural Society, for the Year 1891. Brenham, Texas: Banner Steam Book and Job Printing House, 1891. (Texas State Archives.)

Rodgers, Anthony Denny, III:

• *American Botany, 1873–1892: Decades of Transition.* New York: Hafner Publishing Company, 1944, reprinted 1968.

• *Liberty Hyde Bailey: A Story of American Plant Sciences.* Princeton, New Jersey: Princeton University Press, 1949.

Roth, Darlene R., and Andy Ambrose. *Metropolitan Frontiers: A Short History of Atlanta.* Atlanta, Georgia: Longstreet Press, Inc., n.d.

Scheef, Robert F. *Vintage Missouri: A Guide to Missouri Wineries.* St. Louis: Patrice Press, 1991.

Serlis, Harry G. *Wine in America.* New York: Newcomen Society in North America, 1972.

Spearing, Darwin. *Roadside Geology of Texas.* Missoula, Montana: Mountain Press Publishing Company, 1991.

Stanislawski, Dan. *Landscapes of Bacchus: The Vine in Portugal.* Austin, Texas: University of Texas Press, 1970.

Texas and Texans: Special Limited Edition. Chicago: The American Historical Society, 1914.

Texas State Horticultural Society: First Annual Convention and Exhibition. Dallas, Texas: W. A. Shaw & Co., 1887.

Tonkin, Lunley M. *A Daughter of Gardens.* Chicago: Libby Company, n.d.

Unwin, Tim. *Wine and the Vine: An Historical Geography of Viticulture and the Wine Trade.* London: Routledge, 1991.

Van Ravenswaay, Charles. *Drawn from Nature: Joseph Prestele and His Sons.* Washington, D.C.: Smithsonian Institution Press, 1984.

Viala, (M.) Pierre:

• *The Texas State Geological Scientific Association: Bulletin No. 3. The French Viticultural Mission to the United States.* Houston, Texas: A. C. Gray, 1888.

• *Une Mission Viticole en Amerique.* Montpellier: Camille Coulet, 1889.

• *Les Maladies de la Vigne.* Montpellier: Camille Coulet, 1893.

Viala, P(ierre), and V(ictor) Vermorel. *Ampelographie.* Tome VII. Paris: Masson et Cie, Editeurs, 1909.

Visitor's Guide to the World's Industrial and Cotton Centennial Exposition. Louisville, Kentucky: Courier-Journal Job Printing Co., 1884.

Volney, C. F. *The Ruins, or Meditations on the Revolutions of Empires: And the Law of Nature.* New York: Peter Eckler, 1890.

Wagner, Philip M.:

• *Wine Grapes: Their Selection, Cultivation and Enjoyment.* New York: Harcourt, Brace and Company, 1937.

• *A Wine-Grower's Guide.* New York: Alfred A. Knopf, 1945.

• *Grapes Into Wine: The Art of Winemaking in America.* New York: Alfred A. Knopf, 1974, 1976, and 1981.

Warner, Charles K. *The Winegrowers of France and the Government Since 1875.* New York: Columbia University Press, 1960.

Whetzel, Herbert Hice. *An Outline of the History of Phytopathology.* Philadelphia: W. B. Saunders Company, 1918.

Who Was Who in America.

Wright, John D., Jr. *Lexington: Heart of the Bluegrass.* Lexington-Fayette County Historic Commission, 1982.

Zielinski, Quentin Bliss. *Modern Systematic Pomology.* Dubuque, Iowa: Wm. C. Brown Company, n.d.

Index

Abilene, Texas, 122
Academy of Science of Texas, 50
Academy of Sciences, 49, 79
Acheson, Archibald, 146
Acheson, Fern, 146
Acheson, Marcus, 63, 91, 186
Acheson, Volney, 186
Acoma Indian Reservation, 103
Ada grape, 42
Adams Fund grant, 154
agricultural experiment stations, 51, 96, 135, 136, 144, 181, 191; *see also* Texas A&M
Agricultural College of Mississippi, 58
Albania, 202
Albuquerque, New Mexico, 103
Allen, John Fisk, 42
Altus, Arkansas, 199
America grape, 131, 152, 200
American Agriculturist, 137
American Association of Nurserymen and Florists, 52, 114, 124, 137
American Association for the Advancement of Science, 114
American Breeder's Association, 145, 153, 175, 177
American Cyclopaedia, The, 43
American Entomologist, The, 48, 71
American Forestry Association, 49
American Gardening, 135
American Genetics Association, 145
American Horticultural Society, 49, 56, 96, 99, 136
American Journal of Forestry, 52

American Plant and Stock Breeders' Association, 145
American Pomological Society, 56, 57, 101, 114, 154, 162, 175, 179
American Seed-Trade Association, 114
American Theosophical Society, 131
American vitis, 58, 60
American Wine Depot, 48
American Wine Growers Association, 162
Anderson, Dr. John, 203
Anderson, Mrs. John, 203
Annual Report of the Ohio State Horticultural Society, 56
apple varieties, 56, 142, 175
Aransas Pass, 116
Ardmore, Oklahoma, 181
Argentina, 74
arizonica, 54
Arnold Arboretum, 154
Ashland, 7-8, 12, 18, 19
Ashland Institute, 8
Astoria, 6
Augusta, Missouri, 199
Austin, Stephen F., 38
Austin, Texas, 50, 82
Australia, 71, 74, 77, 79, 87
automatic dumping chart, 32-33
Avignon, 79

Bacon, Francis, 63
Bailey grape, 122, 197, 199, 200, 202
Bailey, Liberty Hyde, vii, 53, 57, 58, 114, 118, 122, 123, 136, 141, 144, 165, 175, 186

Balcones Fault, 82
Bateson, William, 177
Batjer, W.F.D., 122
Baussan, A., 133
Bavarian State Institute for Viticulture, 87
Beacon grape, 131, 138, 152, 199
Bear Canyon, 103
Bell grape, 202
Bell, Charles Stewart, 16, 18, 19, 20, 22, 142, 146
Bell, Charlie, 19
Bell County, Texas, 82, 84, 133
Bell, Ellen Scott ("Nellie"), 16-20; *see also* Munson, Nellie
Bell, Eudora, 20
Bell, George, 19, 20, 30
Bell, John M., 179
Bell, Margaret Bunyan Smith, 16, 18, 19, 22
Beltsville, Maryland, 112
Bessey, Charles, 26, 91
Big Bend, 203
Biltmore Herbarium, 148
black rot, 48, 64, 65, 67
Black Hamburg, 42
Blackland Prairies, 82
Blacklands, 40
Blanco County, Texas, 133
Bliss, Don, 149
Blue Lick Battlefield, 16
Blue Mountains Vineyard, 203
Bluffton Wine Company, 48
Bocking, E. A., 111-112
boll weevil, 137
Bonham, Texas, 197
Bordeaux, 64, 65, 80, 81, 87, 191
Bostetter, Mr., 35
botany, development of, 12
Bouquet, Alain, 193

Bowles, M. W., 149
Bradford's Best apple, 142
Bradshaw Mountains, 66
Brass Button, 9
Brilliant grape, 101, 131, 136, 152
Broadway Christian Church, 19
Brown Valley, 104
Bryant & Stratton Business College, 6
Buckley, S. B., 39
Buerri pears, 152
Bulletin of the Texas Department of Agriculture, 111, 135, 153, 179
bunch grapes, 184
Burbank, Luther, vii, 104, 119, 120-121, 132, 136, 139, 142, 144, 146, 177, 186
Burgundy, 68, 73, 90
Burlington & Missouri, 25
Burnet County, Texas, 133
Bush, Isidor, 48, 76-77, 132-133
Bushberg Nursery, 48, 132-133
Bushberg vineyards, 76

California, 71, 75, 77, 87, 139-140, 148, 190
California Wine Industry, 1830–1895, The, 140
Calvert, Col, 183
Calvert, Eleanor, 184
Calvert, N. C., 146
Calvert, Olita Munson, 146, 183, 184, 188, 191, 204
Camp Mabry, 137
Campbell, George W., 108, 132, 136
Canada, 54
Captivator grape, 199, 202
carbon disulphide, 73
Carman, E. S., 119, 136
Carman grape, 119, 122, 132, 146, 184, 194
Carosso, Vincent P., 140
Carson, J. W., 154
Carthage, Missouri, 35
Catawba, 64
Center for American History, 152
champagne, 86, 191, 199
Champanel grape, 199, 200, 202
Charente, 71, 78, 80, 86
Charlottesville, Virginia, 64
Chasselas, 86
Chateau Lafitte, 71

Chateau Pavie, 87
Chevalier du Merite Agricole, 89, 90, 162
Chicago, 114, 123
Chicago Exposition, 109
Chile, 74
chlorosis, 78, 83, 86
Clark, B. L., 9
Classical Revival style, 114
Clay, Henry, 7-8
Cliff Dwellers Gulch, 103
Clift, John, 192-193, 197, 205
Cloeta grape, 200
Cognac, 71, 75, 80, 86, 202
Cognac Research Center, 196
Coleman, Norman J., 65
College Station, Texas, 97
Colobel, 199
Columbian Exposition, *see* World's Columbian Exposition
Columbus, Christopher, 114, 115
"Combined Cultivator and Subsoiler," 32
Commercial House, 25
Committee on Nomenclature, 56
Comstock, Theo B., 164, 170
Concord grape, 24, 42, 47, 64, 200
Congress on Horticulture, 114, 124
"Continuous Stream, Chain force Pump," 20
Cooke County, 192
Cordifolia, 54, 55, 85
Cornell, 96
Cornu, Paul, 167
Corsica, Switzerland, 71
Cortes, 38
Cote d'Or, 71, 75
Cotton Exposition, 55, 56, 99
Cotton States and International Exposition, 136-137
Crooked Creek, 4
Crosby County, Texas, 104
cross-breeding, 86, 122, 123, 131, 145, 162, 175, 199
Culberson, Charles Allen, 136
Cummings, Nina, 146
Cuvee T. V. Munson champagne, 199

da Vinci, Leonardo, 167
Dactylasphaera vitifoliae, 69-70

Dallas Historical Society, 183, 186, 188, 204
Dallas Morning News, 89, 96, 125
Dallas-Fort Worth International Airport, 202
Dallas Opera, 200
Dancy, W. E., 192, 194, 205
Darwin, Charles, 12, 57, 58
Darwin's Variation of Animals and Plants, 57
Davis, Ben, 51
de Vries, Hugo, 177
DeBow's Review, 39
Del Rio, Texas, 52, 103
Deland, —, 35
Delaware, 64
Delicatessen, 199
Dempsey, Jack, 198
Denison Canning Company, 116
Denison City Directroy, 102
Denison/Cognac Sister City program, 196-197, 198
Denison Cotton Mill, 129, 142
Denison Herald, 192
Denison High School, 99, 128, 129
Denison Iron Rolling Mill Company, 116
Denison Light & Power Company, 102
Denison Nursery, 35, 37, 44, 45, 47, 49, 52, 58, 60, 61, 62, 63, 64, 82, 101, 102, 106, 118, 119, 142; *see also* Munson Nursery
Denison Opera House, 171
Denison Post Office, 181
Denison Public Library, 165
Denison Rod and Gun Club, 199
Denison School Board, 148, 178
Denison Sunday Gazetteer, 52, 118, 148
Denison, Texas, 20, 22, 43, 46, 81, 96, 101-102, 109, 116, 129, 143, 148, 184, 196
Denison Wine Renaissance Festival, 196, 198
Descent of Man, The, 58
Devonian formation, 16
"Diamond Scuffler Hoe," 20, 49, 52
Diana grape, 42
Dickinson, Texas, 137
Dixie grape, 199

Doan, Judge J., 66
Doan's Grape, 67
Dog Ridge, 83, 84 200, 202
Domain Cordier, 91
downy mildew, 79
Dr. Collier grape, 200
drought, 71
dry-land farming, 23, 25
Dubois, Wilber, 179

Earle, Parker, 48, 53, 56, 99
East Texas State University, 197
Eastern Cross Timbers, 40
Ecole Nationale Superieure
    Agronomique, 193
Edison, Thomas Alva, 89
Edmondson, P. C., 178
Edward, C. L., 45
Eisenhower, Dwight D., 184, 196
El Paso grapes, 38
El Paso, Texas, 38, 39, 82
Ellen Scott grape, 152, 199,
    200, 202
Ellis County, Texas, 82
Ellis, Dr. John, 178
Engelmann, George, 43, 49, 50,
    54, 105, 132
erosion, 51
Erwin, Anne Clay, 8
Eureka Springs, Arkansas, 192
Exposition Universelle de Paris,
    88, 89, 106, 136
Extra grape, 200

Fairview Cemetery, 127, 178
Fall Creek Winery, 191
Fannin County, Texas, 203
Farmer-Stockman Magazine, 186
Farmers Home Journal, 32
Farmers Institutes, 137
Fayetteville, North Carolina, 64
Fern Munson grape, 199
Fincastle, Tennessee, 118
First National Bank, Denison, 58
Fisher, D. V., 200
Flagstaff, Arizona, 103
fleshy-rooted, 57
Florida, 184
Florists' Exchange, 146, 179
Floyd County, Texas, 104
Foex, Gustave, 73, 76
Fort Davis, 203
Fort Hood, 83
Fort Worth Gazette, 89

41B rootstock, 86
Foster, Horace, 193, 201
Foster, L. L., 137
Foundations of American Grape
    Culture, 52, 95, 113, 135,
    161-165, 175, 179, 186, 191,
    192, 199, 202
France, American vines sent to,
    48, 60, 64, 74-90, 133, 196,
    202, 205
France, viticulture in, 68-90
France, wine industry in, 68, 69,
    75, 79, 86
Fraser, Samuel, 65
Freethinkers Association, 131
French Legion of Honor, 87, 88,
    89, 136, 162, 186
Frost grapes, 40, 57
Fruit Investigation Lab, 190
Fuller, Andrew S., 136, 139
Fulton County, Illinois, 4, 5, 6
fungal diseases, 47, 64

Galet, Pierre, 86, 113, 193, 195,
    197
Galloway, B. T., 135
Galton, Sir Francis, 177
Galveston County, Texas, 119
Galveston, Texas, 156-157, 159
Gard region, 71, 72, 73, 78, 82-
    83
Garden and Forest, 122-123,
    136, 154
Geiser, Samuel Wood, 119, 141
Gentlemen's World's Fair
    Association of Texas, 125
Germany, 71
Geyser Peak Winery, 200
Gill, Richardson, 191
Giordano, Frank, 200
Gird, Mr., 104
Gironde district, 71, 75
Gloire de Montpellier, 139
Goff, Emmett Stull, 136
Golden Chasselas, 42
Gonzales pecan, 143
Gonzales, Texas, 103
Goree strawberry, 142
Government Printing Office,
    109, 111
grafting, 69, 73-77, 78-79, 85-
    86, 87, 139
Grand Rapids, Michigan, 56-57
grape classification, 42-43, 47,
    48-49, 50, 51, 52, 54-56, 57,

58, 65-67, 81, 96, 107-108,
    123, 132-133, 139, 140, 213-
    235
Grape Culturist, The, 48, 136
grape, evolution of, 54, 145
Grape Fest, 202
Grape Growers Association,
    190
Grapes and Wines from Home
    Vineyards, 188
Grapes of New York, The, 148
Grapevine, Texas, 202
grasshoppers, 30, 32, 35, 45
Gray, Asa, 15, 54, 57
Grayson County College, viii,
    20, 90, 193-194, 197, 198
Grayson County College
    Foundation, 203
Grayson County, Texas, 22, 40,
    148-149, 151
Greater Kansas City Cellarmaster
    competition, 199
Greece, 75
Green, E. C., 151
Green, Viala Munson, 162, 171;
    see also Munson, Viala
Green, W. C. "Billy," 171, 183
Grimes, Henry H., 36
Groom, Daryl, 200
Guadalajara, Mexico, 95
Gulbeau, Francis, 76
Gulf of Mexico, 54
Gunter & Munson, 45
Gunter, Jot, 37

hard-wooded, 57
Harrington, Henry, 137
Harshbarger, W. A., 179
Harvard, 49, 96, 105, 123, 154,
    155
Harwood, 55
Hawkins, Anthony, 202
Hayne, Arthur, 139-140
Hays County, Texas, 82
Hays, W. M., 175
Headlight grape, 148, 152, 153,
    202
Hearn, Lafcadio, 160
Hedrick, U. P., 148, 188
helicopter, 167
Herault region, 68, 79, 80
Herbemont grape, 39, 55, 132
Hermann Jaeger grape, 101, 131
High Commission on
    Phylloxera, 74

Hill Country, 82, 143
History of Fruit Growing, 200
History of Wine in America, A, 200
Holland, Frank P., 137
Holland's Magazine, 137
Holley, Mary Austin, 38
Hooker, Sir Joseph, 50
Hopkins, C. L., 103-105, 109, 141
Horticultural Hall, 53
Horticultural Society of New York, 144
horticulture, development of, 12
Houston & Texas Central Railroad, 101
Hungary, 87
Husmann grape, 122, 202
Husmann, George, 42, 47, 50, 56, 77, 90, 122, 139
Husmann, George, Jr., 139, 147
Husmann Nurseries, 48
hybrids, 41-42, 47-48, 80, 83, 86, 132, 133, 135, 148, 151, 152, 162, 175, 184, 186, 189, 190, 192, 193, 194, 197, 199, 199, 203

Imperial College, 133
Independent Literary Society, 30
India, 87
International Commission of Viticulture, 175
International Conference of Hybridizers, 144
International Conference on Plant Hardiness and Acclimatization, 161
International Jury of Awards, 151
International Phylloxera Conference, 80
International Symposium in Viticulture and Enology, 197
Investigation and Improvement of American Grapes at the Munson Experimental Grounds near Denison, Texas, 135
Irvine, Kentucky, 16
Isabella grape, 42, 69
Israel, 87
Italy, 71, 72, 75, 86
Ivanhoe, 203
Ivy, H. A., 149

Jaeger, Hermann, 42, 47, 50, 52, 53, 55, 56-57, 58, 59, 60, 64, 77, 81, 85, 86, 87, 89, 90, 101, 108, 128, 132, 144
Jefferson, Thomas, 149
Johnson County, Texas, 82, 83
Johnson, J. R., 82
Johnson, Mrs. J. R., 99
Johnson, Sidney, 184
Johnson, Mrs. Sidney, 184

Karchner, Mr., 90
Katy Hotel, 172
Kawakami persimmon, 152
Kendall, George Wilkins, 39
Kennedy, John F., 130
Kentucky A&M, 7-8, 12, 15, 16, 18, 19, 52, 97, 128, 157
Kentucky Military Institute, 9
Kentucky River, 16
Kentucky University, 7, 18
Kerr, John S., 179
Kittery, Maine, 3
Klamath Valley, 105
Kniffin, William, 138
Knox, Matthew, 76
Kohout, Joseph, 118
Krause, E. J., 178
Kyle, E. J., 154, 178

L'Ecole Nationale Superieure Agronomique, 73, 76
l'Institut National Agronomique, 60, 80
La Bodega Winery, 202
La Buena Vida Winery, 191
La Moine River, 4
La Reunion (Dallas), 39
La societe de la Charente-Inferieure, 80
labruscas, 42, 55, 69, 78, 85
Lake Belton, 83
Lampasas County, Texas, 66, 82, 103
Lamson-Scribner, Frank, 64, 65, 81, 82, 87, 89, 90
Languedoc, 68
Law, Oscar, 26, 35
Laws, 30
Le Noir, 55
Le Progres Agricole et Viticole, 56, 122
Leet, John Edward, 16, 156
Legate, R. S., 178

Lenoir grape, 38, 139
LeNotre Culinary Institute, 203
Leon River, 83
Lewis County, Kentucky, 7
Lewis, J. H., 103
Lexington Cemetery, 18, 20, 146
Lexington, Kentucky, 7-8, 20
Lexington pear, 142
Leyendecker Collection, 152
lightning conductor, 26, 28
"Lightning Weeder and Cultivator," 20, 34
Lincoln, Mary Todd, 7
Lincoln, Nebraska, 25, 30, 36
Linley, Dr. Thomas, 4
Linley, Isaac, 4
Linley, Joseph, 3, 5
Linley, Louisa, 4
Linley, Maria, 3, 7; *see also* Munson, Maria
Linley, Sibilla Benjamin, 3, 5
Linnaeus, Carl von, 12, 43
Llano Estacado Winery, 190, 191, 202
Loire Valley, 71
Lomanto, 199
Los Angeles, California, 82, 103
Loubere, Leo, 189
Louisiana, 55
Louisiana Purchase Exposition, 145, 149-152
Louisiana Territory, 151
Lubbock, Texas, 104, 203
Lucky grape, 52

Mabry, Woodford, 137
Maki, Joyce Acheson, 181
Mally, Fred W., 137
Manson, John, Jr., 3
Marble Falls, Texas, 82
Marcus, Stanley, 200
Margate, England, 69
Martin grape, 140
Massachusetts Agricultural College, 179
Massachusetts Horticultural Society, 155, 179
Masson, Paul, 75
McCarty's Station, 103
McClenahan, Georges A., 184
McEachern, George Ray, 154, 193, 200
McGinnis, N. M., 147
McPherson, Clinton, 190
meaty-fruited, 55, 57

Meissner, Gustave Edward, 48
Mendel, 153
Messina Hof Winery, 191, 202
Mexico, 87, 133
Michigan State University, 202
mildew, 48, 65
Millardet, A., 55, 108
Miller, Samuel, 46, 50
Mission grape, 38, 140
Mission Valley Nursery, 47
Mississippi River, 54
Mississippi Valley Horticultural Society, 48, 49, 51, 52, 53, 54, 56, 97
Missouri, 76
Missouri Botanical Gardens, 50, 105
Missouri, Kansas & Texas Railroad, 22, 46
Missouri State Fruit Experiment Station, 194, 199-200
Monism, 159
Monson, Richard, 3
Monson, Thomas, 3
Montague County, 190, 193
Montpellier, France, 48, 74, 76, 84, 93, 96, 133
Moore, G. W., 161
Morris, Justin, 199
Mount Pleasant Vineyards, 199
Mount Shasta, 105
Mount Sterling, Kentucky, 16
Moyer Texas Champagne, 191, 199
Muench, Frederick, 199, 202
Muench grape, 146, 199
mulberry, 142
Munson & Son, 129, 136
Munson, Ben, 4, 6, 7, 9, 15, 16, 22, 36-37, 44, 58, 63, 101, 116, 117, 118, 131, 146, 156, 171, 172, 178, 179, 184, 191
Munson, Ben, III, 193
Munson Cling Peach, 141
Munson, David M., 195
Munson, Ella Newton, 44, 172
Munson, Elmerwin, 20, 22
Munson, Eloise, 184, 191-192
Munson, Fern, 36, 62, 102, 128, 146, 177, 181, 184, 191
Munson, Huxley, 20
Munson, Jennie, 5, 6, 44, 101, 116, 129, 130-131, 146, 171, 172, 177-178, 184
Munson (Manson), John, Jr., 3
Munson, Joseph Theodore

(Theo), 4, 5-6, 9, 15, 44, 52, 58, 62, 101, 117, 118, 129, 131, 146, 149, 171, 172, 178, 183, 184
Munson, Louisa, 4, 5, 23, 30, 35, 131, 171, 172, 178, 184
Munson, Lydia Philbrook, 3
Munson, Marguerite (Margaret), 111, 146, 177, 179, 181, 183
Munson, Maria Linley, 3-4, 5, 7, 30, 36, 44, 111, 130
Munson, Mary Ginevra, see Munson, Jennie
Munson, Minnie Secoy, 90, 165, 186, 187, 188
Munson, Nellie, 22, 25, 32, 36, 49, 62, 82, 99, 101, 127, 146, 152, 171, 179, 181, 183, 204
Munson, Neva, 49, 62, 129, 146, 181, 183, 184, 191
Munson, Nina Cummings, 146, 171
Munson Nursery, 119, 123, 142, 147, 148, 152, 161, 165, 175, 181, 183, 184-186; see also Denison Nursery
Munson, Olita, 49, 62, 129, 146, 183; see also Calvert, Olita
Munson Realty Company, 118, 146, 184
Munson (Monson), Richard, 3
Munson, Rupert Scott, 127, 129, 179
Munson, Theodore (son of Richard), 3
Munson, Thomas Volney
addition in Denison, 101-102, 116, 146, 181
ancestry of, 1-4
apples of, 142, 175
archives of, vii, 59, 118, 183, 186, 188
Asiatic varieties of, 143
associates/correspondents of, 47-48, 50, 53, 56-57, 58, 60, 77, 80, 81-82, 90, 95, 96, 99, 108, 132, 139, 154, 164-165, 186
association/organization membership of, 8, 30, 36, 49-50, 52, 53, 56, 60, 99, 114, 116, 133, 137, 144, 145, 151, 154, 155, 162, 172, 175, 177, 179

attitude toward women, 98-99, 130, 159
awards/honors, 55, 56, 57, 87-90, 99, 106, 115, 123, 136, 152, 154, 162, 175, 180, 186, 191, 195, 200, 204
as bank director, 58, 116
Bible of, 8, 14
birth of, 4
book by, 52, 65-67, 95, 107-113, 128, 133, 135, 152, 156, 161-165, 175, 179, 191, 192, 202
as cadet, 9
Campbellite heritage of, 8, 14, 157
catalogs of, 49, 52, 73, 88, 90, 96, 109, 122, 131, 141, 142, 152, 156, 161, 180, 185, 187
childhood of, 5
children of, 20, 25, 36, 49, 58, 82, 90, 99, 111, 127, 128-129, 146, 181, 183
civic activities of, 116, 148, 172, 175
climbs Mount Shasta, 105
crepe myrtle of, 143
on death, 160
death of, 177
defines "species," 54
described, 1, 5, 55, 97, 98, 118, 133, 148, 178-179, 188, 200
diaries/journals of, 26-30, 31, 35, 36, 45, 65, 183, 204
drawings by, 20, 63, 135
on education, 129
education of, 5-6, 7-19, 52, 97, 154
educational sets offered by, 91-96
evergreens of, 152
family crest used by, 2, 63
as farmer, 5, 8, 26, 30, 35, 36, 45, 97
funeral of, 177-178
funeral oration written by, 160-161, 178
international sales by, 73, 85-86, 87, 133, 147
field reports of, vii, 103
as French speaker, 122, 153
graduates from Kentucky A&M, 19

on grafting, 73
grape classification by, 42-43, 47, 48-49, 50, 51, 52, 54-56, 57, 58, 65-67, 81, 96, 107-108, 123, 132-133, 139, 140, 213-235
and grape evolution, 50, 54-56
grape exhibits of, 53-54, 56, 99, 106, 109, 114, 115-116, 123-125, 151, 152
grape hybridization/experimentation of, 23-24, 32, 40, 41-43, 46-47, 49-52, 54-56, 57-58, 60, 65, 80-90, 91-93, 96, 101, 118, 119-122, 123, 131-133, 135, 144-146, 147, 148, 152-155, 162, 175
grape varieties created by, 213-235 *and see* grape hybdridization/experimentation of
honorary doctorate of, 154
horticultural training/study, 5, 12, 97
illness of, 19, 36, 44, 65, 95, 106, 111, 129, 161, 165, 170, 177
Internet sites about, 202
inventions of, 15, 20, 21, 26, 28, 32, 33-34, 45, 49, 52, 137-139, 165-170
on isolationism, 97
in land sales, 45
library of, 63, 81, 113, 186, 188
in literary society, 8, 36
and machinery, 5, 15, 20, 32, 97, 165-170
marriage of, 19
mentors of, 16, 18, 20, 23, 128, 146
mulberry of, 142
name/surname of, 2, 3, 4
as nominee for president, Texas A&M, 137
as nominee for Texas legislator, 148, 151
nursery of, *see* Denison Nursery and Munson Nursery
nut varieties of, 142 *and see* pecans of
opponents of, 118-119

orchards of, 26, 30, 32, 35, 43-44, 45-46, 58, 175
on overproduction, 97
peaches of, 141-142, 145, 152
pears of, 142, 152
pecans of, 99, 143, 147, 151
persimmons of, 99, 142, 152
plum species named for, 155
plums of, 99, 141, 155
potato experiments of, 32
property owned by, 20-21, 26-30, 35, 36, 43-44, 52, 62-64, 101, 146, 175, 181; *see also* Vinita Home
pseudonym of, 15, 156
publicity about, 58-59, 67, 81, 86, 122, 136, 144, 146, 152, 164
religious views of, 8-9, 12, 14-15, 57, 97, 98, 156-161, 178
sends cuttings (to France), 73, 85-86, 87, 196, 205 (to Japan), 133
siblings of, 4-5, 130, 171-172, 184
specimens collected by, 16, 46-47, 52, 54, 56, 57, 65-67, 73, 81, 85, 91-96, 97, 103-106, 116, 142, 151, 155
speeches by, 51, 97-98, 124, 129-130, 137, 144, 145, 153, 161, 212-213
strawberries of, 46, 99-101, 119, 142
tax records of, 32
as teacher, 5, 6, 8, 15-16, 19, 91
on temperance, 125, 148-149
and theosophy, 131
thesis of, 52
tombstone of, 177, 183
trellis designed by, 137-139, 140, 165, 193
tributes to, 178-179, 196
and USDA projects, 64-67, 81, 82, 91, 96, 103-106, 107-113, 142, 143, 147-148, 151
vineyards of, 26, 30, 32, 39, 44, 46, 48, 52, 85, 93, 142, 153, 188
and whole root products, 118-119, 120-121
will of, 179, 181

winemaking by, 63, 127
wines using hybrids of, 199
writings of, 8, 9-15, 32, 50, 51, 52, 54, 56, 57, 58, 60, 65-67, 91, 97, 98, 104, 107-113, 122-123, 124, 125, 131, 133, 135, 143, 144, 145, 153, 156-165, 175, 186, 188, 191, 204, 207-213
"Munson Three-Wire Trough Trellis," 137-139, 140, 165, 193
Munson Trail, 202
Munson, Triphena Mary (Trite), 4, 6, 44, 101, 116, 130-131, 146, 171, 172, 177-178, 184
Munson, Viala Laussel ("Vee"), 62, 63, 82, 146, 161, 162, 171, 181, 183
Munson, Volney Earle, 58, 127
Munson, W. B., 118, 161
Munson, W. B., III, 196, 203
Munson, W. H., 62
Munson, Warder, 49, 61, 62, 129, 146, 177, 181, 183
Munson, William Bell (Will), 32, 36, 61, 62, 90, 101, 102, 106, 119, 124, 128-129, 136, 146, 152, 162, 171, 181, 182, 183, 184, 186
Munson, William (son of Richard), 3
Munson, William (T.V.'s father), 3-5, 25, 44, 111, 130
Munson's Pink, 143
Munsoniana, 57
Murray, B. C., 156
Murray's Stream Printing House, 52
muscadine grapes, 38
Museum of Natural History at A&M, 16
Mustang grapes, 38, 40, 55, 76, 78
Mustang wine, 40

Napa Valley, 104
Napoleon III, 72, 89
National Agricultural Library, 112
National Archives, 59
National Archives II, 103
National Bank of Denison, 116
National City, California, 104

National Commercial College of Denison, 116
National Cyclopaedia of American Biography, 85, 111
National Herbarium at the Smithsonian, 66, 103, 105, 124, 142
National Institute for Agronomic Research, 193
National Museum, 49, 93, 95, 104, 124
National Nurseryman, The, 161
National Plant Germplasm System, 200
native grapes of California, 103-106
native grapes of Texas, 40-41, 52, 82-83, 84, 103, 119, 122
native grapes of U.S., 46, 41-43, 47-48, 54-56, 57, 80-81, 96, 132, 140
Navasota, Texas, 103
Nebraska, 25, 30, 37
Nebraska Farmer, 153
Nebraska State Journal, 32, 35
Neosho, Missouri, 47, 52
New Braunfels, Texas, 82
New Haven, Connecticut, 3
New Orleans Cotton Exposition, 55, 56, 99
New Revelation, The, 15, 156-161
New York Agricultural Experiment Station, 147
New York Botanical Garden, 144
New York State Agricultural Experiment Station, 181
New York Times, 89
New Zealand, 71
Newton, Ella, 44
Nimon, James, 52, 99-101, 116
No. 3, 99
No. 43, 60
No. 70, 47
nomenclature of American grapes, *see* grape classification
North American Fruit Explorers, 200
North Carolina Department of Agriculture and Consumer Services, 202
North Central Texas, 203
North Texas Horticultural Society, 49, 60

Number 333 School of Montpellier rootstocks, 86

Ohio State Horticultural Society, 56
O'Mara, Patrick, 146
oidium, 68-69, 70
Ojai Valley, 104
Oliver Dewey Mayor Foundation, 195
olives, 75
Olmo, Harold P., 193
Onderdonk, Gilbert, 47, 145-146, 164
140 Ruggeri, 87
1103 Paulsen, 86-87
110 Richter, 86
orange trees, 119, 148
Orange Judd Company, 161
Origin of Species, 12
ornamental trees, 52
overplanting, 51
Owingsville, Kentucky, 16
Oxford Companion to Wine, 87
Ozark Prize grape, 200

P. australis, 141
Pacific Rural Press, 56
Pacific Tree and Vine, 139, 199
Pacific Wine and Spirit Review, 140
paddle churn, 32
Palace of Agriculture, 152
Palace of Horticulture, 149, 150
Palestine, Texas, 50
Panhandle-Plains Historical Museum at Canyon, 118
Panic of 1907, 161
Paris, Kentucky, 16
Paris, Texas, 118
Parker Earle strawberry, 99-101, 142
Parras, Mexico, 133
Pauline, 55
PawPaw Creek, 54
peaches, 45, 47, 52, 141, 145, 152
pears, 119, 142, 152
pecans, 143, 147
Pecos County, Texas, 191
"Perfection" stove, 20, 21
peritymbia vitisana, 69
Perry grape, 146
Perry, Oklahoma, 137
persimmons, 142, 152

Peter, Robert, 18, 23-24, 30, 128, 146
Pfeffer, William, 139, 164, 199
Pheasant Ridge Winery, 191
Philosophius, Theophilus, 15, 156-161
phylloxera, 48, 64, 66, 69-77, 79-80, 81, 86, 87, 89-90, 91, 108, 133, 139-140, 196, 205
Phylloxera Congress, 79-80
Pickett, Joseph D., 8
Pierce's Disease, 190, 199
Pilot Knob, Texas, 16
Pinney, Thomas, 200
Planchon, Jules-Emile, 48, 54, 56, 60, 69, 70, 74, 75, 76, 108, 122, 162
Planchon, Jules-Emile, memorial to, 133-134
plums, 47, 58, 141, 155
Point Loma, California, 131
Pomology Division, USDA, 64, 67, 103, 105, 112, 142, 143
Portugal, 71, 75
Post Familie Vineyards, 199
Post Oak grapes, 38, 40, 47, 123, 131, 199
President grape, 202
President port, 199
Prestele, Wilhelm (William) Heinrich, 67, 110, 111, 112, 115, 133
Preston Anticline, 40-41
prohibition, 125, 127, 144, 148-149, 184, 186, 189
Prohibition Party, 148-149
Prunus munsoniana, 155
Prunus Simonii, 58
Prussia, 72, 90

R. W. Munson grape, 131, 200
Ramsey, F. T., 178
Randolf, U.A., 193
Ray peach, 152
Red River, 16, 40, 54, 192
Reed, Bob, 190
Renfro, Roy E., Jr., 90, 193, 194, 195, 197, 200, 203, 205
Report of the Pomologist for 1890, 143
Report on the Experiments Made in 1887 in the Treatment of the Downy Mildew and Black Rot of the Grape Vine, 65
Resistant Vines, 140

revolving shuttle sewing machine, 32
Revue de Viticulture, 133
Revue des Hybrides Franco-Americains, 135
Rhone River, 79
Richardson, T. C., 186, 188
Richter family, 60, 61, 85, 86, 133, 191
Riley, C. V., 48, 70, 71, 75, 76-77, 90
Riverside grape, 40
Rives, Max, 90, 191, 193, 196
Rogers, E. S., 42
Rogers hybrids, 24
Rogue River Valley, 105
Rolla, Missouri, 35
Rommel, Jacob, 42, 101, 144
Rommel grape, 101
Ross, Lawrence Sullivan "Sul," 137
"Rotary Dasher Churn," 20
Rothschild, Baron Charles, 71
Rotundifolia, 55
Royal Horticultural Society of England, 144
Ruidoso, New Mexico, 183
Rupestris-Candicans, 83
Rural New Yorker, 58, 119
Russia, 184
Rutledge apple, 142

Sacramento River, 104
Safety Flying Car, 165-170
Salt Creek, 32, 36, 87, 200
Sam Houston State Univeristy, 129
Sam Rayburn Foundation, 197
San Antonio Regional Wine Guild Competition, 199
San Antonio, Texas, 76
San Bernardino Mountains, 104
San Diego, California, 104
San Francisco, California, 156-157, 159
San Jacinto grape, 152
Sand grape, 40
Sandia Mountains, 103
Santa Rosa, California, 104
Saragosse, Spain, 80
Sargent, C. S., vii, 123, 154-155
Satsuma orange trees, 119
Saunders, William, 177
Sayers, Joseph D., 137
Sayre Female Institute, 18

Scarlet Oaks, 43, 44, 58
Schoppaul Hill Winery, 203
Schuyler County, Illinois, 4
Science, 164, 170
Scientific Society of San Antonio, 111-112
scuffle hoe, 20, 49, 52
Scuppernong grapes, 54, 132, 145, 152
Secoy, Minerva, 181
Sedalia, Missouri, 16
Serbia, 77
Shakespeare, 63
Shaw's Garden, 105
Sherburn Mills, 16
Sheridan grape, 200
Sherman, Texas, 22
Shirley apple, 142
shrubs, 52
Sicily, 87
Sikorsky, Igor, 167
Simpson, J. H., 57
"Single-hand Drill," 32
Sissons, California, 105
Sister City Movement, 196-197
6666 Ranch, 184
Smith, Clara Blackford, 195
Smithsonian Institution, 50, 91, 93, 95, 103, 104, 124
Snodgrass, Eldon, 35
Societe Centrale d'Agriculture de l'Herault, 133
Societe des Viticulteurs de France, 133, 162
Societe Nationale d'Agriculture de France, 133, 162
Society for (the Promotion of) Horticultural Science, 145, 175
Society of American Florists, 114
soft-rooted, 55
soil conditions, 78-79, 80-81, 83, 85, 87, 140
Sonoma, 71
Sour Winder grapes, 40
South Africa, 71
South America, 74
Southern Horticultural Journal, 104
Southern Methodist University, 119
Spain, 71
Springfield, Missouri, 35
St. George grape, 139, 140
St. Louis, Missouri, 48, 149
St-Emilion, France, 69

Standard Cyclopaedia of Horticulture, 175
Stark Brothers nursery, 118-119
State Agricultural College of Mississippi, 52
State Fair of Texas, 200
State House Rock, 16
Ste. Genevieve Vineyards, 191, 202
Stell, William Wynne, 118, 119
Stephenville, Texas, 186
Stevens, Linda, 90
Stiles, —, 122-123
strawberries, 46, 99-101, 119, 142
Stringfellow, Henry Martyn, 119, 153
Sturtevant, E. Lewis, 95
Sullivan, Charles L., 139
sulphides, 73
Swartz Studio, 109
Sweet Winter grapes, 40
Sweetey grape, 146

T. G. Bright & Company, 199
T. V. Munson Award, 200
T. V. Munson Viticulture and Enology Center, viii, 20, 90, 106, 123, 152, 165, 194-198, 203-204
T. V. Munson Instructional Winery, 203
T. V. Munson Memorial Vineyard, 193-198, 203
T-Anchor Ranch, 184
table grapes, 189, 190, 200
Tarrant County, Texas, 82
taxonomy, 12
Taylor, William A., 111-112, 148
Tecumseh, Nebraska, 23
Temple-Belton, 82, 85
Terroir, 200, 204
Texas 4-H, 137
Texas A&M, 97, 130, 135, 137, 139, 148, 151, 154, 190, 200
Texas Academy of Science, 50
Texas Agricultural Experiment Station, 135, 136, 151, 153
Texas Agricultural Extension Service, 137, 154
Texas Agricultural History Committee, 186
Texas Almanac, 149
Texas Association of

Community Services and
Continuing Education, 198
Texas Department of Agriculture,
111, 135, 153, 179
Texas Farm and Ranch, 65, 99,
118, 119, 125, 136, 137, 167
Texas Farm Winery Act, 191
Texas Farmers' Congress, 179
Texas Fruits at the World's Fair,
152
Texas Fruits, Nuts, Berries and
Flowers, 143
Texas Grand Vin Brut, 199
Texas Heritage Hall of Honor,
200
Texas Historical Markers, 184,
191
Texas Horticulturist, 179
Texas Industrial Congress, 172,
175
Texas Legislature, 125
Texas Local Option
Association, 149
Texas Pavilion, 125, 151
Texas Post Oak grape, 131
Texas Power and Light, 171
Texas State Fair, 137
Texas State Horticultural
Society, 60, 99, 115, 152
Texas State Teachers College,
129
Texas Tech University, 190
Texas Vineyards Winery, 191
Texas, viticulture in, 38-41,
189-204
Texas, wine industry in, 39,
149, 189, 190-191, 198, 199,
200, 202-203
Texas wine trail, 202
Texas Wine and Grape Growers
Association, 200
"Texas Wine Month," 202
Texas Women's World's Fair
Association, 125
Texas World's Fair Association,
151
Texas-French Shootout, 203
Texoma Chapter of the
American Wine Society, 197
theosophy, 131, 146, 184
Theosophy Center, 184
Thomas Volney Munson
Symposium, 197
Thompson, J. W., 183
Thompson, Margaret, 183
Thompson, S. R., 30

Throckmorton, J. W., 58
Tingley, Katharine, 131
Tobin, Bernadette, 125
Tomas grape, 132
Tone & Munson, 44
Tonkin, Linley M., 17
Traite General de Viticulture:
Ampelographie, 153
Transactions, 52
Transylvania University, 7, 18
Travis County, Texas, 82, 142
Trelease, William, 50, 105
trellis system, 137-139, 165, 193
Troup, Texas, 148
Tucson, Arizona, 82
XXI Club, 178
TXU Electric, 102

U.S. Department of
Agriculture, 39, 58, 59, 60,
64-65, 80, 85, 91, 103, 107-
113, 123-124, 135, 143, 147,
151, 161; *see also* Pomology
Division, USDA
U.S. Geological Survey, 80, 85
U.S. Patent Office, 39
Une Mission Viticole en Amerique,
108
Union Pacific, 25
United States Bureau of Alcohol,
Tobacco, and Firearms, 199
Universidad de la Rioja, 202
Universite de Montpellier, 72,
73
University of Arkansas, 194,
199
University of California, 139
University of California/Davis,
193, 194
University of Kentucky, 16, 19,
154
University of Missouri at
Columbia, 56
University of Nebraska, 26, 91
University of Texas at Austin,
128, 129, 191
Upshall, W. H., 200

Val Verde Winery, 189, 190, 202
Valk, William W., 42
Valley of the Dordogne, 69
Van Deman, H. E., 58-60, 64,
66-67, 81, 82, 85, 96, 103,
105, 106, 107, 108, 109-110,

111, 112, 115, 123, 141,
142, 164, 178
Vasey, George, 91, 93, 95, 96, 142
Vaucluse Agricultural Society,
70
Vaucluse region, 71
Vermorel, Victor, 61, 153
vermouth, 75
Viala, M. Pierre, 48, 56, 58, 80-
87, 89, 108, 122, 133, 153,
162, 196
Victoria County, Texas, 47
Vienna Exposition, 48
Vinita Home, 62-64, 81, 91,
101, 103, 126, 127, 146,
147, 171, 181, 183, 184,
191, 195
Vintinto V65115, 199
Virginia grapes, 47
viticulture in U.S., 41-43, 48,
64, 75, 96, 162
Viticulture and Enology
Research Center in
California, 200
Vitis candicans, 40, 78
Vitis Aestivalis, 54, 55, 57, 153
Vitis berlandieri, 82, 83, 85-87,
133
Vitis californica, 113, 139
Vitis Champini, 113, 200
Vitis cinerea, 40, 54, 55, 83, 85,
124
Vitis Doaniana, 67
Vitis Doanii, 66
Vitis Floridiana, 107
Vitis Lincecumii, 40, 123, 131, 152
Vitis Linsecommi, 144
Vitis Monticola, 112, 133
Vitis Nuevo Mexicana, 54, 66
Vitis palmata, 122
Vitis riparia, 40, 42, 54, 55, 57,
86, 87
Vitis rubra, 122
Vitis rupestris, 40, 54, 81, 86,
139-140
Vitis Simpsoniana, 107
Vitis Spaldingi, 103
Vitis Texana, 66
Vitis vinifera, 24, 38, 40, 42-43,
47, 49, 53, 55, 68, 69, 70,
71, 73, 81, 86, 105, 123, 139
Volney, Comte de Constantin F.
C., 4
Vulpina, 55

287

W. Aubrey Smith Charitable Foundation, 195
W. B. Munson Foundation, viii, 165, 192, 193, 194, 195, 205
Waco Searchlight, 153, 178
Waelder, Texas, 103
War Department, 178
Warder, John A., 49
Warren, 55
warty-wood, 57
Washburn College, 179
Washington University, 50
Waugh, Frank A., 179
Waxahachie, Texas, 82
Welch, William C., 36
West Texas, 190, 191
Western Fruit Grower, 136
White, H. H., 15

White Sulpher Springs Hotel, 16
Wichita River, 39
Wilbarger County, Texas, 66
Wilder Award, 57, 88, 136
Williams, H. C., 39
Williamson County, Texas, 133
Willits, USDA Secretary, 109, 111, 115
Wilson, James, 200, 204
Wine and Fruit Grower, The, 56
Wine Society of Texas, 203
Winter Garden Station, 190
Winton, 23
Witcher, Mr., 149
Wolfe Nursery, 186
Wolfe, Ross R., 186
Woodbine sands, 40
Woodland Township, Illinois, 5

Woodruff, J., 118
wooly-leaved, 57
World War I, 184
World's Columbian Exposition, 53, 103, 114, 123-125, 180
World's Industrial and Cotton Exposition, 53-54
Wright, Orville, 167
Wright, Wilbur, 167
Wu 83-82-4, 87

Xlanta grape, 200

Yokohama, Japan, 133